The Logic of Entailment and its History

What follows from what, and how do we make statements (whether true or false) about which inferences are correct? In this book, Edwin Mares provides a new philosophical, semantical, and historical analysis of and justification for the relevant logic of entailment. In the first half of the book he examines some key ideas in the historical development of the logic of entailment, looking in particular at the notion 'is derivable from' and at how symbolic logic has attempted to capture this notion. In the second half of the book he develops his own theory connecting ideas from the traditions in mathematical logic with some ideas in the philosophy of science. The book's fresh and original perspective on the logic of entailment will be valuable for all who want to know more about the historical and philosophical origins of modern symbolic logic.

Edwin Mares is Professor of Philosophy at Victoria University of Wellington. His publications include *Relevant Logic: A Philosophical Interpretation* (Cambridge University Press, 2004).

The Logic of Entailment and its History

EDWIN MARES

Victoria University of Wellington

CAMBRIDGE
UNIVERSITY PRESS

Shaftesbury Road, Cambridge CB2 8EA, United Kingdom

One Liberty Plaza, 20th Floor, New York, NY 10006, USA

477 Williamstown Road, Port Melbourne, VIC 3207, Australia

314–321, 3rd Floor, Plot 3, Splendor Forum, Jasola District Centre, New Delhi – 110025, India

103 Penang Road, #05–06/07, Visioncrest Commercial, Singapore 238467

Cambridge University Press is part of Cambridge University Press & Assessment,
a department of the University of Cambridge.

We share the University's mission to contribute to society through the pursuit of
education, learning and research at the highest international levels of excellence.

www.cambridge.org
Information on this title: www.cambridge.org/9781009375313

DOI: 10.1017/9781009375283

First published 2024

A catalogue record for this publication is available from the British Library.

A Cataloging-in-Publication data record for this book is available from the Library of Congress.

ISBN 978-1-009-37531-3 Hardback

To my mother and the memory of my father.

Contents

Preface

A logic of entailment is a logic that represents the notion of valid deductive inference by means of a connective. The problem of developing an adequate logic of entailment was a central issue in the literature on philosophical logic of the first half of the twentieth century. This was made a central issue by C. I. Lewis.[1] Lewis constructed a logic of strict implication in order to represent the notion of deducibility and, until mid-century, modal logic was understood largely as a way of representing a notion of deducibility. The debate was about how exactly to formulate a logic of entailment. In the work of G. H. von Wright, Arthur Prior, and others, there was a realisation that modal logics could be used to represent a wide range of modal notions and to represent notions that were not previously understood as modal, such as tense operators (past and future operators), deontic operators (obligation and permission operators), epistemic and doxastic operators (knowledge and belief operators), and so on. During this later period, the problem of entailment became just one of a wide variety of issues to do with modal logic and lost its prominence. In addition, the very attempt to construct a logic of entailment was strongly criticised by philosophers such as W. V. O. Quine and Casimir Lewy who questioned whether the very notion of a logic of entailment was coherent.

The problem of entailment, however, did not disappear altogether from the philosophical logic literature. Building on the work of Wilhelm Ackermann and others, Alan Anderson and Nuel Belnap constructed their logic E of entailment in the early 1960s and a school developed around them. Early in their work, however, Anderson and Belnap also constructed the logic R of relevant implication. R was not a logic of entailment. Its implication was not meant to express what is deducible from what. Rather, it formulated a form of contingent implication that captured a connection of relevance between antecedents and consequents.

In the early 1970s, a Kripke-style semantics for relevant logics was constructed by Richard Routley and Robert Meyer, building on the work of Alasdair Urquhart. Routley–Meyer semantics showed that a huge number of new and interesting relevant logics could be constructed, by placing different conditions on the accessibility relation of models. In this semantic framework, E was seen as just one among many logics, but its Routley–Meyer semantics was clumsy and unintuitive.

---

[1] Although Lewis's influence in contemporary philosophy is relatively minor, during the 1920s and 1930s, he was a towering figure in American philosophy. Bruce Kuklick says of Lewis that "his "conceptual pragmatism" made him the most influential American philosopher of his generation" [95, p 533].

The Routley–Meyer semantics also showed that traditional modal notions – possibility and necessity – could quite easily be combined with the vocabulary of relevant logics. Very recently, this idea has been extended to create deontic, epistemic, and doxastic relevant logics, all of which have interesting uses in representing the notions of obligation, permission, belief, and knowledge.

All this has made most researchers in relevant logic more interested in logics other than E. I was among them. In my book, *Relevant Logic*, I championed a modal version of the system R as a logic of entailment. I still stand by most of what I said in that book about the interpretation of R, but I now think that E has been unfairly neglected. In this book, I give E an interpretation and try to secure its place not just in the history of relevant logic but also in the future of this subject.

My approach in the first half of the book is historical. I examine some key ideas in the development of the logic of entailment. I do not, however, attempt to give a comprehensive history of the subject. There are some interesting ideas that I only mention in passing, such as Timothy Smiley's construction of a non-transitive logic of entailment, and others that I do not deal with at all, such as Mark Lance and Philip Kremer's interpretation of the logic E as a theory of linguistic commitment [98]. I regret that I could not fit these ideas into the pages of this book in any natural way. The discussion, especially of Lance and Kremer's view, would take me away from the topics that I wish to address and cause me to talk about, for example, Brandom's theory of linguistic commitment, which plays no role in the interpretation of E that I put forward in the second half of this book.

In the second half of the book, I use the lessons extracted from this history to justify a semantical approach to the logic E based on Kit Fine's "Models for Entailment" [61]. I use Fine's semantics to interpret E as a theory of theory closure. Some of the theories that it can be used to examine are also theories of entailment (mostly, incorrect theories of entailment). Using this idea, I explain the nature of nested entailments and the relationship between the semantic consequence relation and the entailment connective.

This book is not a sequel to my earlier *Relevant Logic*. It does, however, sit nicely beside that book, I think. The two books together give interpretations of E and R, that are rather different, despite the similarities of their proof theory (discussed in Chapter 4). I think the relationship between E and R is rather like the relationship between the logics of indicative and counterfactual conditionals. The two sorts of conditionals are very similar in what they allow and disallow in terms of inferences. But they have very different semantics and very different interpretations. Counterfactual conditionals are often thought to represent dispositional features of reality and indicative conditionals are usually taken to represent doxastic or epistemic attitudes of agents. On my interpretations of E and R, the implication of R represents facts about what information is carried by propositions, whereas the entailment connective of E represents the closure of theories. Both sorts of logics play useful roles in reasoning and do not clash with one another in any way.

Before I end this little preface, I would like to discuss one issue that is often closely connected with relevant logics. One of the referees pointed out to me that I rarely use the expression 'paraconsistent' in this book and do not discuss inconsistent theories at any length. This is because I am no longer certain about the best approach to inconsistent theories. Certainly, there are inconsistent theories that are interesting in and of themselves. But theories that are made up of various inconsistent elements, like Bohr's old quantum theory that combined an orbital view of electrons with a classical theory of electrodynamics, might be best understood in terms of collections of classical theories instead of a single unified theory. At any rate, that is not what this book is about and I try to avoid the issue as much as possible. The principle of explosion – that contradictions entail everything – is to be avoided mainly because it is the contraposition of the principle of implosion – that everything implies any theorem. And implosion is a principle that any reasonable theory of entailment must avoid.

Acknowledgements

My greatest intellectual debt is to Robert Goldblatt. At our weekly lunches, I talked to Rob about the technical problems of the various logics that I dealt with in this book and he would often tell me of techniques that would help. I also talked to Max Cresswell and Adriane Rini often, especially about the historical material that I discussed in the book, and they made some extremely useful observations. I gave various talks at a number of philosophy departments about the material in Chapters 6 and 7, and I'm grateful to the audiences at those universities – the Graduate Center at CUNY, the University of Connecticut, the University of Cagliari, UNAM in Mexico City, and the University of Alberta – as well as the audiences at the Conditionals Conference at National Taiwan University in 2018 and at the joint ASL/APA meeting in Denver in 2019. In particular, I'm grateful to Axel Barcelo, Catarina Dutilh Noves, Nicholas Ferenz, Hartry Field, Melvin Fitting, Ole Hjortland, João Marcos, Graham Priest, Greg Restall, Lionel Shapiro, Andrew Tedder, Heinrich Wansing, and Yale Weiss for very insightful discussions. I am especially indebted to Francesco Paoli, with whom I have been engaged in a project on the history of the problem of entailment, and Shawn Standefer, with whom I wrote a paper about the interpretation of E and nearby logics [137]. Although my interpretation in this book is rather different than the one that Shawn and I gave in that paper, the latter influenced me strongly in constructing the former. The two readers for Cambridge University Press both gave extremely helpful reports and implementing the changes that they recommended made the book more readable. Of course, my greatest personal debt is to my partner, Susan Wild, who encouraged me in my work and discussed grammatical issues with me. Some of the research for this book has been funded by the Marsden Fund of the Royal Society of New Zealand (grants "The Logic of Natural Language" and "A Natural History of Necessity" both jointly held with Cresswell and Rini).

Table of Symbols

Name	Symbol	Logical System	Chapter
Entailment	$\rightarrow$	Almost all	All
Relevant implication	$\Rightarrow$	R	4
Strict implication	$\dashv3$	Modal logics	2
Material implication	$\supset$	Classical logic	All
Fusion consequence	$\vdash$	E	4 onwards
Conjunctive consequence	$\Vdash$	E	4 onwards
Satisfaction/semantic consequence	$\vDash$	Almost all	3 onwards
R-fusion	$\circ$	R	5
E-fusion	$\otimes$	E	5
Conjunction	$\wedge$	All	All
Disjunction	$\vee$	All	9
Negation	$\neg$	All	9
Universal quantifier	$\forall$	QE	10
Existential quantifier	$\exists$	QE	10
Identity	$=$	QE	10
Consequence operators	C, C_Σ, C_a	TL, GTL, E	6 onwards

1   Why Entailment?

1.1   Planning Proofs

I want to prove that a statement B holds in a particular theory. I think I can show that A follows from the axioms of that theory and I think that I can prove that A entails B. In order to prove B, I formulate a **proof plan**: *If A entails B, then if Ax (the conjoined axioms of the theory) entails A, then Ax entails B*. The 'if ... then's in this statement of the proof plan themselves do not express a standard indicative conditional. Rather, they express logical entailments. Formalising logical entailment with an arrow, $\rightarrow$, this proof plan can be written

$$(A \rightarrow B) \rightarrow ((Ax \rightarrow A) \rightarrow (Ax \rightarrow B)).$$

What makes this a good proof plan? Part of the answer to this is that a proof plan is good only when it is a theorem of the logic of entailment. Proofs have to be logically valid arguments and the salient notion of validity is captured by the logic of entailment. This book is about this logic and its philosophical and formal semantics.

Of course, there are also non-logical criteria that go into determining whether a proof plan is good. As Saunders Mac Lane[1] said:

A proof for a given theorem is not a haphazard collection of individual steps, taken arbitrarily one after another, as the classical logic might easily lead us to believe. On the contrary, there is some definite reason for the inclusion of each one of these steps in the proof; that is, each individual step is taken for some specific purpose. [99, p 125]

The rationale for the ordering of the steps in a proof may be logical. It may have to do with how they fit together logically into the proof. But it may have to do with psychological factors that govern how people understand proofs. The construction of logic of proof plans involves the abstraction away from these psychological factors to produce a theory of the formal requirements of a good proof.

One might object that we could attribute every feature that makes a proof good that is not available in classical logic, say, to the contingencies of human psychology and merely say that classical logic is the theory of proof plans. In this book, I suggest that classical logic and modal logics based on classical logic do not make certain distinctions or treat the connectives in ways that make a logic useful in distinguishing between good and bad proof plans.

---

[1] I take the terminology of proof plans from Mac Lane, who used the very similar "plans of proof".

Suppose that B really does follow from the axioms of the theory (Ax). Then, if we treat $\rightarrow$ as the classical conditional or as a classical form of strict implication (see Chapters 2 and 3), the following inference is valid:

$$\frac{\vdash Ax \rightarrow B}{\vdash C \rightarrow (Ax \rightarrow B)}$$

for arbitrary formulas, C. Treating this $\rightarrow$ as entailment means that we treat $C \rightarrow (Ax \rightarrow B)$ as a good proof plan, at least as far as the formal logic of the proof plan is concerned. The acceptance of this rule and its classical kin severely limit the usefulness of a logic of entailment. It is not the case that mathematicians or others would accept this $C \rightarrow (Ax \rightarrow B)$ as a good proof plan when C is unrelated to Ax or B. Thus, accepting classical logic or one of its modal extensions would force us to give an almost entirely psychological and pragmatic analysis of proof plans concerning logical theorems. Surely, a formal logic of entailment can contribute a lot more than this.

As the above indicates, in order to formulate proof plans, I use an *entailment connective*. I need to be able to nest expressions about proving one formula from another within an entailment. Thus, I have entailments nested within entailments. All logical systems give rise to one or more entailment relations. A *logic of entailment* contains an entailment connective.

In order to develop and justify a logic of entailment, I have to delve very deeply into the problem of nested entailments. In this book, I adopt G. E. Moore's definition of 'entailment' as the 'converse of deducibility' [147, p 291]. On this definition,

'$A \rightarrow B$' is true if and only if $A \Vdash B$.

This truth condition does not help when it comes to the meanings of nested entailments. For it does not make sense in our standard logical languages to write, say,

$$(Ax \Vdash A) \Vdash (Ax \Vdash B).$$

It is the central aim of this book to develop a proof theory and a semantic theory that both make sense of the claim that the entailment connective expresses a reasonable deducibility relation and characterises a logic of entailment that is robust and useful in formulating proof plans. The proof theory that I adopt is Alan Anderson and Nuel Belnap's natural deduction system for their logic E of relevant entailment [3, 4] together with a labelled natural deduction system which is closely tied to the semantics. The semantics that I employ is a version of Kit Fine's "Models for Entailment" [61] modified to incorporate Robert Goldblatt's and my semantics for quantification [134].

1.2   Logic and Intuition

When trying to create a formal logic of any concept we need some criteria of success and of failure. We need to be able to tell, at least within some boundaries, whether we have captured the concept that we want to represent. The most common benchmark used to justify or reject theories of entailment employed over the last century was whether this notion captured the intuitive notion of a deduction. Some philosophers have claimed, however, that our intuitions regarding deduction conflict with one another. One philosopher who said this was Casimir Lewy:

> [T]here is in any case no theory of entailment, however complex, which would enable us to accept all those entailment propositions which on an intuitive level we wish to accept and reject all those which on an intuitive level we wish to reject. And, so far as I can see, there is no *hope* of such a theory. [116, pp 133–134]

Timothy Smiley agreed with Lewy [194]. To support this view, Smiley gave the following list of theses about entailment, the first four are strongly intuitive, but the last is strongly counter-intuitive:

1. A entails $A \vee B$.
2. If A entails B, then $A \wedge C$ entails $B \wedge C$.
3. $(A \vee B) \wedge \neg A$ entails B.
4. If A entails B and B entails C, then A entails C.
5. $A \wedge \neg A$ entails B.

He said about this list that "if [we are] uncorrupted by acquaintance with the polemical literature", we should accept 1–4 all hold but reject 5. The problem, however, was that 5 followed from 1–4.

Smiley did try to construct a theory that almost captured our intuitions regarding entailment. It did reject one thesis – the fourth. His logic did not accept the transitivity of entailment.[2] I argue for the acceptance of transitivity of entailment in Section 1.11 and in Chapter 7. But I set aside my disagreement with Smiley for a moment to discuss a methodological issue. Lewy and Smiley are right that there are certain intuitive theses and rules of inference that cannot be included in a logic of entailment without that logic's thereby containing some counter-intuitive theses or rules. Among these counter-intuitive theses are the so-called paradoxes of strict implication that are discussed at length in this book.

Lewy and Smiley attempted to weigh intuitions against one another to justify the logics that they accept. Smiley argued for his non-transitive logic on this basis and Lewy argued for C. I. Lewis's theory of strict implication, saying that its counter-intuitive consequences were "less counter-intuitive than the consequences of accepting any of the alternative theories" [116, p 134].

I find, however, that this weighing of intuitions, taken alone, is very unsatisfying. It may be that in any justification of a logical system, one has to make use of some intuitions. The problem with the use of intuitions, however, is that they can be malleable or ephemeral. As Lewy pointed out [116, p 108], Lewis had originally rejected the theses that all propositions entailed every tautology and that every contradiction entailed every proposition, but after producing his so-called independent proofs (see

---

[2] Others have also rejected the transitivity of entailment for similar reasons, in particular Neil Tennant [204] and David Ripley [181].

Chapter 2) he came to accept these supposed entailments. Lewis's argument was that the premises that he used in the independent proofs were so intuitive that they overruled his previous rejection of the paradoxes of strict implication. But as I also discuss in Chapter 2, Lewis's students Everett Nelson and William Parry rejected certain of Lewis's intuitions and created alternative logical systems that did not contain these paradoxical theses. This example illustrates the limitations of the method of playing off intuitions against one another. There seems to be no reason why someone else, who sees things slightly differently, will not have different weightings attributed to these various supposed logical principles and reject some of the premises of the independent arguments instead.

There are, however, other sorts of evidence that can be used. In this book, I appeal quite often to the jobs that I want a logic of entailment to do. If the logic is suited well to doing the tasks that are set out for it, this is good evidence for its acceptability.

1.3　　Using Proof Plans to Justify the Choice of Logic

The project of representing proof plans is one of the central motivators for creating a theory of entailment. The notion of a proof, as something that is certain and settled, justifies the two principles that I discuss in Sections 1.9 and 1.10.

The first of these is the principle of inclusion, which says that if a proposition A is in a theory, even after inferences are made, A remains in that theory. The point of making inferences from, say, axioms of a theory is to discover the propositions other than the axioms that are in the theory. But these inferences do not change the fact that the axioms are in the theory. I discuss this feature of entailment at greater length in Section 1.9, but here I wish to discuss a methodological issue. This motivation for the principle of inclusion is not a mere appeal to intuitions about proof, but an appeal to the reason for proofs and the structure of proofs as they are commonly understood. We commonly understand proofs about theories as a way of increasing our understanding of proofs by discovering more propositions that are in those theories.

When we appeal to the tasks that a logic must be able to accomplish, rather than just intuitions, we have a better (or at least a second) measure of the viability of that logic. It may be that a logic cannot do everything that seems to be included in its job description, but we can certainly talk about what makes a logic overall the best at doing its job. With regard to the formulation of proof plans, a logic's having a good proof theory is one of those things that makes it good at being the logic of proof plans.

By 'proof theory', I do not mean merely a good set of axioms, all of which seem intuitive, but rather a structure, like a natural deduction system or a Gentzen sequent system, which gives a basic structure into which particular rules fit (or fail to fit). And this structure (and the rules) must be able to be explained in terms of possible proof plans. Consider again the plan to prove that $A \rightarrow B$ and $Ax \rightarrow B$ in order to show that $Ax \rightarrow B$. Here is a Fitch-style natural deduction proof:[3]

---

[3] For a brief introduction to natural deduction of this sort, see Chapter 4.

1	$A \rightarrow B$	hypothesis
2	$Ax \rightarrow A$	hypothesis
3	Ax	hypothesis
4	$Ax \rightarrow A$	2, reiteration
5	A	3,4, $\rightarrow$E
6	$A \rightarrow B$	1, reiteration
7	B	5,6, $\rightarrow$E
8	$Ax \rightarrow B$	3–7, $\rightarrow$I
9	$(Ax \rightarrow A) \rightarrow (Ax \rightarrow B)$	2–8, $\rightarrow$I
10	$(A \rightarrow B) \rightarrow ((Ax \rightarrow A) \rightarrow (Ax \rightarrow B))$	1–9, $\rightarrow$I

The structure of this derivation both shows what needs further proof ($A \rightarrow B$ and $Ax \rightarrow A$) and what is assumed, as well as those that are to be used in a proof of B and, through its use of the entailment introduction and elimination rules it shows how the entailment connective represents the structure of the derivation. In this way, a good proof theory can show how an entailment connective can represent deducibility and show how the various principles of the logic are generated and show that they make up a coherent whole.

As I said above, the classical rule that makes any formula entail a theorem makes the logic of entailment of limited use. Thus, a proof theory that does not make this rule valid has, prima facie, an advantage over theories that do make it valid. It is for this reason that I adopt Anderson and Belnap's logic E and their natural deduction system for it. On a classically based natural deduction system, the following derivation might be valid:

1	$Ax \rightarrow B$	hypothesis
2	C	hypothesis
3	$Ax \rightarrow B$	1, reiteration
4	$C \rightarrow (Ax \rightarrow B)$	2–3, $\rightarrow$I
5	$(Ax \rightarrow B) \rightarrow (C \rightarrow (Ax \rightarrow C))$	1–4, $\rightarrow$I

If this is a theorem of the logic of entailment, then we have to accept the rule that $C \rightarrow (Ax \rightarrow B)$ is derivable from $Ax \rightarrow B$. The problem with this derivation is that the second hypothesis, C, is not used to derive $Ax \rightarrow B$. C is merely hypothesised and then in line 4 its hypothesis is discharged. It has not been used and therefore it has not earned the right to be discharged.

The mechanism by which Anderson and Belnap keep track of which hypotheses are really used in derivations is explained in detail in Chapter 4. That mechanism, however, requires further interpretation and explanation. This interpretation and explanation is given by the semantic theory that I present in Chapters 8–10.

1.4   Theory Closure

Another role for the logic of entailment is as a theory of theory closure. In this book, I claim that this function of the logic of entailment is not only one of its central functions, but understanding the relationship between entailment and theories is a key to understanding the meaning of the entailment connective itself.

Proofs and proof plans are often about theories. Logicians are often concerned with mathematical theories. Arithmetic and set theory are philosophers' favourites, but logicians have formalised other mathematical theories, such as algebraic and geometrical theories. The logical positivists and structuralist philosophers of science also had the goal to formalise the theories of physical and even social sciences.

I follow C. I. Lewis [106] and Richard Routley [182] in holding that theories, such as scientific theories, are closed under the principles of entailment. As I have said, logicians and others often prove things about the contents of theories. Among other things, I think that reflecting on the nature of theories allows us to find important restrictions to place on a theory of entailment. For example, in pre-1750s chemical or alchemical theories, there is no mention of oxygen or hydrogen. We cannot attribute to them the idea that 'lakes are filled with water' means the same thing as 'lakes are filled with H_2O', nor can we even attribute the latter sentence to theories of this period.

This example shows that theories cannot be understood as sets of propositions, at least not in the contemporary view of propositions. Moreover, it shows that a logic of entailment is not closed under replacement for identical propositions. This means that just because p and q happen to express the same proposition (in the contemporary sense), it may not be that the formula $p \leftrightarrow q$ is a theorem of the logic, where the double-ended arrow is co-entailment. I deal with this issue in Chapter 10 by introducing a metaphysically thin notion of proposition, as opposed to the metaphysically thick conception of proposition that is discussed in much of the contemporary literature.

I use a double turnstile, $\Vdash$, to represent the relation under which theories are closed. This is one of three consequence relations I associate with a logic. The turnstile, $\vDash$, is used to represent semantic consequence. And the single turnstile $\vdash$ is used to indicate theorems (as in '$\vdash A$', which means 'A is a theorem') and is such that '$A_1; \ldots; A_n \vdash B$' means '$\vdash A_1 \rightarrow (\ldots (A_n \rightarrow B) \ldots)$'. Where Γ is a set of formulas, $\Gamma \Vdash C$ if and only if there are some formulas $G_1, \ldots, G_n$ such that $G_1 \wedge \ldots \wedge G_n \vdash C$. In classically based logics (such as the standard modal logics) and intuitionist logic, $\vdash$ and $\Vdash$ are the same. But this is not true for relevant logics, as I show in Chapter 4.

For my project, it is crucial that the consequence relation (understood syntactically or semantically) is compact. That is, if C is derivable from a collection of premises Γ, then there is a finite sub-collection Γ', such that C is derivable from Γ'. In Chapters 6 and 10, I reply to arguments due to Alfred Tarski and Alonzo Church for the thesis that the consequence relation should not be compact. Their arguments are meant to show that any theory of theory closure must be essentially infinitary, that some inferences with infinitely many premises must be included in this theory that cannot be truncated in any way to finitary inferences.

I find the connection between theories and the logic of entailment to be particularly fruitful. I use the job of theory closure to indicate which properties a logic of entailment needs to have. And I use Tarski's concept of a *theory closure operator* to help provide a semantics of nested entailments. Each theory determines a consequence operator. For example, if T_1 is a theory that contains the formula $A \rightarrow B$, then the consequence operator C_1 is such that, if T_2 contains A, then B is in $C_1(T_2)$. The relationships between these consequence operators and theories can be seen to characterise a logical system. In Chapters 7 and 8, I show how natural conditions on theories and consequence operators can produce semantics for two relevant logics: one which I call *Generalised Tarski Logic* (GTL) and the other is Anderson and Belnap's logic E. Postulates governing the behaviour of consequence operators in relation to one another (and to themselves) give us different properties of nested entailments.

1.5 Entailment and Metaphysics

Like other theories, metaphysical theories are closed under true entailments. But there are other uses of the logic of entailment in metaphysics. The entailment connective has been used to formulate important theses in metaphysical theories. For example, G. E. Moore coined the term 'entailment' with its current philosophical meaning in order to formulate a theory of internal properties and relations. J. M. Dunn modified Moore's theory by using a relevant logic of entailment to develop a theory of "relevant predication" [3, 54, 55]. The concept of entailment has also been used in the contemporary theory of truthmakers [118, §1.1]. A truthmaker is something that makes a sentence or a proposition (i.e. a truth-bearer) true. Truthmakers have been said to entail the truth of their corresponding truth-bearers.

I don't think that any of these three uses are compatible with the use of entailment to capture the structure of proof or the structure of theories. I suggest, rather, that a different semantics should be used to understand entailment as it is used to capture these metaphysical notions. Hence, I claim that a different logical notion should be used than the notion of entailment that is used to understand proofs and theories.

My reason for this is quite simple. For any of these metaphysical purposes, one needs more metaphysically substantial connections between antecedent and consequent than when thinking about theories or proofs. For example, both Moore and Dunn tried to develop theories of what is necessarily contained in the concept of a thing. They both formalise the idea that the property P is essentially internal to the concept of a thing i if and only if the following formula is true:

$$\forall x(x = a \rightarrow Px),$$

where $\rightarrow$ is some form of entailment. I think most philosophers would now think that it is essentially internal to the notion of being a platypus that it is a monotreme (a mammal that lays eggs). In particular, it is essentially internal to Pauline the Platypus that she is a monotreme, and so

$$\forall x(x = pauline \rightarrow Mx).$$

It was, however, an empirical discovery that Platypuses were monotremes. A theory that predates this discovery should not be closed under this entailment. Hence, whatever sort of logic of entailment that formalises the notion of essential containment in this sense is different from the one that is a universal tool for the closure of theories.

The case with regard to truthmakers is rather similar. In formulating his version of the so-called truthmaker principle, John Bigelow said:

> Whenever something is true, there must be something whose existence entails that it is true. The 'making' in 'making true' is essentially logical entailment. [15, p 125]

Formalising this notion, what Bigelow said is that m is a truthmaker for p, then

$$m \text{ exists} \rightarrow p \text{ is true.}$$

Clearly, we should have a truthmaker version of Tarski's principle T, that is, for example,

$$\text{The fact exists that water is wet} \rightarrow water\ is\ wet \text{ is true.}$$

I use 'fact' and 'truthmaker' here as synonyms. Now, let's consider the following facts:

$$\text{water is wet} \qquad H_2O \text{ is wet}$$

On most current philosophical accounts, I think, these are the same fact. Hence, we should have

$$\text{The fact that } H_2O \text{ is wet exists} \rightarrow water\ is\ wet \text{ is true.}$$

Once again, the sort of entailment needed here is too metaphysically loaded to provide an adequate treatment of the closure of theories. For that task, an entailment that can make much more fine-grained distinctions is required.

There are, therefore, at least two distinct notions of entailment: one that treats theory closure and proof plans and another that deals with metaphysical closures of various kinds.

1.6   Problem: Use and Mention

Some philosophers have claimed that treating entailment as a connective is a mistake. W. V. O. Quine was perhaps the most prominent of those who made this objection. He said (using 'implication' in the same way that I am using 'entailment'),

> [C. I.] Lewis, [H. B.] Smith, and others have undertaken systematic revision of '⊃' with a view to preserving just the properties appropriate to a satisfactory relation of implication; but what the resulting systems describe are actually modes of statement composition – revised conditionals of a non-truth functional sort – rather than implication relations between statements. [171, p 32]

Quine had a good point. To say that a statement B follows from a statement A is to posit an entailment relation that holds between A and B. The expressions '$A \rightarrow B$' and '$A \Vdash B$' have very different logical forms. In '$A \rightarrow B$', the statements 'A' and 'B' are subformulas and there is no reference to them as sentences. In '$A \Vdash B$', there is reference to the sentences 'A' and 'B' themselves and the notion of entailment represented is a binary property of sentences (see also [169]).

For example, according to Quine, it was correct to say

'Every dog is mortal' entails 'all Spitz dogs are mortal',

but it was incorrect to say

Every dog is mortal entails that all Spitz dogs are mortal.

Thus, it is a grammatical mistake to try to construct a logic of entailment [171, p 28].

In an appendix to the first volume of their book, *Entailment*, Alan Anderson and Nuel Belnap reply to Quine by arguing that there is no grammatical error in treating 'entails' as a connective. They point out that a relation on sentences can be converted in ordinary English to a sentence operator rather easily. Their example of a monadic sentence predicate that can be converted to a sentence operator (i.e. to a connective) is quite clear.

'Tom is tall' is true.

In this sentence, 'is true' is a predicate, which holds of the sentence 'Tom is tall'. The string of words, 'Tom is tall is true', is not a sentence. For one, it has two main verbs. The expression 'is true', however, can be converted rather easily into a connective:

That Tom is tall is true

This is a well-formed sentence. In the expression, 'that _ is true', one can replace the blank with a sentence and obtain a sentence. Thus, it is an operator that takes sentences to sentences, that is to say, it is a connective [4, pp 479–480]. Even if 'entails' is a relation between sentences, in the same manner it can easily be converted into a connective:

That every dog is mortal entails that all Spitz dogs are mortal.

The expression 'that _ entails that _' is a binary connective.

Another approach to dealing with Quine's complaint is to claim that entailment is a relation between *propositions* instead of sentences. Thus, if we have a relation Ent such that '$Ent(A, B)$' means that the proposition that A is entailed by the proposition that B, then '$Ent(A, B)$' itself can express a proposition that can all be expressed by '$A \rightarrow B$'. As Quine said, 'entails' "would come to enjoy simultaneously the status of a binary predicate and the status of a binary sentence connective" [171, p 32]. Quine rejected this idea on the grounds that the entities needed (i.e. propositions) are too obscure and should not be postulated.

Alasdair Urquhart replied to this latter worry in his article, "Intensional Languages via Nominalization" [206]. In that paper, Urquhart shows how to construct a semantics for a modal logic in which '$\Box$' is taken to mean 'it is a tautology that'. The resulting logic is called TS for "Tautology System". In the syntax and semantics of TS, sentence names are *nominalised*. This means that, when they are arguments of the representations of expressions like 'it is necessary that' and 'entails', they are numerals that represent the Gödel numbers of the expressions that they nominalise. So, 'It is necessary that A' is represented as $\Box g(A)$, where $g(A)$ is the (numeral representation of the) Gödel number of A. Unless Quine wished to claim that numbers are obscure entities, he would have had to accept that Urquhart's construction allowed 'entails' and other intensional idioms to enter into our logical language.

I think, however, that Quine had another worry in mind in accusing intensional logicians of confusing use and mention. The logicians whom Quine was attacking, C. I. Lewis and Rudolf Carnap in particular, thought of their necessity operators as representing logical truth. For them (in more modern notation), '$\Box A$' meant 'A is logically true' or, to follow Anderson and Belnap, 'It is logically true that A'.

For Quine, logical truth was just generality.[4] Quine said that "a logical truth is a statement which is true and remains true under all reinterpretations of its components other than the logical particles" [168]. I think that, for Quine, logical truth was not primarily a property of sentences, but rather of schemes. A scheme is a formula that contains metavariables representing grammatical particles such as statements. Consider the scheme, $A \equiv B$. We cannot say that '$A \equiv B$' is logically true, logically false, or logically contingent, because it does not have any truth value at all unless 'A' and 'B' are given an interpretation. By themselves, the expressions A and B have no truth values. To deal with this problem, Quine introduced his quasi-quotation marks. The expression, $\ulcorner A \equiv A \urcorner$, for example, is true for all its substitution instances and so is logically necessary, and $\ulcorner A \equiv B \urcorner$ is logically contingent because some of its substitution instances are true and others are false [171, p 35].

In order to use Urquhart's device to reply to Quine's worry, we would have to introduce Gödel numbers for schemes. This would require a formalised metalanguage (with a finite primitive vocabulary). It would be possible to do this. Instead of having infinitely many "atomic" schematic letters, $A, B, C, \ldots$, as we usually do in informal metalanguage presentations, we could have a single basic letter A with superscripted primes – $A_{\prime}, A_{\prime\prime}, A_{\prime\prime\prime}$, etc. Then what we get, if the appropriate semantics is given to this language, is a logic with some implicit propositional quantification. The sentence $\Box g(\ulcorner A \urcorner)$ says that the scheme A is logically true, and hence that all instances of A are logically true. This seems rather complicated, but it seems doable.

Another, and I think better, way to avoid Quine's problem is to reject his view that schemes, rather than sentences, are the real logical truths. We can still maintain that a sentence is logically true only if all of its substitution instances (for its so-called non-logical particles) are true. Schemes are convenient means of expressing logical truths, but they are not the primary bearers of logical truth. In his "Syntactic Construction of Systems of Modal Logic", J. C. C. McKinsey treated modality in just this way. He evaluated formulas relative to a set of substitutions. A substitution associates with each propositional variable a formula of classical propositional calculus (i.e. one that

---

[4] Russell held this too (see Section 2.2).

does not contain any modal operators). The formula $\Diamond p$ is true because there is at least one formula that could be substituted for p that is true. And $\Box(p \lor \neg p)$ is true because no matter what formula is substituted for p, $p \lor \neg p$ comes out true [46, 140]. We could interpret entailment in this way too. We could say that $A \rightarrow B$ is true if and only if for every substitution that makes A true also makes B true. The result is a normal classical modal logic. I argue in Chapter 3 that this sort of logic is inadequate to represent logical entailment, but not because it runs afoul of any strictures governing use and mention.

I do not think, however, that all this formal machinery is really necessary. Quine is right that '$A \Vdash B$' mentions the formulas A and B in a way that '$A \rightarrow B$' does not. But, given the right semantics we can see that there is a correlation between true entailments and valid deductions, and this semantics should explain why there are these correlations. In this way, entailments do express the validity of sequents and the use-mention problem is a pseudo-problem.

This representation of elements of the metalanguage in the object language raises other interesting issues. We can make explicit that we are talking about sentences if we add names of sentences to the language and represent entailment as a predicate rather than a connective. '$E(x, y)$' could mean 'x entails y', where x and y range over sentences. As I discuss in Chapters 5 and 10, the explicit addition of names for sentences together with certain widely held logical theses, can lead to difficulties such as Jc Beall and Julian Murzi's "validity Curry paradox" [12].

1.7   Logicality

As I say in Section 1.6, Quine thought of logical necessity as generality. Although I do not think that logical necessity is merely generality nor do I think that entailment is reducible to universally true implication, I do think that generality has something to do with logical entailment. A logic of entailment, in my view, has to be *formal*, but what exactly this means needs to be spelled out in some detail.

In his PhD dissertation, John MacFarlane set out three different senses of the claim that logic is formal. This taxonomy has become quite influential and so I think it will be helpful to use it to situate my own view. These were MacFarlane's three senses of 'formal' [121, ch 3]:

1. Logic is formal in the sense that its norms are "constitutive of thought as such";
2. Logic is formal in the sense that it is "indifferent to the particular identities of objects";
3. Logic is formal in the sense that it "abstracts entirely from the semantic content of thought".

The first sense of 'formal' is to be found in Kant's writings. Kant said that, for example, a thought that violated the law of non-contradiction was not actually a thought. Thinking would cease to be thinking if it violated any laws of logic. The third sense could be found in Kant and also in the Wittgenstein and the logical positivists.

Wittgenstein and the Positivists held that the laws of logic had no semantic content. This view is incorporated into Carnap and Bar Hillel's theory of information too. On that theory, the amount of information carried by a proposition was proportional to how implausible it is. A contradiction carried the maximum amount of information and a tautology carried no information at all.

The second sense of 'formal' is the one that I attribute to the logic of entailment. The idea is very old. It was held by several mediaeval philosophers, in particular by Peter Abelard and Jean Buridan. Both Abelard and Buridan were trying to elucidate Aristotle's distinction between perfect and imperfect deductions [6, 24b23]. Aristotle called a deduction perfect if nothing had to be added to it to make evident that the conclusion followed from the premises. Abelard and Buridan give similar examples of imperfect deductions. One of Abelard's examples was "If every man is an animal, every man is alive" [121, p 281] and one of Buridan's was "A human is running, so an animal is running" [27, ch 14].

Abelard's gloss on the imperfection of these sorts of inferences was that they relied on the nature of the subjects and predicates involved rather than just on the logical form of the expressions. Buridan said that an imperfect deduction was valid in a sense because it was impossible for its premise to be true and conclusion false, but it is flawed in that not every deduction of the same form is valid [27]. If we substitute 'rock' for 'human' in Buridan's example, the argument is no longer valid even in the weaker sense.

I think that the logic of entailment should be topic neutral in this way for pragmatic reasons. I do not wish to treat conditionals such as 'if x is water, then it is H_2O' as logical entailments. Taking them to be entailments can interfere with one of the jobs I think is important for the logic. A logic of entailment is, among other things, a theory of how theories should be closed. Forcing all theories to be closed under these sorts of analytic conditionals may interfere with the content of theories. We do not want to close a theory about phlogiston in such a way that it talks about oxygen. For this reason, I think that the logic of entailment is formal in MacFarlane's second sense.

As John Etchemendy [59] pointed out, it was the hope of some logicians that by designating a set of "logical constants" (the usual connectives and quantifiers), they could give a reductive analysis of logical truth and entailment in terms of actual general truth. For example, 'every cat is either a cat or a dog' is logically true because its generalisation, $\forall F \forall G \forall x (Fx \supset (Fx \lor Gx))$ is true. I suggest this was true of Quine (see Section 1.6) and Russell (Section 2.2), and Etchemendy pointed out that it was in Tarski as well. Etchemendy argued, successfully I think, that demarcating logical from non-logical vocabulary alone could not provide a successful reductive account of logical as generality, nor could it provide the basis for a logic of entailment.

Etchemendy suggested, moreover, that the choice of certain pieces of vocabulary as logical was arbitrary. I think that Etchemendy was right that, if we rely on our intuitions about logical necessity and derivability alone, then it seems there is nothing special about a *formal* necessity such as '$\forall F \forall G \forall x ((Fx \land Gx) \supset Gx)$' as against an informal necessity such as 'every red thing is coloured'. I think the problem here, however, is that Etchemendy uses as his only benchmark of truth the compliance of a theory with a set of intuitions about logical truth and logical consequence. I use as evidence for the correctness of my view its ability to do certain tasks that a logic of entailment is supposed to do.

These tasks are to act as a general theory of theory closure and to provide a logical tool to analyse proof plans. To do the latter, I argue, we need an entailment connective. To do the former, we need to represent not just what each sentence in a theory entails, but also what a collection of sentences jointly entail. As we shall see, representing the closure of theories requires that we have a conjunction that has certain properties. Thus, the logic has to have an entailment and a conjunction, and both of these need to have certain properties (discussed in the proceeding sections). I put forward a logic with just these two connectives and I call it the "core logic of entailment" (see Chapter 8).

I hold that all theories should be thought of as closed under the logic. In order to do so, the logic has to be indifferent to actual and even some metaphysically necessary features of things. So, given the need for substantive principles governing entailment and conjunction (at least) in order to provide an adequate logic of theory closure and the need to be indifferent to the metaphysics of the subject matter, there is a distinction that we can make between the logical and non-logical vocabulary in statements.

1.8   Entailment and Necessity

Since the dawn of the study of logic in the Western tradition at least,[5] logicians have thought of the notion of logical consequence as incorporating some notion of necessity. In the *Prior Analytics* Aristotle wrote that

A deduction is a discourse in which certain things being stated, something other than that which is stated follows of necessity from their being so. [6, 24b18]

Apart from those logicians who question the very coherence of the notion of necessity, such as Russell and Quine, it has been accepted universally that logical consequence incorporates some notion of logical necessity. Even in introductory lectures on logic we appeal to some form of logical necessity. We typically tell students that a valid argument is one in which it is impossible for the premises to be true and the conclusion false. I follow the mainstream in the theory of entailment by claiming that true entailment sentences reflect the necessity of the corresponding valid deductions, that the entailment connective embodies this sort of necessity. What this necessity consists in, however, is quite a difficult issue.

Many philosophers equate logical necessity with metaphysical necessity. Perhaps, the most vocal modern proponents of this view were Ludwig Wittgenstein (in the *Tractatus*), David Lewis [115], and Frank Jackson [88]. There are, however, other views of the necessity involved in logical consequence. Michael Dummett viewed

---

[5] There is also a debate about the existence and nature of logical necessity in Classical Indian philosophy but I am certainly not enough of an expert to contribute in any way to that debate. So I do not discuss it here.

logical consequence in terms of the transmission of justified assertion. We can understand the sort of necessity that he postulated as a form of epistemic necessity. Another interpretation of logical necessity has been suggested by Greg Restall. On Restall's view, a classical sequent,

$$A_1, \ldots, A_n \Vdash B_1, \ldots, B_m,$$

is to be understood as saying that one should not simultaneously assert all of $A_1, \ldots, A_n$ and deny all of $B_1, \ldots, B_m$ [179]. In the sort of inferences that I am concerned with in this book, where there is only a single conclusion, a sequent $A_1, \ldots, A_n \Vdash B$ is said to be valid if and only if one should not simultaneously assert all of $A_1, \ldots, A_n$ and deny B. In this view, norms that are represented by valid sequents express a notion of deontic necessity. I discuss this view in more depth in Section 3.8.

In Chapter 3, I discuss attempts to base a theory of entailment on a theory of logical necessity and possibility. In particular, I discuss attempts to give a possible worlds semantics for logical necessity and treat entailment as truth preservation on all logically possible worlds. I argue there that such approaches do not give an adequate treatment of nested entailments. Even though I find this sort of worlds treatment of entailment wanting, I do think that any adequate theory of entailment (like any adequate theory of logical consequence) does need to capture some notion of logical necessity. In Chapter 8, I discuss the relationship between the logic of entailment and logical necessity and argue that the rather weak sense of necessity captured in the logic is adequate.

1.9   The Principle of Inclusion

I hold that there are four core principles of entailment. They are the principles of inclusion, monotonicity, transitivity, and compactness. Each of these is controversial, so I will spend a few words formulating and defending them. I start with the principle of inclusion. (This principle is sometimes called the "principle of reflexivity", but I use "reflexivity" for another property, see Chapter 5.)

The principle of inclusion says that any sentence in a set of sentences is a logical consequence of that set, follows from that set. In formal notation, the principle of inclusion is

$$\frac{A \in \Gamma}{\Gamma \Vdash A}.$$

The reason for accepting inclusion is quite obvious. In a proof about a theory, one often makes use of the axioms or other known theorems of that theory. In so doing, one is in effect appealing to the principle of inclusion.

There have been, however, logicians who have rejected inclusion. Robert Meyer and Errol Martin said:

Which arguments are valid? This has been the central question of logic. "Reasoning is an argument in which, certain things being laid down, something *other than these* necessarily comes about through them," said Aristotle (*Topics* 100a 25–27). The emphasis is ours. "He who repeats himself does not reason," as Strawson correctly notes. The fallacy of concluding what one has assumed is almost universally condemned. Some of the rubrics under which it is condemned are the following: *circular reasoning, begging the question, petitio principii.* [145]

If we accept inclusion, we also have to accept that the proof plan, $A \rightarrow A$, is valid. Meyer and Martin said that we should not accept a proof plan of this form. This form is the form of what has traditionally been thought of as a logical fallacy – begging the question.

At this point, it may seem that I have stumbled upon a place in which the logic of proof plans diverges sharply from the logic of theory closure. It is clear that in thinking about theory closure the principle of inclusion is required. Given some statements from a theory, such as its axioms (if it has any), one can infer various things about what that theory contains, including those statements themselves. But a proof plan of the form $A \rightarrow A$ might seem useless. In fact, such proof plans may seem harmful. We think it is important that when proving something the conclusion should not appear in the premises. One who sneaks his conclusion into his premises would be called a charlatan of some sort, or at best declared to be sloppy. I think, however, that this divergence between the logics of proof plans and theory closure is merely apparent rather than real.

Consider the following example. Suppose that there is a logical conflict between A and B, that is, it is true that $\neg(A \wedge B)$. Suppose also that we know that it is a theorem that A entails B. Then, we know that A cannot be true, that is $\neg A$ obtains. One way of representing the inference this is:

$$\frac{A \Vdash B}{\neg(A \wedge B) \Vdash \neg A} \ .$$

This seems a good inference. What is going on here, I maintain, is that once we know that B follows from A, we also know that $A \wedge B$ follows from A, that is

$$\frac{\dfrac{A \Vdash B}{A \Vdash A \wedge B}}{\neg(A \wedge B) \Vdash \neg A} \ .$$

How do we know that the argument above is valid? One good reason is because the following inference is obviously valid:

$$\frac{\dfrac{A \Vdash A \qquad A \Vdash B}{A \Vdash A \wedge B}}{\neg(A \wedge B) \Vdash \neg A} \ .$$

We tacitly assume the validity of $A \Vdash A$ in this and many other inferences. We do not often use inclusion overtly, but we do sometimes use it tacitly.[6]

---

[6] A referee correctly pointed out that there may be many ways to derive that $A \Vdash A \wedge B$. But, note that if we accept that any of these derivations is valid, then we can infer that $A \Vdash A$ from $A \Vdash A \wedge B$ and conjunction elimination. Thus, if every formula implies some formula or other, then inclusion is valid. The point here is that it really is very difficult to avoid making inclusion valid.

1.10   Monotonicity

The principle of monotonicity says that if a statement follows from a set of premises, then it still follows if we add further premises. Formally stated, this is what the principle of monotonicity says:

$$\frac{\Gamma \Vdash A \qquad \Gamma \subseteq \Delta}{\Delta \Vdash A}.$$

Recall that in '$\Gamma \Vdash A$', we think of the premises (the members of Γ) as being conjoined to one another. As I say in Chapter 4, in relevant logic there is a sense in which the premises of an inference are not treated as being conjoined to one another and on this notion of inference, monotonicity fails. For the present discussion, however, I put aside this relevant notion of consequence ($\vdash$) and discuss only the relation, $\Vdash$.

There is an important challenge to monotonicity that takes the most general form of reasoning to be a type of counter-factual reasoning. Counter-factuals are not monotonic. It does not follow from 'If Nova had been good, then Sue would have been happy' that 'If Nova had been good and there were a major earthquake in Wellington, then Sue would have been happy'. Daniel Nolan puts forward a view of this sort [154, 155]. He suggests that reasoning about, for example, inconsistent situations is like reasoning counter-factually. He says:

imagine I reasoned by saying "suppose naive set theory were true. Then the Russell set is a member of itself and not a member of itself. So at least one set is a member of itself and not a member of itself". One way to understand what has happened is that the effect of the reasoning is much the same as before: I have not categorically supported the conclusion that at least one set is a member of itself and not a member of itself: instead, the force of my conclusion is captured by "if naive set theory were true, at least one set is a member of itself and not a member of itself". [155, p 422]

On Nolan's view, there is no general logic of entailment. If one considers the hypothesis that some alternative logic holds, then one cannot use any of her usual logical principles that are not contained within this alternative system. This view of deduction and the correlated view of entailment do not allow for a single logic of entailment (except for the logic with no generally valid principles).

Thinking about deduction in this non-monotonic fashion, however, clashes with our ordinary notion of proof. Let's say that one proves something from some (but not all) of the axioms of a theory. If this is an actual *proof*, then the mention of further axioms cannot show that the conclusion is false. But, if the relation of provability is non-monotonic, then we can undermine conclusions in this manner. The notion of proof that I am discussing, then, requires that entailment be monotonic.

I return to the subject of non-standard logical theories at the end of Chapter 8. There, I suggest that we can retrench our logic in terms of the negation and disjunction principles that we accept when faced with certain non-standard theories. In *Relevant*

Logic, I suggest that when faced with the need to reason in a very non-standard way, we should think of this in a metalinguistic manner. That is, our reasoning in such situations is in effect in the scope of an "according to the logic such-and-such ..." operator. While reasoning in the scope of such an operator, only the inferential moves licensed by the logic mentioned are legitimate. But such reasoning is, in a sense, playing a formal game – one manipulates the symbols of the language in the way that is set down by the proof theory or by the model theory of the logic. In different situations, I think that it is appropriate to choose one and not the other of these solutions, but I cannot see that there are situations in which neither is adequate. Thus, I accept a monotonic notion of proof.

1.11 Transitivity

The third constraint that I place on a consequence relation is *transitivity*:[7]

$$\frac{\Gamma \Vdash A \qquad A \Vdash B}{\Gamma \Vdash B}.$$

In Chapter 6, I give an argument for transitivity in this form and in Chapters 7 and 8 I give arguments for stronger forms of transitivity. These arguments have to do with closure operators used to construct theories. This is a technical topic and is better left until after the required formal background is discussed. Here, instead, I briefly present an intuitive and informal motivation in favour of transitivity broadly construed.

Our notion of proof seems to include a notion of steps towards a proof. In reasoning about empirical theories, we might have an entailment between a theoretical postulate and a possible observation. In mathematical proof plans, steps in reasoning are axioms or lemmas. When we derive a consequence from a set of axioms, say, we add it to the axioms and allow it to be used in further proofs. A theory is what we construct, or rather what could be constructed in ideal circumstances, out of all the consequences drawn in this way. A consequence that is unwanted is a purported counter-example to the theory. The fact that it is done in one step or in two hundred steps makes no difference to its status as a counter-example.

The concept of a step in a proof seems integral to the structure of proof plans. We think of these steps to be taken one after another in a logical progression towards the desired conclusion. Sometimes the way in which lemmas are put together to derive the conclusion is not sequential or the proof has a more complicated structure than the transitivity rule given above, but that does not undermine the claim that this simple rule is required by our ordinary notion of proof.

---

[7] David Ripley claims that the property that I set out is not really a transitivity property, but a "linking property" that he calls "KS$_{SF}$" [181]. It does not matter for my purposes whether this is referred to as transitivity rule or a linking property.

1.12   Compactness

One of the most important properties a consequence relation must have in the context of the current project is *compactness*. A consequence relation, $\Vdash$, is compact if and only if for every set of formulas Γ and every formula A, if

$$\Gamma \Vdash A,$$

then there is some finite subset of Γ, Γ', such that

$$\Gamma' \Vdash A.$$

The reason that this is so important is that the connectives of the language that I use are finitary. I represent the derivability of A from the set Γ by the entailment of A by some conjunction of formulas of Γ. If there is no finite set of formulas in Γ from which A can be derived, then I cannot represent the derivation of A from Γ in this way. I discuss the possibility of constructing an infinitary logic of entailment in Chapter 6, but the central point is that human inference is in an essential sense, finitary. I do think we can *appeal to* infinitary rules, but we cannot use them. We might discuss a version of the theory of arithmetic that is closed under the infinitary omega rule, for example. We cannot use the omega rule, but we can appeal to it to say, for example, that Gödel's theorem is not provable of this theory. I make clear this distinction between using a rule and appealing to it in Chapters 6 and 11.[8]

Another way of dealing with the issue of the failure of compactness is to add names for sets of formulas to the language. Let γ be a name for a set of formulas, Γ. Then, we could represent $\Gamma \Vdash A$ by $K(\gamma) \rightarrow A$, which means that the conjunction of formulas in Γ entails A. Although this avoids the move to an infinitary language, it still has problems. If there are $\aleph_0$ many formulas, then (assuming the continuum hypothesis) there are $\aleph_1$ many sets of formulas, and hence we would need at least $\aleph_1$ many names in the language. If there are $\aleph_1$ many names, then there are at least $\aleph_1$ many formulas that can be constructed from them and so there would then be $\aleph_2$ many sets of formulas. Then, we would need $\aleph_2$ many names. And so on, up the hierarchy of the transfinite cardinal numbers. Now, suppose that $\Gamma \Vdash A$ if and only if $K(\gamma) \rightarrow A$ is a theorem of the logic, for all sets of formulas Γ. Then, there is something interestingly finitary about the logic: for every valid sequent, there is a finite sequent (and a valid formula) that represents it. This is an interesting approach, but difficult to formulate and it is essentially unaxiomatisable. And so I set it aside.

For these reasons, I place the constraint consequence relations that are to be considered as the basis for a theory of entailment that they be compact.

---

[8] Sometimes this property is called "finitariness" rather than "compactness", where the latter term is reserved for the property that a model or class of models have according to which a set of formulas is satisfiable if and only if every finite subset is satisfiable. Classically, finitariness and compactness are equivalent, but they are not equivalent in all non-classical frameworks [86, ch 1].

1.13 Method

An entailment $A \rightarrow B$ means that B is deducible from A. So, $A \rightarrow B$ is true if and only if B is deducible from A. Any adequate theory of entailment has to make plausible that this biconditional obtains. It may seem very straightforward to do this, but in fact it can be rather tricky. Consider, for example, a theory that claims that the deducibility relation of the modal logic S5 is the one true deducibility relation. On this theory, $A \rightarrow B$ is true if and only if B is deducible from A is S5. What we need here is a model in which all and only the S5 valid inferences are reflected in true entailments. We cannot use any of the usual models for S5. Consider one such model and pick a random world w to be the actual world. Let p be a formula that is true in w. Then, at w, the entailment

$$\Box \neg p \rightarrow B$$

holds for every formula B. The corresponding deduction,

$$\Box \neg p \Vdash B,$$

however, is not valid in S5 for all formulas B. So this is not a model in which every true entailment expresses an S5-valid deduction. As I explain in Chapter 3, those who treat entailment in terms of a possible worlds semantics have tended to accept extensions of S5 as the logic of entailment.

In Chapters 7–10 and the Appendix, I argue that there is a model (the "intended model") that makes all and only true entailments true. And this is a model for the logic E.

The development of a model theory for the logic also has another goal. In my theory, the model theory is a semantics for the proof theory. The model theory explains the proof theory. In explaining the proof theory, the semantics also justifies the proof theory. The semantics demonstrates the rationale for the various rules of proof. But the fact that the proof theory is elegant and to a large extent intuitive also gives some justification to the model theory. It is justified by being a semantics for that proof theory.

Although intuitiveness is an important criterion for the choice of a logic of entailment, it is not the only criterion. Other criteria include:

1. The entailment connective of the logic must express a deducibility relation that has plausibility independent of being expressed by the connective;
2. The treatment of nested entailments must be reasonable;
3. The logic must prove as theorems the formal representations of paradigm cases of good proof plans.

These three criteria have to do with the logic being able to do the job of a logic of entailment. As such, our choice of a logic of entailment cannot fail to have any of these three properties. If a logic does fail to do any of these jobs, then it is not really a logic of entailment. A logic of entailment has jobs to do and it had better do them.

One might find my adherence to criterion three surprising, because I am an advocate of relevant logic. No relevant logic contains all instances of the following scheme:

$$((A \lor B) \land \neg A) \to B.$$

But, surely (an objector might say), the following is a paradigm of a good form of a proof plan: if I can prove $A \lor B$ and $\neg A$, then I can prove B. I argue in Chapter 11 that this proof form, that is disjunctive syllogism, is to be understood not as a universal proof form but only as a rule used in the construction of some theories. In making inferences about those theories, we can appeal to disjunctive syllogism, but we cannot do so in making inferences about theories in general. I leave the argument for that view for later, but here I would like to make a point about method. Denying that a particular rule – even a widely accepted rule – is a good deductive principle is allowed in the search for a logic of entailment as long as one can explain why that rule seems falsely to be a good rule. In the case of disjunctive syllogism, the fact that the creation of consistent theories is a widely accepted goal of theory creation in the formal and empirical sciences makes the appeal to disjunctive syllogism so common that it seems to some to be a generally applicable rule of inference.

Satisfying the criteria listed above does not uniquely determine a logic of entailment. We need to appeal also to the theoretical virtues such as elegance, strength, and, of course, intuitiveness, in deciding what logic to accept. This may leave some room for a form of logical pluralism about entailment. I have chosen, however, to avoid discussing the topic of pluralism in this book because the literature on pluralism is now very extensive and engaging it would mean writing a very different book than the one I have in mind.

1.14   Plan of the Book

The book is divided into three parts. Part I gives a history of logics of entailments. It is not meant to be a complete history of the subject. Rather, it looks at some of the more important and influential approaches to entailment and in particular those that influenced me in my adoption and interpretation of Anderson and Belnap's logic E. Anderson and Belnap's logic is itself the subject of one of the chapters in Part I. This inclusion, of course, introduces the logic, but it also argues that their logic was, despite their proof theory, in need of further interpretation. In Part II of the book, I introduce Tarski's consequence operators and then employ them to construct a model for a simple logic that I call Generalised Entailment Logic (GTL). The idea is to use the notion of theory closure, which is what is produced by a consequence operator, to act as the basis for the semantics of entailment. Part III begins with an argument that GTL is too weak to be treated as the logic of entailment. In Chapter 8, a semantics for the implication and conjunction fragment of E is presented, and in Chapters 9 and 10, this semantics is extended to treat all of propositional and first-order E. In the final chapter, I look at the integration of E and its semantics into a more general view of deductive inference.

Part I

Entailment in the Twentieth Century

2   C. I. Lewis and His School

A Guide to Part I

This is the first chapter of Part I of this book. In Part I, I provide a historical background for the theory that I present and defend in Parts II and especially III. In Part I, I look at four approaches to the construction of a logic of entailment. I find three of these approaches wanting and one promising, but rather incomplete.

I first examine the difficult history of the construction of C. I. Lewis's logics of strict implication. I consider Lewis's 1912 article as the first attempt to produce a formal logic of entailment. One thing that is interesting about Lewis's method is that he justified or rejected individual theses about entailment largely on the basis of his own intuitions. This led him in his early writings to formulate a logic that was not a real candidate for the logic of entailment because it collapsed into classical propositional logic. The logics he eventually accepted (S3 in 1920 and S2 from 1932) still made valid the paradoxes of strict implication, and this seemed unsatisfactory even to Lewis's own students, Everett Nelson and William Parry. Their logics, however, also had serious problems, and I suggest it was partly because of their adoption of Lewis's axiomatic method of logic construction.

Perhaps the best-known and most widely accepted alternative to the pure axiomatic method is to start with a model theory that seems to represent the concept one has in mind and then discover what logic is characterised by the model theory. This was the approach that Rudolf Carnap adopted in the mid-1940s as the basis for his modal logic. In Section 3.1, I examine this approach and the ones that followed on from it, such as Cocchiarella's. The logic that is characterised by their semantics is an extension of S5. While I have no quarrel with the idea that S5 gives a good treatment of the notion of logical necessity, I do not think that its strict implication can be taken to be logical entailment. Its treatment of nested conditionals is inadequate.

The standard alternative to the model-theoretic treatment of logics is the proof-theoretic treatment. In Chapter 4, I look at Anderson and Belnap's use of a system of natural deduction to formulate and justify a logic of entailment. This system has the virtue of giving a subtle and useful treatment of nested conditionals. The natural deduction system, however, requires further interpretation. This interpretation is the task of Part III of the book.

Chapter 5 is the final chapter of Part I. In this chapter, I examine another proof-theoretic approach to entailment. This one is primarily due to Dana Scott. According to this, a logic of entailment is one that contains an implication connective that represents the internal and external rules of the logic itself. In this logic, one can recreate and to a certain extent justify the reasoning that one does about the logic. This approach to entailment is highly influential and it can be adapted to a wide range of logical frameworks, including to relevant logic. But I argue that it is inadequate because it does not take into account the useful and intuitive distinction between those rules that are supposed to hold of theories in general and those that are used in particular cases to construct theories of a special nature. Therefore, despite the fact that this approach to entailment is elegant, I reject it. And I move on to Part II, where I present an alternative semantic framework in which to interpret the logic of entailment.

2.1   Historical Introduction

I begin my historical investigation with the work of C. I. Lewis. Lewis was, as far as I can tell, the first person self-consciously to formulate a logic in which there is a connective that expresses deducibility. The reasons for looking at Lewis is not that he produced the right logic, nor was he just an example of a logician who got a lot wrong, but rather that we can learn from his work certain criteria of what a correct theory of entailment would look like as well as some mistakes to avoid.

Lewis studied logic with Josiah Royce in the very early years of the twentieth century. He learned Boole–Schröder algebra and Boole–Schröder set theory from Royce, who taught his students how to calculate complicated equivalences between descriptions of sets using Euler–Venn diagrams.[1] Lewis, however, read *Principia Mathematica* very soon after its first volume was published in 1910, and he quickly became a convert to the logistic method of presenting a logic. One difference between the algebraic method and the logistic method, as Lewis saw it, was that an algebraic logic was presented by stating some basic equations and then the theorems were derived assumed logical rules, whereas a logistic systems contained axioms and explicitly stated rules of inference.

After reading *Principia*, however, Lewis quickly became dissatisfied with Whitehead and Russell's formulation of logic. There was no device in their logic to express the necessary connection between premises and conclusions of valid deductions. Lewis formulated his theory of strict implication to rectify this deficiency. In this chapter, I examine what Lewis thought was wrong with the classical logic of *Principia*, and how he thought his system superior to it. I also look at the paradoxes of strict implication, which he originally wished to avoid, but eventually came to accept. The avoidance of these paradoxes motivated his students, Everett Nelson and William Parry, to formulate radically different logics. I also examine their attempts and find them deficient as theories of logical entailment.

---

[1] Lewis's notes on Royce's lectures are in the Stanford archives. Also, Royce's lecture notes and his notes on logic, which contain hundreds of pages of Euler–Venn diagrams, are in the Harvard archives.

Before I begin my detailed discussion of Lewis and his school, I should say something about the logic of Hugh MacColl. Lewis took McColl's logic to be his predecessor as a modal logician. MacColl set out a modal logic in various articles in the 1870s and 1880s and then again in writings in the early 1900s, including his 1906 book, *Symbolic Logic and Its Applications* [119]. That book was hardly a systematic work in logic, and it is very difficult to determine the exact system he was presenting. Stephen Read has argued that MacColl's logic was the normal modal logic T (perhaps *re*discovered by Robert Feys and G. H. von Wright) [176]. If Read is right about this, then MacColl deserves credit for discovering an important system of modal logic. But I am not concerned here about modal logic per se. Rather, I am interested in whether MacColl thought of his logic as a logic of entailment.

I suggest that MacColl did not think that he was developing a logic of entailment. He defined implication (:) as follows:

$$A : B \ =_{df} \ (AB')^{\eta}$$

'η' was an impossibility operator and $\prime$ was negation. The definition said that 'A implies B' meant that it is impossible that both A and not-B. This was the same definition that Lewis used for strict implication in his *Survey of Symbolic Logic* (1918). In contrasting his own notion of implication with Russell's concept of material implication, MacColl considered the following example: Let W be 'This year there will be a great war in Europe' and E be 'This year there will be a disastrous earthquake in Europe'. An astrological almanac states that both E and W will occur. Russell held that it is true that 'either W implies E or E implies W', but for MacColl, for it to be true that W implies E, it must have been impossible that both W and not-E, and similarly for E to imply W, it must have been impossible that E and not-W. MacColl said:

> Our hypothetical predictions of the astrological almanac may, and probably will, turn out false; but as they contradict no certain or admitted data – no law of nature or of linguistic consistency – they can hardly be called *impossible*. [120, pp 453–455]

MacColl gave connections according to laws of nature, linguistic consistency, and "certain or admitted data" as examples of the sort of modality he meant. But these were not all examples of the same form of necessity. They were examples of nomological necessity, epistemic necessity, and logical necessity (broadly construed). I suggest that what MacColl was doing was giving a logic of modality that could be applied to any form of modality. It was a formal logic with many different interpretations. I do not think MacColl thought of himself as giving a specialised system of logical entailment, unlike Lewis. Nor did MacColl even seem to consider the problem of entailment at all.

2.2 Frege, Russell, and Lewis on Logical Truth

Lewis's logic of strict implication was from its very origins supposed to be a logic of entailment. In a 1911 letter to Royce, Lewis said:

I am quite convinced now of the possibility of modifying the calculus of propositions so as to bring its meaning of implication into accord with that of ordinary inference and proof. I worked up a preliminary paper, and would trouble you with it if I had a legible copy. Since it only costs postage, I have sent it to "Mind."[2]

He stated this goal in the paper itself – "Implication and the Algebra of Logic" [102, p 359]. Thus, it seems quite clear that Lewis did think of himself as putting forward a theory of entailment, although he rarely used the term 'entailment' and preferred to call it 'implication'.

In commenting on the origin of modal logic, Quine criticised Lewis's treatment of entailment (or implication) as a connective:

Lewis founded modern modal logic, but Russell provoked him to it. For whereas there is much to be said for the material conditional as a version of 'if-then', there is nothing to be said for it as a version of 'implies'; and Russell called it implication, thus apparently leaving no place open for genuine deductive connections between sentences. Lewis moved to save the connections. But his way was not, as one could have wished, to sort out Russell's confusion of 'implies' with 'if-then'. Instead, preserving that confusion, he propounded a strict conditional and called it implication. [169, p 323]

Quine was right in claiming that Lewis based a theory of "genuine deductive connections" on his strict implication, but I do not think that Lewis was confused in using a connective to represent implication.[3]

Lewis was providing an alternative to the logics of Frege and, in particular, Russell. I examine Quine's complaints about the logic of entailment in Section 1.6, and I do not repeat this discussion here. Rather, I look at Lewis's reason for thinking that he had to formulate a logic that contained an entailment connective.

In Russell and Frege's logic, there did seem to be a way to represent a valid deduction. If there was a valid deduction from the premises $A_1, \ldots, A_n$ to C, then the material implication $(A_1 \wedge \ldots \wedge A_n) \supset C$ was provable. This was to become known in logic as the "deduction theorem". What it showed, many think, is that an entailment is expressible as a provable material implication. Thus, 'A entails B' really means '$A \supset B$ is provable'.

The deduction theorem was not proven until Jacques Herbrand's PhD thesis in 1930, and so Lewis did not know about it when he penned his attacks in the 1910s. But even if he had known of the deduction theorem, I think it would not have changed his view. To see why this was so, I discuss the notion of provability in Frege and Russell.

The key idea here, for Lewis, was that deducibility and provability should be represented as a necessary connection. The notion of provability is represented formally in Frege's notation by his assertion sign. My interpretation of Frege on the assertion sign closely follows that of Nicholas Smith [196]. Smith has argued that Frege's logic was not a theory of consequence. It was not designed to tell us which propositions follow

---

[2] Letter to Josiah Royce, 15 October 1911, Josiah Royce Archives, Harvard University Libraries, Box 122, Folder 44.

[3] Note that Quine did not think that *every* interpretation of modal logic required that one confuse use and mention, but rather that Lewis's interpretation did [169].

from which, but rather which judgements follow from other propositions [196, p 645]. Frege said:

An inference simply does not belong to the realm of signs; rather, it is the pronouncement of a judgement made in accordance with logical laws on the basis of previously passed judgements. Each of the premises is a determinate Thought recognised as true; in the conclusion too, a determinate Thought is recognised as true. [68, p 175]

The idea here is that an inference is a way of justifying the move from a set of judgements to another judgement. A judgement was a proposition (a Thought, in Frege's terminology) that is taken to be true.

This may sound like a logic of the closure of belief sets, but I do not think that is what Frege wanted to construct. He was attempting to produce a logic that took truths to truths and, in particular, that generated analytically true judgements from analytically true premises, especially from axioms. This view of logic, and of mathematical inference, may seem extremely problematic. After all, mathematicians make inferences about theories they believe to be false and those that are really neither true nor false, such as those geometries that do not accurately describe space and algebraic theories like group theory. For Frege, these "theories" did not become real theories until their axioms were given interpretations and then could be said to express propositions.[4] Not only did real theories have to express collections of thoughts, but the axioms of a theory, to be considered axioms, have to be true [58, 68].

Frege's understanding of independence proofs in geometry illustrated his view well. Consider a formal geometrical theory, such as Euclidean geometry without the parallel postulate. The question is how to show that the parallel postulate is independent of this theory. On Frege's view, what we should do is to interpret this formal theory in terms of some real theory in which each of the axioms is to be understood as expressing a truth. Then it can be shown, say, that the parallel postulate, as understood on this interpretation, is false. If this can be accomplished, then we have proof of the independence of the parallel postulate (see [68]). Mathematicians give this sort of proof all the time when they give concrete models for mathematical theories.

There was a slight irony here. Frege's axiom V of his *Grundgesetze* was not true, and so by his own lights, was never really an axiom. His derivations from it were in some sense not real derivations, and there was no real theory of arithmetic, for an inconsistent theory was not a real theory.

This view of logic contrasted starkly with Lewis's conception of logic. Lewis wanted a logic that could treat inference from hypotheses, and these hypotheses did not have to be even remotely plausible, let alone true. In his 1917 paper, "Issues Concerning Material Implication", Lewis said that it is a flaw in Russell's logic that it considers 'Socrates was a solar myth implies $2 + 2 = 5$' to be true because of the falsehood of the antecedent [103, p 355]. At this stage of his development, Lewis wanted a theory that could treat implications seriously even if they had conceptually impossible antecedents. In the *Survey*, he came to accept that all implications with

---

[4] See Frege's correspondence with Hilbert [67].

logically impossible antecedents are true (see Section 2.9), but he still wanted a logic that could treat any possible hypotheses, even very implausible ones, and not just hypotheses that were taken to be true.

Like Frege, Russell thought that the statements of his logical theory were mere truths (although of a very general sort) and that there was no vertebrate sense of necessity that held between the premises and conclusions of valid inferences. For the most part, Russell rejected any sort of metaphysical or linguistic necessity. He thought that logical necessity, insofar as he could make sense of the notion, could be reduced to a form of generality [186, p 170]. Necessity was to be understood as holding not of propositions but of propositional functions. For example, where p is a propositional variable, $(p \lor p) \supset p$ is a propositional function. Russell stated that '$\vdash (p \lor p) \supset p$' means that $(p \lor p) \supset p$ is true when any elementary proposition is substituted for p [218, vol. 1, p 97]. The propositional function $(p \lor p) \supset p$ is a logical truth (and is logically necessary) because every substitution instance of it is true.[5]

Russell's rejection of any substantive notion of necessity precluded him from postulating any important sort of necessary connection between premises and conclusions of valid inferences. In formulating his logic of strict implication, Lewis was trying to remedy this. Like Frege and Russell, Lewis treated his logical axioms and theorems as merely true and as being general in the same way as Frege and Russell. The provable strict implications of the logic then were supposed to express the universally valid inferences. Lewis's logic stated in its theorems that there was a necessary connection between premises and conclusions of good inferences.

2.3   Lewis and the *Tractatus*

From the mid-1920s onwards, Lewis was influenced by Wittgenstein's treatment of logical truths as truth-table tautologies. To understand the nature of this influence, I first explain two ways of understanding the truth tables for classical logic. Consider the table for the material conditional:

p	q	$p \supset q$
T	T	T
T	F	F
F	T	T
F	F	T

On the first interpretation of the truth table, we think of the rows of this truth table giving the result of interpreting the variables p and q in different ways. If we take p

[5] On the other hand, Russell did write to Lewis to thank the latter for sending him a copy of the *Survey*, and Russell said that he believed in strict implication but left it out of *Principia* because it was not needed to do what he wanted to do in that book: "I am glad to see your chapter on "Strict Implication." I have never felt that there was any very vital difference between you and me on this subject, since I fully recognize that there is such a thing as "strict implication", and have only doubted its practical importance in logistic". Letter from Russell to Lewis, 15 May 1919, Stanford University Archives, C.I. Lewis Collection, M0174 Box 1, Folder 2.

and q to represent actual truths, such as 'The earth orbits the sun' and 'There are no talking donkeys', then the material conditional is true. If p represents a truth and q represents a falsehood, such as 'The moon is made of green cheese', then the material conditional is false. And so on. On the second interpretation of the truth tables, each row of the table is a different *possible situation*. The meaning of p and q is assumed to be fixed and each row tells us what else would be true or false given the truth values stated for p and q. Thus, for example, let us say that p is fixed as meaning 'The earth orbits the sun' and q as There are no talking donkeys'. Then, for example, in a situation in which there are talking donkeys and the earth orbits the sun, $p \supset q$ is false. Let us call the first interpretation the *generality* interpretation and the second the *possibility* interpretation.

In his earlier writings, as I say above, Russell equated logical necessity with generality, but in the *Philosophy of Logical Atomism* (1917–1918) and *Introduction to Mathematical Philosophy* (1919), Russell adopted a possibility reading of the truth tables and a view of logical truth as truth in all possible worlds [187, p 72]. He said that "[p]ure logic, and pure mathematics (which is the same thing), aims at being true, in Leibnizian phraseology, in all possible worlds, not only in this higgledy-piggledy job-lot of a world in which chance has imprisoned us" [185, p 192]. But in the *Analysis of Matter* (1927), he returned to his old view that there were no necessary propositions and that logical necessity was just a form of generality properly attributed to propositional functions [186, pp 169–170]. I suggest that the interlude of two or three years in which Russell adopted the possibility reading may have been a period in which Wittgenstein's influence on him was quite strong. When that influence abated, Russell returned to his old actualism and to the generality reading of the truth tables.

Another logician who held the generality reading to the truth tables was one of the inventors of the truth table (in modern tabular form), Emil Post. He treated rows of truth tables as tools for calculating the truth of complex formulas given the values of propositional variables [161, p 167].

In the *Tractatus*, Wittgenstein advanced a version of the possibility reading of the truth tables. A row of a truth table was, for Wittgenstein, a function from elementary statements (*Elementarsätze*) to truth values. Each such function is a *possibility*. An elementary statement was, to borrow Ramsey's explication, the name of a quality or n-place relation R joined to the names of individuals $i_1, \ldots, i_n$, to get $Ri_1 \ldots i_n$ [172, p 168]. A logically necessary statement is one that is true in all possibilities. On this view, rather than treat each row of a truth table as what would happen given certain sorts of substitutions for each proposition letter in a schema or propositional function, a fully determinate statement is said to be either possible, impossible, or necessary [219, 4.31–4.46]. These functions are, perhaps, infinitary and determine the logical status of quantified statements as well as of statements containing the classical connectives.[6]

---

[6] Wittgenstein and Ramsey were involved in giving what John Etchemendy called a "representational semantics" for logical truth [59]. They thought of the meanings of statements as fixed and the rows of the truth table determined different combinations of the propositions that the statements represent. Tarski and Quine, on the other hand, were engaged in producing an "interpretational semantics" for

In "Facts, Systems, and the Unity of the World", Lewis put forward his own construction of possibilities. A system was for Lewis a consistent set of "facts", which we would now call "states of affairs" (since they can be false), that is closed under conjunction and entailment. Correlates to the logical connectives, conjunction and negation, were elements of facts. For example, if A and B were facts in a system Σ, then there is a fact $A \wedge B$, and $\neg A$ and $\neg B$ are not in Σ [106, p 385]. A possible world is a system that is negation complete, that is, a system Σ is a possible world if and only if for every fact A, either A or $\neg A$ is in Σ [106, p 387].

In *Symbolic Logic*, Lewis suggested that Wittgenstein's construction enhanced the intelligibility of the notion of logical truth (and logical necessity):

the nature of logical truth itself has become more definitely understood through the discussions of Wittgenstein. It is 'tautological' – such that any law of logic is equivalent to some statement which exhausts the possibilities; whatever is affirmed in logic is a truth to which no alternative is conceivable. [113, p 24]

Following Wittgenstein, Lewis thought of logical necessity as truth in all possibilities. His use of "law of logic", however, is rather misleading. One might naturally think that any axiom or theorem of a logical system is a law of logic. This was not the case for Lewis. Most of the axioms that he stated were not necessarily true according to his preferred logic.

From approximately 1930, Lewis adopted the logic S2 as his preferred system. The theorems of S2 are not closed under a rule of necessitation. Consider, for example, Lewis's axiom A3:

$$p \prec (p \wedge p).$$

Written in more modern notation, this is $\Box(p \supset (p \wedge p))$. On Lewis's view, $p \supset (p \wedge p)$ was a law of logic, but A3 was not. What Lewis says is misleading since it seems that every logical axiom is "affirmed in logic". In a later paper, "Notes on the Logic of Intension" (1951), Lewis stated his view more accurately as saying that every theorem of logic was either of the form $\Box A$, was equivalent to a formula of the form $\Box A$, or was of the form A, where $\Box A$ was also a theorem [112, p 420]. For Lewis, it seems, it was the job of a logical theory to determine a set of logical truths (that are true in all possibilities), but it is not necessary for all the statements of a logical theory to be themselves logically true.

2.4   Lewis, Carnap, and Reichenbach

Lewis was attacked by Carnap, Reichenbach, and Quine for confusing object and metalinguistic issues in constructing his logic of strict implication. The formal background

---

logical truth. A statement is deemed to be logically true if and only if it is true on any interpretation of the non-logical terms that occur within it. It is extremely difficult to tell whether Russell was a representationalist or interpretationalist in the *Lecturers on Logical Atomism*. In his earlier works, such as the first edition of *Principia*, he was certainly thinking of logical truth in an interpretational manner.

to their complaint was made up of two results about classical propositional logic. The first was Emil Post's completeness theorem. He showed that a formula is provable in classical propositional logic if and only if it is a truth-functional tautology [162]. This proves that a formula is provable in classical logic if and only if it is necessarily true in the sense of the *Tractatus*. Second, there was Jacques Herbrand's 1930 proof of the deduction theorem for classical logic. As it was stated in Herbrand's thesis, the deduction theorem said that for a finitely axiomatised theory, T, a formula P is provable in T if and only if $H \supset P$ is a theorem of classical logic, where H is the conjunction of the axioms of T [79, ch 3.2]. This shows that for formulas of propositional logic, provable material implication is equivalent to provability. This second result is particularly important since what it says is that from the standpoint of a metalanguage that talks about what is provable in classical logic, classical logic determines a class of entailments.

Lewis would have almost certainly known about Post's result, since he knew Post. Lewis probably did not know about Herbrand's theorem,[7] but Lewis did think that "to know that $[p \supset q]$ is a tautology [is] to know that $p \dashv 3\ q$" [113, p 262], and he thought that strict implication represented valid deduction. Let us add a little bit of terminology. A *first-degree entailment* is a formula of the form $A \dashv 3\ B$, where A and B do not contain any modal connectives, including entailment. A first-degree entailment was a theorem of Lewis's system if and only if the corresponding material implication was a tautology. And so his view coincides with Herbrand's deduction theorem in what they consider to be first-degree entailments.

Carnap and Reichenbach argued that the fact that classical logic itself determines the first-degree entailments shows that there was no need for a logic of strict implication. In his *Logical Syntax of Language* (1937), Carnap said that for this reason, "a special logic of meaning [was] superfluous" [29, p 259]. Reichenbach held exactly the same position as Carnap, saying that the job of strict implication could be done by extensional statements in the metalanguage for classical logic [177, p 42].[8]

Lewis disagreed vehemently with Carnap and Reichenbach on this point. In a seminar presented to the Harvard Philosophy Club in early 1948, Lewis said:

I expect you to comment at this point – if you have not done so already – then why all the fuss? And you will probably have added that the fact that the analytic character of postulates and theorems, and the deducibility of theorems from postulates cannot be expressed in the symbolism of *Principia*, is likewise an unimportant consideration, since these are readily formulatable and can be certified in terms of a metalogic or metamathematics which would codify considerations appearing in the unofficial part of *Principia* written in plain English.

My answer is that I am concerned over the point that the explicitly extensional character of all symbolised functions, together with the above facts, makes it appear that a logic of extension exclusively is sufficient for mathematics, and as an easy and plausible corollary, probably sufficient for every purpose for which logic is needed. Whereas in fact, the logic of extension is only half of logic and is subordinate to the logic of the necessary relations of

---

[7] The deduction theorem was already shown by Bernard Bolzano in 1837 ([19] II §224.2) but this would not have been known to Lewis or to Herbrand.

[8] Carnap reversed his opinion on modal logic (and the need for a logic of meaning) in the 1940s. For more about this, see Chapter 3.

concepts; indeed there is nothing which belongs to logic or which logic alone can certify except that can be certified by reference to the necessary relations of concepts.

Logic can certify only what is analytic, and the character of analytic truth is something which is neither expressible nor certifiable in terms exclusively of extension, but only in terms of concepts and their relations of intension. [111]

The first paragraph of this quotation stated exactly the point that Carnap and Reichenbach made. An extensional logic (i.e. classical logic) was all the formal logic needed. In order to express its deducibility relation, one turned to an informal metatheory. For Lewis, hiding the deducibility relation in the metatheory gave the appearance of a purely extensional theory. This impression, he thought, was false. The sources of logical truth are in the intensional relationships between concepts. Only by expressing these relations and making clear that they are intensional can a logic become a true theory of deductive inference.

2.5   Nested Modalities

I wish that Lewis had explicitly cited the fact that his logics could treat nested entailments, whereas classical logic could not, as a reason for preferring his logics, but he did not do so. The issue of nested entailments and nested modalities, however, does play an interesting, but not always clear, role in Lewis's philosophy of logic. As I discuss in Section 2.6, Lewis preferred logics that did not have *collapse theorems*. A collapse theorem says that a string of modal operators can be collapsed (in an equivalent formula) to a single operator. For example, in S5, the scheme $\Box \Diamond A$ is equivalent to $\Diamond A$, so $\Box \Diamond A \equiv \Diamond A$ is a collapse theorem of S5.

From the early 1920s, Lewis developed and adopted a *pragmatic* conception of a priori knowledge, analyticity, and necessity. Lewis equated a priori propositions with analytic propositions, and with necessary ones [108, ch 7 and 8]. The pragmatic conception of necessity was a form of conventionalism. What was necessary, analytic, and a priori was what followed from one's conceptual scheme. Lewis held that the choice of one's conceptual scheme was done by "fiat" [108, p 213] [7, p 190]. The choice of a conceptual scheme, moreover, was to be done on pragmatic grounds. One should choose the scheme that fits best with the sort of needs one has in categorising and making inferences about the world.

One example that Lewis discussed was the application of geometry to physical space [108, pp 253–259]. The choice between the use of Euclidian and non-Euclidean geometry to construe the structure of space was seen in terms of the way in which it enabled the confirmation of various purported empirical laws. Empirical laws are to be understood as regularities, but also as ways of relating various concepts in one's conceptual scheme to one another.[9] The laws of geometry are necessary, by virtue of the fact that they are true by the choices of agents.

---

[9] For Lewis, some fundamental laws are not empirical, but a priori and necessary. These only relate concepts to one another [108, p 254] (and see [87].)

In his earlier work (in particular, written before 1930), Lewis seems to flirt with some form of linguistic conventionalism, that 'it is necessary that *A*' means, in certain cases, that *A* is true by linguistic convention. If Lewis were to hold that position, then he would be guilty of the sort of confusion between use and mention that Quine warns against. But, in his 1946 book, *Analysis of Knowledge and Valuation*, Lewis explicitly rejects linguistic conventionalism. He says that whereas the "verbal expression "All squares are rectangles" conveys a logically necessary fact, could not be determined in entire independence of what the constituent expressions 'square' and 'rectangle' convey" but the proposition that this sentence expresses is logically true independently of linguistic conventions [110, p 153].

Lewis's positive view – his pragmatic theory of the a priori and analyticity – in *Analysis of Knowledge and Valuation* was expressed in the following passage:

> That what is asserted by an analytic statement is knowable in advance of particular occasions and incapable of being adversely affected by any empirical finding, reflects the fact that it is a relation of classifications and their criteria which is in question. If the test for the property of character *A* and that for the property or character *B* are so related that in satisfying the former anything must satisfy the latter, their relationship can be discovered without raising any question of what in particular satisfies or fails to satisfy either test. [110, pp 153–154]

In this passage, Lewis was referring to his notion of a sense meaning. An expression has a linguistic meaning and a sense meaning. The linguistic meaning of 'cat', for example, contains the expression 'animal', because "nothing is nameable by 'cat' that is not nameable by 'animal'"[110, p 94]. The sense meaning of an expression is a way (or ways) of relating that expression to the members of its extension. For example, the sense meaning of 'chiliagon' is a method (or methods) for counting the sides of a figure and coming to find that there are one thousand of them [110, p 134]. In choosing a conceptual scheme, one chooses a coherent collection of sense meanings, or ways to apply expressions to things. In choosing a geometry to describe space-time, one chooses a way of relating expressions like 'co-linearity' and 'simultaneity' to events in space-time. The analyticity of a sentence is not determined just by the linguistic meanings of its constituents, but also by their sense meanings. Since one is free to choose the package of sense meanings that one is to use, the analyticity of an expression is to some extent a matter of choice.

Analyticity, on Lewis's theory, did not iterate. It was part of the theory of relativity that two events were simultaneous relative to a point if light from those events hit the point at the same moment. In choosing the theory of relativity, one made this claim analytic and logically necessary. But it was not analytic that the theory of relativity was chosen. It was contingent. Lewis thus rejected the 4 principle of modal logic, that is $\Box A \dashv 3 \Box \Box A$.

The same was true for classical logic. The choice of the two-valued truth tables yielded a sense meaning for the truth-functional connectives – negation, disjunction, and conjunction. One could have chosen, for example, a three-valued logic [109]. Consider the law of excluded middle, $A \lor \neg A$. It was analytic, because Lewis had chosen classical logic as his basic system. This made $\Box(A \lor \neg A)$ true. But $\Box\Box(A \lor \neg A)$

did not hold because it was not analytic that classical logic was chosen. Hence Lewis rejected the rule of necessitation:

$$\frac{\vdash A}{\vdash \Box A}$$

Lewis's weaker logics, S1, S2, and S3 were not closed under necessitation. It is to these systems (and S4 and S5) that I now turn.

2.6   Lewis's Modal Logics

In *Symbolic Logic* and its appendices, Lewis developed five modal logics. In the *Survey*, he had presented a single logic of strict implication, which was badly flawed (see Section 2.7). He corrected that system in 1920 [105], but was then shown by Mordechai Wajsberg and William Parry that the resulting system, which became known as S3, still included a postulate that Lewis wished to reject (which is also discussed in Section 2.7). And so Lewis proposed two weaker logics – S1 and S2. He also developed two stronger logical systems – S4 and S5 – both based on the addition to S3 of principles suggested by Oskar Becker [13] who advocated various collapse principles.

I set out Lewis's logics in this section and then discuss the properties of them that he liked and those that he disliked.

The primitives of Lewis's language were $\Diamond$, $\neg$, and $\wedge$. He said that the possibility operator should be thought of as consistency, that '$\Diamond(A \wedge B)$' meant that A and B were consistent with one another. He defined disjunction using the now-standard De Morgan equivalence:

$$A \vee B =_{df} \neg(\neg A \wedge \neg B).$$

Material implication was defined as usual:

$$A \supset B =_{df} \neg(A \wedge \neg B).$$

He defined strict implication as the inconsistency of the antecedent with the negation of the consequent:

$$A \dashv B =_{df} \neg\Diamond(A \wedge \neg B).$$

And strict equivalence was defined as usual using the conjunction of two strict implications:

$$A = B =_{df} (A \dashv B) \wedge (B \dashv A).$$

His axiomatisation of S2 was the following:

Axioms

1. $(p \wedge q) \dashv (q \wedge p)$;
2. $(p \wedge q) \dashv p$;

2 C. I. Lewis and His School

3. $p \prec (p \land p)$;
4. $(p \land (q \land r)) \prec ((p \land q) \land r)$;
5. $((p \prec q) \land (q \prec r)) \prec (p \prec r)$;
6. $p \prec \neg\neg p$;
7. $(p \land (p \prec q)) \prec q$;
8. $\Diamond(p \land q) \prec \Diamond p$.

Rules

$$\frac{\vdash A \prec B \qquad \vdash A}{\vdash B} \ \text{MP}$$

$$\frac{\vdash A = B}{\vdash C(A) = C(B)} \ \text{Eq}$$

$$\frac{\vdash A \qquad \vdash B}{\vdash A \land B} \ \text{Adj}$$

$$\frac{\vdash A[p]}{\vdash A[B]} \ \text{US}$$

The rule US (for 'uniform substitution') said that if A was a theorem of the logic and p occurred in A, then the result of replacing *all* occurrences of p in A with an arbitrary formula B was also a theorem. By contrast, the rule Eq (for 'equivalents') said that if A and B were provably equivalent, then the occurrence of one could be replaced with the occurrence of the other in any formula to produce a strict equivalent of the original formula.

Lewis's axiomatic basis for S1 was the same as for S2, except that S1 lacked axiom 8. Because of the removal of axiom 8, S1 did not obey the rule of the distribution of possibility over provable strict implications:

$$\frac{\vdash A \prec B}{\vdash \Diamond A \prec \Diamond B}$$

This lack made S1 a very weak logic. It also made it impossible to give a relational Kripke semantics for necessity and possibility for S1 [45]. On the other hand, the axiomatic basis for S3 included all the axioms and rules of S2 and the axiom:

$$(3) \ (p \prec q) \prec (\neg\Diamond q \prec \neg\Diamond p).$$

This is equivalent to adding $(p \prec q) \prec (\Box p \prec \Box q)$. To get S4, Lewis added to S3:

$$(4) \ \Box p \prec \Box\Box p.$$

And to obtain S5, he added to S4 the axiom:

$$(5) \ \Diamond p \prec \Box\Diamond p.$$

Simpler, and more widely used, axiomatisations of Lewis's logics were produced by E. J. Lemmon [100], but since they did not play a role in the history that I am discussing, I do not present them here.

The last two sentences of the first edition of *Symbolic Logic* were:

Those interested in the merely mathematical properties of such systems of symbolic logic tend to prefer the more comprehensive and less 'strict' systems, such as S5 and material implication. The interests of logical study would probably be best served by an exactly opposite tendency. [113, p 502]

A modal logic was strict if it does not contain collapse theorems, that is, if $\Box\Diamond A$ is not equivalent to $\Diamond A$, $\Diamond\Diamond A$ is not equivalent to $\Diamond A$, and so on. Strictness came in degrees. The more collapse theorems a logic had, the less strict it was. S1 and S2 were very strict, and S4 and, especially S5, were not very strict at all. Lewis preferred strict systems because their treatment of necessity was consistent with his views on the meaning of necessity. As I say above, necessity for Lewis was the same as analyticity and analyticity did not iterate.

2.7   The Axiomatic Method

Lewis's method was purely axiomatic. He had a very informal notion of deduction in mind, and perhaps an informal philosophy of the modalities, although he did not articulate a real theory of modality until a decade after he first presented his logic of strict implication.

The axiomatic method was not the only method used in formal logic until that point. Boole, Schröder, and Peirce all had classes of structures in mind when they set out their logics. They thought of their algebras as characterising classes of spaces of sets. Boole's logic of primary propositions (his theory of classes) was an attempt to represent syllogisms in terms of sets of objects, and his logic of secondary propositions (his propositional logic) was supposed to characterise relationships between sets of times [20].

Having a structure or class of structures in mind when creating a logic gives one something against which to check one's creation. Boole Schröder, and Peirce could, in effect, prove the soundness of their axioms in the structures that they were describing. The same is true for Frege, who described truth tables in a non-tabular way, and Post and Wittgenstein who produced truth tables in the modern form.

Lewis, however, did not have a complete picture of a semantic structure for his logic. He had partial pictures. In lecture notes, he gave a truth condition for necessity first in terms of systems and then later in terms of possible worlds. But there is no clear indication of how classes of systems or classes of worlds were to characterise his logics. In his earlier systems (produced before 1920), Lewis seems to have found each of the axioms that he put forward independently intuitive. And this led him astray, and not just once.

Three of his early axiom systems generated results that he did not want. In his first published article on strict implication, "Implication and the Algebra of Logic" (1912), Lewis used an intensional disjunction as his primitive modal connective. I use '$\oplus$' for this connective.[10] Lewis defined $A \multimap B$ as $\neg A \oplus B$.

---

[10] This might be a bit confusing. Lewis uses $+$ as extensional disjunction and $\vee$ as intensional disjunction. But the standard usage in the substructural logic literature matches the conventions used here.

Lewis accepted a principle of permutation for intensional disjunction:

$$(A \oplus (B \oplus C)) \mathbin{⥽} (B \oplus (A \oplus C)).$$

This principle probably seemed quite reasonable, taken by itself. It turned out, however, that it had terrible consequences. By the definition of strict implication, and a little bit of fiddling, this principle implied a principle of permutation for strict implication:

$$(A \mathbin{⥽} (B \mathbin{⥽} C)) \mathbin{⥽} (B \mathbin{⥽} (A \mathbin{⥽} C)).$$

The principle of self-identity $(A \mathbin{⥽} A)$ was also a theorem of this system, and an instance of that is

$$(A \mathbin{⥽} B) \mathbin{⥽} (A \mathbin{⥽} B).$$

Applying permutation to the scheme given immediately above, we obtain

$$(!)\ A \mathbin{⥽} ((A \mathbin{⥽} B) \mathbin{⥽} B).$$

The thesis(!) was something Lewis definitely did not want in a theory of strict implication. To see why, substitute any true proposition p for both A and B. Then we obtain, $p \mathbin{⥽} ((p \mathbin{⥽} p) \mathbin{⥽} p)$. By the truth of p and modus ponens, we get $(p \mathbin{⥽} p) \mathbin{⥽} p$. Thus, every true proposition is entailed by a logical truth. If strict implication really represents derivability, then every true proposition is said to be derivable from a logical truth, and this was clearly not something that Lewis wanted.

Even after Lewis removed the principle of permutation, problems persisted with the theory of strict implication. In the *Survey* (1918), Lewis constructed a system that included the following principle:

$$(NC)\ (A \mathbin{⥽} B) = (\sim B \mathbin{⥽} \sim A),$$

where $=$ is strict equivalence and $\sim$ is an impossibility operator. The name 'NC' stands for 'necessary contraposition'. In the *Survey*, Lewis uses the impossibility operator to define strict implication:

$$A \mathbin{⥽} B =_{df} \sim(A \wedge \neg B).$$

Thus, the formula '$A \mathbin{⥽} B$' means that it is impossible for A to be true when B is false. Unfortunately, Emil Post showed that in the system of the *Survey*,

$$\sim A = \neg A$$

(see [105]). Here is a short version of Lewis's rendition of Post's proof:

1.	$(\sim\neg A \mathbin{⥽} \sim A) \mathbin{⥽} (A \mathbin{⥽} \neg A)$	NC
2.	$\sim(\sim\neg A \wedge \neg\sim A) \mathbin{⥽} \sim(A \wedge \neg\neg A)$	1, def. $⥽$
3.	$(A \wedge \neg\neg A) \mathbin{⥽} (\sim\neg A \wedge \neg\sim A)$	2, NC used as a rule
4.	$(A \wedge A) \mathbin{⥽} (\sim\neg A \wedge \neg\sim A)$	3, PC
5.	$A \mathbin{⥽} \sim\neg A$	4, minor fiddling

Now, we substitute $\neg A$ for A uniformly to obtain

$$6. \ \neg A \prec \sim\neg\neg A.$$

And then we can erase the double negation in the consequent to obtain

$$7. \ \neg A \prec \sim A.$$

The converse, $\sim A \prec \neg A$, is axiom 1.7 of Lewis's system [104, p 295], so $\sim A = \neg A$ is a theorem of Lewis's system. This makes Lewis's system of the *Survey* into a complicated-looking variant of classical logic. For it is provable that

$$A = \neg\neg A = \sim\neg A = \Box\neg\neg A = \Box A$$

and so A is strictly equivalent to $\Box A$, and A can be replaced without changing any consequences by $\Box A$ in any formula and vice versa. This is a very bad result for this logical system. It means that it is not a real alternative to Russell's or Frege's logic.

After Post's discovery, Lewis replaced NC with a weaker principle,

$$(A \prec B) \prec (\sim B \prec \sim A),$$

to create the logic we now know as S3.

But even S3 turned out to be too strong for Lewis. William Parry and Mordechai Wajsberg showed that it contained as a theorem the thesis of suffixing – a strong form of the transitivity of strict implication:

$$(\text{Suffixing}) \ (A \prec B) \prec ((B \prec C) \prec (A \prec C)).$$

Lewis rejected suffixing in a famous passage in *Symbolic Logic*:

I doubt whether [suffixing] should be regarded as a valid principle of deduction: it would never lead to any inference $p \prec r$ which would be questionable when $p \prec q$ and $q \prec r$ are given premises; but it gives the inference $(q \prec r) \prec (p \prec r)$ whenever $p \prec q$ is a premise. Except as an elliptical statement for "$((p \prec q) \wedge (q \prec r)) \prec (p \prec r)$ and $p \prec q$ is true," this inference seems dubious. [113, p 496]

In this passage, Lewis both rejected suffixing and suggested the acceptance of the weaker thesis CS (for 'conjunctive syllogism'): $((A \prec B) \wedge (B \prec C)) \prec (A \prec C)$. Lewis clearly had a very strong intuition in this regard. He proposed a logic, S2, that is weaker than S3 that does, in fact, fail to contain Suffixing but does have CS as a theorem. When he proposed S2, however, he was uncertain whether it really did reject NT and claimed that if found to contain suffixing, he would reject S2 in favour of an even weaker logic – the very awkward system S1 (see Section 2.7). Luckily for Lewis, Parry showed that suffixing is not a theorem of S2, and after that, S2 became Lewis's official logic [114, p 507].

My point in this section is about the fragility of logical intuitions and its effect on the pure axiomatic method. Note that I am not claiming that the safe creation of a logic requires there to be a model theory that guides the logician. Many logics and logical theories have been successfully constructed without that. And a precise proof-theoretic framework (such as a natural deduction system or a Gentzen sequent system) could be used instead. But I am claiming that logical intuitions about axioms by themselves can be very unreliable. A more systematic philosophical view, preferably backed up by a formal semantics (or proof theory), which allows for the testing of proposed axioms, helps a great deal.

2.8 Truth and Entailment

For Lewis, entailment came in two forms. First, there was logical entailment and, second, there was meaning inclusion. The second of these comprised such entailments as 'x is red entails x has a colour'. I call the second type of entailment, *meaning inclusion entailment*. Logical entailment differed from meaning inclusion entailment, according to Lewis, in that logical entailment was formal. He said:

[L]ogic is formal in the sense that its principles are statements involving variables; and any statement having a variable constituent or constituents is appropriately called a formal statement. We must, however, carefully distinguish formal statements from statement functions. The expression, "Nothing is both A and not A" may be intended as a statement; but if so it is elliptical and means, "For every term 'A', nothing is nameable both by 'A' and by 'not A'." Without the prefix, "For every term 'A'," it would, strictly, make no statement, either true or false, but would be a statement function. [110, pp 113–114]

Lewis went on to say that statement functions (i.e. propositional functions) were not the sort of things that can be asserted. This remark was clearly aimed at Whitehead and Russell's view in the first edition of *Principia*, where most of their theorems were asserted propositional functions. However, in the introduction to the second edition, Russell came to the view that Lewis expressed in this passage. A statement of logic did not contain non-logical vocabulary, but only variables that were understood as bound by universal quantifiers at the head of the statement.

What is important for my present project in this passage is that Lewis claimed that all statements of logic are general. For example, that $\neg(A \wedge \neg A)$ holds when any sentence replaces A. With regard to entailment statements, this is especially interesting. Lewis held that $A \prec B$ is true if and only if B is derivable from A. Moreover, a strict implication is a law of logic only if it is formal in the sense we have just explored. I think that the best way of stating this is that the propositional logic of strict implication is correct and complete on Lewis's view if and only if, for any formulas A and B of propositional logic, the formula

$$A \prec B$$

is a theorem of the logic if and only if the closed formula

$$\forall p_1 \ldots \forall p_n (A \prec B)$$

is true, where $p_1, \ldots, p_n$ are all the propositional variables that occur free in A or B.

One of the problems with the lack of a semantics in Lewis's theory is that it makes it difficult to confirm (or refute) the claim that for every strict implication $A \prec B$, it is true if and only if B is deducible from A. If Lewis had endorsed a model theory, then we might be able to see whether there is a plausible model that makes true only those strict implications that represent "ordinary inference and proof".

Fortunately, it is not that difficult to construct a model that allows us to show that at least some of what Lewis thought was in the ordinary notion of proof is captured by strict implications true in that model. By 1918, Lewis had come to accept what we would now call the classical notion of consequence as part of what constitutes the ordinary notion of proof. In the model that I construct, a modality-free formula C follows classically from modality-free $A_1, \ldots, A_n$ if and only if the first-degree formula $(A_1 \wedge \ldots A_n) \mathbin{-\!3} C$ is true. Here is a construction of such a model for the logic S5. Let $\mathcal{L}$ be a language that includes all the classical connectives (conjunction, disjunction, and negation) and strict implication. The model consists of a set of possible worlds. Each world, w, is a function from the propositional variables of the language into the set $\{0, 1\}$, and every such function is included as a world. For each world w and each propositional variable p, $w \vDash p$ if and only if $w(p) = 1$ and the following recursive definition of satisfaction holds in the model:

- $w \vDash A \wedge B$ if and only if $w \vDash A$ and $w \vDash B$;
- $w \vDash A \vee B$ if and only if $w \vDash A$ or $w \vDash B$;
- $w \vDash \neg A$ if and only if $w \nvDash A$;
- $w \vDash A \mathbin{-\!3} B$ if and only if, for all world w', $w \nvDash A$ or $w \vDash B$.

Clearly, C is classically derivable from $A_1, \ldots, A_n$, if and only if in this model $(A_1 \wedge \ldots A_n) \mathbin{-\!3} C$ is true at every world. Since this model is a model for S5, it also is a model for S1, S2, S3, and S4 (see Section 2.6). Therefore, it was coherent for Lewis to hold that a strict implication was true if and only if the corresponding derivation was valid for formulas of classical logic. We can construct similar Kripke models for S2, S3, and S4, which are not models for stronger Lewis systems that make true only those strict implications that correspond to valid classical inferences.

The construction of such models, however, does not show that the ordinary notion of deduction, when the modal operators are included in the language, is captured fully by any of Lewis's logic. We need to know more about the ordinary notion of proof to construct such an argument. In my opinion, none of Lewis's logics does well in treating nested entailments, and so does not capture the notion of proof or entailment that is needed for my purposes.

2.9   The Paradoxes of Strict Implication

The axiomatic method consists in the writing down of axioms that give a logic the properties that one wants it to have. Lewis also had in mind some principles that he did not want it to have. Among these were the so-called paradoxes of material and strict implication. Among these were that any proposition implied any truth or any tautology, that any falsehood implied any proposition, and that any contradiction implied any proposition.

By 1918, when he published the *Survey of Symbolic Logic* [104], Lewis came to accept the paradoxes of strict implication, in particular that any proposition implied any logical truth and that any contradiction implied any proposition, that is, implosion and explosion. In the *Survey* he gave an argument for both implosion and explosion. His proof began by stating three principles that he took to be unassailable:

I $(p \wedge q) \prec q$

II $((p \wedge q) \prec r) \prec ((p \wedge \neg r) \prec \neg q)$

III $(((p \wedge q) \prec r) \wedge (r \prec s)) \prec ((p \wedge q) \prec s)$

Principle I was, of course, the principle of simplification for conjunction and principle III was a form of transitivity for strict implication.

Principle II was a form of the principle of *antilogism*. The rule form of this principle had been isolated and discussed at length by Christine Ladd-Franklin in her dissertation of 1883 [97], and Lewis was influenced by her work.[11] Ladd-Franklin reconstructed the standard syllogisms as "inconsistent triads" of propositions. For example, the syllogism Barbara was

Every *A* is *B*

Every *B* is *C*

∴ Every *A* is *C*

Construed as an inconsistent triad it became

Every *A* is *B*

Every *B* is *C*

Some *A* is not *C*

Any two of these propositions entailed the negation of the third. Hence, the valid syllogisms that could be generated by antilogism were:

- Every *A* is *B*, every *B* is *C* ∴ every *A* is *C*;
- Every *A* is *B*, some *A* is not *C* ∴ some *B* is not *C*;
- Every *B* is *C*, some *A* is not *C* ∴ some *A* is not *B*.

The fact that antilogism could generate the syllogisms in this way was, for Lewis, definitely a reason to accept it.

Armed with principles I–III, Lewis attacked his own previous rejection of explosion and implosion. This was his argument:

1	$(q \prec r) \prec ((p \wedge q) \prec r)$	I, III
2	$((p \wedge q) \prec r) \prec ((q \wedge \neg r) \prec \neg p)$	II
3	$(q \prec r) \prec ((q \wedge \neg r) \prec \neg p)$	1, 2, transitivity of $\prec$

---

[11] Lewis cited Ladd-Franklin's reasoning about syllogisms with approval both in the *Survey* [104, pp 108–110] and in *Symbolic Logic* [113, pp 60–61].

By substituting q for r in 3 we obtain

$$4 \ (q \dashv 3 \ q) \dashv 3 \ ((q \wedge \neg q) \dashv 3 \ \neg p).$$

Lewis assumed that $q \dashv 3 \ q$ is a theorem, and so by modus ponens, he obtained

$$\vdash (q \wedge \neg q) \dashv 3 \ \neg p.$$

As Lewis said "p itself may be negative" and so "this impossible proposition implies anything". This last step appealed to the equivalence of a proposition and its double negation and the substitutivity of propositions and their logical equivalents.[12] This was the argument for explosion and by contraposition, we get

$$\neg \neg p \dashv 3 \ \neg (q \wedge \neg q),$$

which is clearly equivalent to

$$p \dashv 3 \ (\neg q \vee q).$$

This was an important instance of implosion. Thus, Lewis came to accept both imposition and explosion. As I argue in Chapter 1, implosion is really not a principle one wants in a theory of entailment. If a formula is provable, then every proof plan for that formula becomes valid. And explosion yields at least some instances of implosion in any logic in which contraposition is valid.

2.10   The Deduction Theorem

Another problem for Lewis concerns the deduction theorem. I briefly discuss Herbrand's deduction theorem for classical logic in Section 2.4. The now-standard deduction theorem for classical logic states that if a formula B is derivable from a set of formulas Γ and a formula A, then $A \supset B$ is derivable from Γ. If we replace the material implication in this statement with strict implication, then we obtain a statement that should be false. For $A \wedge B$ is derivable from $\{B\}$ and A, but it is not the case that $A \dashv 3 \ (A \wedge B)$ is derivable from $\{B\}$.

A more reasonable version of the deduction theorem for strict implication is what Anderson and Belnap call the "entailment theorem" [4, §23.6]: If there is a derivation of B from the premises, $A_1, \ldots, A_n$, then $(A_1 \wedge \ldots \wedge A_n) \dashv 3 \ B$ is a theorem of the system. Where the strict implication is replaced by $\rightarrow$, this statement is true of Anderson and Belnap's logic E of relevant entailment (see Chapter 4). The question here, of course, is whether the theorem holds for Lewis's logic S2.

---

[12] Given these assumptions, the argument can be made simpler:

1.	$(q \wedge \neg p) \dashv 3 \ q$	I, US
2.	$(q \wedge \neg q) \dashv 3 \ \neg \neg p$	1, II, MP
3.	$(q \wedge \neg q) \dashv 3 \ p$	2, DN, Eq

Ruth Barcan Marcus proved that the entailment theorem did not hold of S2 [123].[13] Barcan Marcus pointed out that if Lewis's strict implication was to represent deducibility, then it made sense to think that some form of deduction theorem should have held. The derivability of a formula from a set of premises should have been represented in the logic by a provable strict implication [123, 124].

Barcan Marcus used the following definition of a derivation. A derivation of a formula A from a finite set of formulas Γ is a finite list of formulas, $B_1, \ldots, B_n$ such that B_n is A and for each B_i ($1 \leq i \leq n$), either B_i is a member of Γ, B_i is an axiom, or B_i follows from one or more formulas that appear prior to B_i in the derivation by a rule of the logic. It is easy to see that every theorem of the logic is derivable from the empty set.

One counter-example to the deduction theorem for S2 is the following. Let Γ be $\{A \prec B\}$. It is easy to prove that

$$(i)\,(A \prec B) = (A = (A \wedge B))$$

and

$$(ii)\,(B \prec C) \prec ((A \wedge B) \prec C).$$

From (i) and $A \prec B$, we obtain

$$(iii)\,A = (A \wedge B)$$

and from (ii) and (iii) and the rule Eq, we get

$$(iv)\,(B \prec C) \prec (A \prec C).$$

But, as I said in Section 2.7,

$$(A \prec B) \prec ((B \prec C) \prec (A \prec C))$$

is not a theorem of S2. Thus, the deduction theorem fails for S2.

This failure clearly shows that strict implication, as it is formulated in S2, does not express the notion of derivability that is formulated by Barcan Marcus. The question, then, is whether this notion of derivability is the appropriate conception. It is difficult to know what Lewis meant by derivability, other than, perhaps, that conception given by the class of pairs of finite sets of formulas, (Γ, A) such that $\wedge \Gamma \prec A$ is a theorem of S2. Just saying that the notion of deduction captured by the logic is the one determined by the theorems about strict implication is rather disappointing since Lewis had claimed to be capturing the notion of derivability that underlay "ordinary inference and proof". If there is no independent way of characterising this sense of deducibility, then it is extremely difficult to evaluate Lewis's view.

The best route for Lewis to reject Barcan Marcus's definition of a deduction from hypotheses was to claim that one or more of the rules of his logic were not applicable to hypotheses, even though they were applicable to theorems of the logic. Clearly,

---

[13] Barcan Marcus, in fact, showed that it did not hold of quantified S2, but it did not hold for propositional S2 either.

the rule of uniform substitution is of this sort, but we do not need to use it in the definition of a derivation. We can replace the clause that says that we can use axioms with one that says that we can use uniform substitutions of axioms to obtain the same results.[14] The only other rule that Lewis might have restricted in this way is the rule of replacement of equivalents (Eq). I am not sure exactly how to do this in a way that seems adequate.[15]

2.11   Connexive Logic

After Lewis showed that the two paradoxes still hold in his logic of strict implication, others tried to avoid them by formulating alternative logics of entailment. One of the first attempts was by Lewis's student, Everett Nelson. Nelson formulated what has become known as a *connexive logic*. The term was coined by Storrs McCall to indicate that a very strong connection between the antecedent and consequent obtains in provable conditionals in these systems. Nelson explicitly claimed that his logic was a logic of entailment. He even used 'E' (for 'entailment') as his conditional. I use '$\rightarrow$' for this connective.[16]

Connexive logics are characterised by the provability of Aristotle's and Boethius's theses:

(AT) $\neg(\neg A \rightarrow A)$;
(AT') $\neg(A \rightarrow \neg A)$;
(BT) $(A \rightarrow B) \rightarrow \neg(A \rightarrow \neg B)$;
(BT') $(A \rightarrow \neg B) \rightarrow \neg(A \rightarrow B)$.

I explain Nelson's reasons for adopting Aristotle's theses after I present his theoretical framework. But he gave a justification for Boethius's theses that I can explain just by appealing to our intuitions about natural language.

As Nelson pointed out [152], Boethius's theses seemed very natural if one took $\rightarrow$ to be the indicative conditional of English:

Jane: If Manon has gone to the bakery, she will have bought a chocolate cake.
Sue: No, if she has gone to the bakery, she will have only bought an apple pie.

Examples like this made it plausible that in natural language 'if p, then not-q' is a contradictory of 'if p then q'. These are not contradictories if we took the indicative conditional to be the material conditional, nor if we took the conditional to be Lewis's strict implication.

I think that Nelson's justification of Boethius's theses should be taken seriously. This only means, however, that we need to take seriously the claim that the indicative

---

[14] To be fair to Barcan Marcus, she used axiom schemata rather than axioms and did not use a rule of uniform substitution.

[15] In [200], Stanisław Surma proves a version of the deduction theorem for fragments of S2, but the statement of the theorem is very technical and it is unclear how the existence of this theorem really bears on the present discussion.

[16] This section is based on an article that I wrote with Francesco Paoli [136].

conditional is some form of connexive conditional. The notion of entailment, understood by logicians and philosophers, is a technical notion and I do not think that it
is expressed by the 'if ... then' of everyday language. So we need a different sort of
justification to adopt a connexive logic as a logic of entailment.

Nelson attempted to give this sort of justification by taking Lewis's framework and
altering it to avoid the paradoxes. He began with a consistency connective [149, 150],
for which I use the symbol '•'. I suggest that • is better understood not so much as
consistency as a *lack of conflict*. Nelson claimed that, for all propositions A,

$$A \bullet A.$$

This meant that no proposition conflicted with itself. This is a very striking claim. At
first glance, it seems wrong. For consider the following intuitive argument:

1. $(A \wedge \neg A) \to A$ simplification
2. $A \to (A \vee \neg A)$ addition
3. $(A \wedge \neg A) \to (A \vee \neg A)$ 1, 2, transitivity

One of De Morgan's laws (together with the principles of double negation and the rule
of the replacement of equivalents) tells us that

$$\neg(A \vee \neg A) \leftrightarrow (A \wedge \neg A).$$

Therefore, from 3 we obtain,

$$4. \ (A \wedge \neg A) \to \neg(A \wedge \neg A).$$

Therefore, a contradiction implies its own negation and so contradictions conflict with
themselves. Hence, there are propositions that conflict with themselves.

Nelson adopted a radical solution to this problem. He rejected the principle of simplification. In abandoning simplification, conjunction became an intensional operator
in Nelson's logic. He said:

I do not take $p \wedge q$ to mean "p is true and q is true", but simply "p and q", which is a unit or
whole, not simply an aggregate, and expresses the joint force of p and q. [150, p 444]

In an entailment, a conjunction of two propositions had to act as a whole. Whatever
was entailed by a conjunction must be entailed by the whole and not just by one of its
parts:

Naturally, in view of the fact that a conjunction must function as a unity, it cannot be asserted
that the conjunction of p and q entails p, for q may be totally irrelevant to and independent of
p, in which case p and q do not entail p, but it is only p that entails p. I can see no reason for
saying that p and q entail p, when p alone does and q is irrelevant, and hence does not
function as a premise in the entailing. [150, p 447]

In Chapter 4, I discuss the relevant logicians' notion of the real use of premises in
deductions. It seems that Nelson had in mind here a strong version of the real use
criterion. Nelson rejects Lewis's unrestricted form of the transitivity axiom:

$$((A \to B) \wedge (B \to C)) \to (A \to C).$$

Nelson said that this only holds when A, B, and C are distinct propositions [149, 151]. It seems that Nelson was worried that there would be theorems like

$$((A \rightarrow B) \wedge (B \rightarrow B)) \rightarrow (A \rightarrow B).$$

This thesis is an instance of both transitivity and simplification. It might seem that $(A \rightarrow B) \wedge (B \rightarrow B)$ does entail $A \rightarrow B$ because we *can* use both premises in the production of the conclusion. But I think that Nelson had in mind a criterion that says that if not all of the conjuncts were *needed* to derive the conclusion, the conjunction did not entail the conclusion. This is a very strong form of the criterion of real use.

The intensional treatment of conjunction, in part, allowed Nelson to avoid the paradoxes of strict implication. In Lewis's argument for explosion, he appealed to simplification and antilogism. Although Nelson accepted antilogism, he rejected simplification, and his logic did not contain the thesis of explosion.

Since Nelson introduced his logic there has developed an important literature on connexive logic. The idea that natural language conditionals are connexive seems to be very fruitful – semantics resembling, Stalnaker's semantics for counterfactuals, has been proposed by Heinrich Wansing and Yale Weiss [214, 216]. Whether Aristotle's and Boethius's theses, however, are reasonable principles of entailment is another matter, and a case for this still has to be made. One issue that needs to be clarified is what notion of entailment is supposed to be expressed by a connexive logic. Nelson's logic is itself particularly difficult to treat either semantically or proof-theoretically. It avoids the paradoxes of strict implication but to say that it captures an intuitive notion of derivability is difficult to believe.

2.12   Analytic Implication

Another one of Lewis's students who presented an alternative logic to avoid the paradoxes was William Parry. Parry discovered the logic of *analytic implication*. Unlike Nelson's logic, Parry's logic was a subsystem of one of Lewis's modal logics, although oddly not of Lewis's preferred modal logic (S2), but of S4. Parry's idea was to impose a *content containment restriction* on valid implications. He wanted to restrict provable implications so that the content expressed by the consequent had to be part of the content expressed by the antecedent. In formal terms, this meant that is to say, $A \rightarrow B$ was a theorem of Parry's system only if all the propositional variables that occurred in B also occurred in A. This very strong variable sharing principle was called the "proscriptive principle" by Parry [158, 159].

Unlike Nelson, Parry did not reject simplification $((A \wedge B) \rightarrow A)$. Nor did he reject the De Morgan definition of disjunction:

$$A \vee B =_{df} \neg(\neg A \wedge \neg B).$$

But he did reject the principle of addition,

$$A \rightarrow (A \vee B),$$

because in some instances of it the consequent had more content than the antecedent. Parry also rejected contraposition [159, p 103],

$$(A \rightarrow B) \rightarrow (\neg B \rightarrow \neg A),$$

because using it one could take a legitimate implication (e.g. $(p \wedge q) \rightarrow q$) and derive an illegitimate one ($\neg q \rightarrow \neg(p \wedge q)$). For similar reasons, he rejected the following form of the transitivity of implication:

$$(A \rightarrow B) \rightarrow ((B \rightarrow C) \rightarrow (A \rightarrow C)).$$

For it was a theorem of his system, as we have seen, that $(p \wedge q) \rightarrow p$. So, if he had accepted transitivity, then by modus ponens, he would have had to accept

$$(p \rightarrow r) \rightarrow ((p \wedge q) \rightarrow r).$$

Since q is in the consequent but not in the antecedent of this formula, it violates the proscriptive principle.

It appears from these examples that Parry's logic is quite weak, but in a sense it is not. It can be understood as a subsystem of S4, but it can also be thought of as an extension of S4. To see this, let us consider a semantics for analytic implication developed by Kit Fine [63]. Fine's semantics is based on the possible worlds semantics for S4. A model contains a set of worlds with a binary accessibility relation, R, that is reflexive and transitive. At each world w, there is a function γ_w that attributes to each propositional variable p, a set of concepts (or things that p is about at w). The function γ_w extends to treat all formulas A: $\gamma_w(A)$ is the union of the sets of concepts $\gamma_w(p)$ for all the propositional variables p that occur in A. Let us call the concepts that are attributed to a formula A at a world w, the content of A in w. With this background now covered, I can state the truth condition for implication:

$A \rightarrow B$ is true in w if and only if, for all worlds v such that Rwv, (i) if A is true in v, then B is true in v and (ii) $\gamma_v(B) \subseteq \gamma_v(A)$.

In other words, an implication holds in a world w if and only if a strict implication holds in that world (in the sense of the Kripke semantics for normal modal logics) and if, in every world v accessible to w, B's content at v is a subset of the content of A at v. This is an intuitive semantics, and captures Parry's motivation for his logic. His logic is supposed to restrict S4 such that a strict implication is true only if the consequent is about only those things that the antecedent is about.

Parry's logic contains the usual classical connectives: negation, disjunction, and conjunction. Material implication can be defined as usual:

$$A \supset B \ =_{df} \ \neg A \vee B.$$

And necessity can be defined as

$$\Box A \ =_{df} \ (A \rightarrow A) \rightarrow A.$$

Given these definitions, we can also define strict implication:

$$A \prec B =_{df} \Box(A \supset B).$$

This strict implication can be shown to have exactly the same behaviour as strict implication in S4. Moreover, in Fine's semantics, at a world w, $A \prec B$ is true if and only if for all worlds v such that Rwv, either A is false at v or B is true at v.

As Fine's semantics shows so clearly, Parry's logic *vindicates* the principles of inference of S4. It is a theorem of analytic implication that

$$((A \prec B) \wedge A) \to B.$$

This tells us that if we have a true strict implication, we can use it to infer from its antecedent to consequent. Now consider a paradox of strict implication, say,

$$p \prec (q \vee \neg q)$$

and suppose also that p is true. It is a theorem of analytic implication that

$$((p \prec (q \vee \neg q)) \to (q \vee \neg q).$$

Moreover, since the antecedent is true so is

$$q \vee \neg q.$$

Now, Parry accepts the law of excluded middle and all its instances, so this may not seem problematic. But that is not the point. The point is that any inference from a strict implication and its antecedent to its consequent is truth preserving. So, why not cut out the analytic implication and merely accept that S4 strict implications codify valid inferences? I don't think Parry's system can give us any reason not to take S4 strict implication, together with all its paradoxes, to be a form of entailment.

Even if we think of analytic implication and the one and only entailment of Parry's logic, we are left with some rather counter-intuitive results. For example,

$$(p \vee q) \to (p \to p)$$
$$\neg p \to (p \to p)$$
$$(p \wedge (q \wedge \neg q)) \to (p \wedge \neg p)$$

are all theorems of his logic. In addition, the following rule is valid:

$$\frac{\vdash A}{\vdash B \to A}$$

where the propositional variables that occur in A all occur in B. It does not seem that $(p \vee q \vee r) \to A$ is always a reasonable proof plan, where A is some complicated theorem with only p, q, and r as propositional variables. Thus, it does not seem that Parry's restricted version of strict implication is restricted enough to constitute an adequate theory of entailment. (I return to the topic of analytic implication and discuss the possibility of combining it with relevant logic in Chapter 4.)

2.13   Lessons from Lewis

In a historical context, Lewis's achievement was impressive. The logics that he developed were interesting in their own right and their creation led to the modern study of modal logic, which has made important contributions to philosophy and computer science, as well as being a field that has been shown to have interesting and useful relationships to fields in mathematics, such as topology.

What are at least as interesting for me, however, are Lewis's failures. His original logic collapsed into classical propositional logic even though each of the axioms of that logic seemed to be reasonable in itself. This shows that there is a serious problem with the method that Lewis used of justifying each axiom and rule by intuition. In this sense, logic needs to be treated *holistically*. A logical system needs to be justified as a whole. The justification of each axiom and rule does not suffice to justify a logical system. Thus, the first lesson we can learn from Lewis's failures is that justification of a system by the independent justification of each axiom (or even of each modal axiom) is inadequate.

Following on from this first lesson is a second lesson. It is better to have a formal theory that allows us to justify the logical system as a whole. Soundness and completeness theorems over a model-theoretic or proof-theoretic semantics prove the formal and perhaps even the philosophical coherence of the system. By itself, the coherence of an axiom system can be extremely difficult to see. Lewis did use finite matrices to prove that certain formulas were not provable in his logics [114], but these did not capture the meaning of strict implication. In fact, it is difficult to know what sort of formal semantics would garner Lewis's approval, since he thought that logic should be understood in an irreducibly intensional manner. Most contemporary semantics are based in classical set theory, which is extensional.

The third lesson from Lewis is that it is difficult to see whether he (or Nelson or Parry) had created a logic that did what it was supposed to do. The logic was supposed to capture an intuitive notion of deducibility, yet what Lewis meant by this was never itself clearly explained. It would be better to have a clear (and perhaps formal) characterisation of deducibility to see whether the logic is successful in capturing this notion. In Nelson's criticism of Lewis's logic of strict implication, Nelson relies on intuitions about natural language conditionals. Nelson, however, was trying to construct a logic of entailment. But he did not make clear what the relationship ordinary conditionals have to entailment. It helps to have a clearly formulated theory of deduction in order to see whether the logics proposed capture that theory. Considerations about natural language particles like 'if ... then' were secondary, at best.

To take these lessons as serious criticisms of Lewis (or Nelson or Parry) would be unfair. Lewis's work in logic largely took place before the creation of model theory and formal proof theory (in Gentzen's sense), and so it would be anachronistic or unfair to require Lewis to have been an extraordinary genius to construct either for his logic of strict implication. Rather, given that we now live in the post-Tarski and post-Gentzen world, Lewis's example shows us how important it is to have either a clear and coherent proof theory or model theory for a logic.

In fact, I think it is important to have both a proof theory and a model theory for a logic of entailment. As I say in Chapter 1, I use the proof theory and model theory to provide epistemic and pragmatic support for one another. In addition, although there are semantic deducibility relations (see Chapters 3 and 6), it is most natural to think of deducibility as a proof-theoretic notion. On the other hand, for a logic of entailment to be correct, it has to be that for any statements $A_1, \ldots, A_n$ and C, $A_1 \wedge \ldots \wedge A_n) \to C$ is *true* if and only if C is deducible from $A_1, \ldots, A_n$. Constructing a model of the language that plausibly represents the actual world (and perhaps the other possible worlds (and perhaps impossible worlds)) is a good way of demonstrating the plausibility of the theory of entailment as a whole.

3 Entailment and Possible Worlds

3.1 Historical Introduction

In Chapter 2, I argue that Lewis's logic needed a proof theory or model theory to give it a firm justification. Lewis relied on intuitions about each of his axioms and rules to justify his logic. He gave no justification of his system as a whole. In the late 1920s and early 1930s, Lewis came to use simple matrices that showed, to a certain extent, that his logic was formally coherent, but this method was inadequate to show that the logic achieved its purpose of representing an intuitive notion of deduction.

In this chapter, I examine an approach to entailment that puts the model-theoretic semantics of the logic first. In particular, I discuss attempts to use possible worlds models for modal logic to define entailment.

The history of the worlds approach to entailment reaches back almost to the origins of the modern debate about entailment. In "External and Internal Relations" (1919–1920), G. E. Moore interpreted the claim that some properties were internal to objects in terms of entailment. The property P was internal to the object a if and only if, for any x, $x = a$ entails Px. He said that '$x = a$ entails Px'

is, as far as I can see, either identical with or logically equivalent to the propositions expressed by "anything which were identical to a would, *in any conceivable universe*, necessarily have P" or by "a could not have existed in any possible world without having P". [147, p 293] (my italics)

Although this statement of the position appeared in Moore in 1920, it was not until Carnap's work in the mid-1940s that the possible worlds semantics for entailment was explored rigorously in a formal manner.

In "On Inductive Logic" [30], Carnap characterised a probabilistic relationship between a hypothesis and evidence as a partial entailment. A statement 'A entails B' means that in every world in which A is true, so is B. His worlds were what he called "state descriptions", which were sets of atomic statements and negated atomic statements. It might seem strange that Carnap had come to accept modal logic. As I mentioned in Section 2.3, in *Logical Syntax of Language*, Carnap rejected Lewis's "special logic of meanings", that is, Lewis's logic of strict implication. By the mid-1940s, however, Carnap had changed his mind about this. He came to think of modal logic and possible world semantics as a framework for understanding the Fregean distinction between sense and reference [34, p 63]. The reference of an expression in a world is given by its semantic value in that world. The sense of an expression is given by a truth-conditional schema that states how the truth value of sentences containing that expression are determined at any world. Carnap had rejected Lewis's logic because it seemed irredeemably intensional. Carnap was committed to an extensional metalanguage to interpret the intensional notion of meaning, and in his theory of state descriptions, he thought he had found such a metalanguage.

In his work on the foundations of probability and on modality, Carnap thought that by using *state descriptions* he could give clear, extensional truth conditions for claims about meanings and for sentences containing various modal idioms, such as 'necessity', 'possibility', and those expressing probabilities. An entailment expressed what occurs in all state descriptions, and by assigning real numbers to individual state descriptions, one could determine the conditional probability of H given E by the sum of these measures of the state descriptions at which $H \wedge E$ holds, divided by the sum of the measures of the state descriptions at which E holds. The use of extensional truth conditions like these could be used to give a compositional semantics for the language, which would show how we can understand complex formulas.

A state description, in his 1946 article "Modalities and Quantification" [31], was a finite conjunction of atomic formulas and negations of atomic formulas. In his 1947 book, *Meaning and Necessity* [33], a state description was a class of (perhaps infinitely many) atomic formulas and negated atomic formulas. For each atomic statement of the language, $Ra_1 \ldots a_n$, either it or its negation, and not both, occurred in a state description. A state description, to use David Lewis's famous phrase, was an ersatz possible world.[1]

Carnap gave a pragmatic argument in favour of his logic and semantics. The semantics enabled one to determine philosophically important properties of sentences of the language. For example, using the truth conditions for the various connectives, one could determine the truth conditions for any sentence and hence determine whether that requires empirical data or merely an understanding of these truth conditions to decide whether the sentence was true. For example, one could determine using the truth conditions for negation and disjunction to determine that $A \vee \neg A$ is logically true (L-true) regardless of the meaning of A.

Although his semantical theory had its origins in the theories of Wittgenstein and Ramsey, Carnap's view differed from theirs in that Carnap treated logical necessity as a connective, whereas Wittgenstein and Ramsey were interested in providing a semantics only for classical higher-order logic. Carnap's rule 41-1 stated that $\Box A$ was true in every state description if A was true in every state description; otherwise, $\Box A$ was false in every state description [33, p 182]. In this way, Carnap thought he could give a semantics of $\Box$ that would have that connective mean 'logical truth' or "L-truth" in his terminology. Moreover, he thought of entailment as L-true material implication [30, p 72, 33, p 11].

---

[1] The theory of state descriptions was developed first by Ramsey in his "Foundations of Mathematics" (1925). Carnap did not refer to Ramsey's construction, but rather attributed the basic idea to the *Tractatus* [33, p 9]. Ramsey explicitly stated that possibilities were sets of atomic statements. This did not appear in the *Tractatus*.

Carnap agreed with Lewis that entailment was some form of strict implication. Carnap's methodology was, however, model-theoretic, whereas Lewis appealed to intuition and an informal understanding of derivability to motivate the acceptance of various axioms. The model- theoretic approach had two advantages over the axiomatic method. First, it used well-understood mathematical notions, like that of a class, which had been in use in mathematics for at least a century, as bases for model structures. Having a structure like this, constructed in a simple way as Carnap's was, gave one greater confidence that the logic it characterised would have the sort of properties that it was desired to have. We see, in Section 3.3, however, that Carnap's logic did have certain surprising and unpleasant properties.

In this chapter, I examine various versions of the worlds semantics approach to entailment. The theory of meaning underlying this approach to entailment is the *truth-conditional theory of meaning*. On the truth-theoretic theory of meaning, the meaning of sentence is given by its truth conditions. On the worlds version of the truth-theoretic theory of meaning, the truth condition for a sentence is just the set of worlds in which the sentence is true. This view of meaning motivates the treatment of propositions (which are the meanings of sentences) as sets of worlds. The meaning of a sentence is just the set of worlds in which it is true, and the proposition that it expresses is also just this set of worlds. However, as I argue in the later sections of this chapter, there is a serious tension between the truth-conditional theory of meaning and the project of creating an adequate theory of entailment.

In this chapter, I examine a variety of possible worlds accounts of entailment. What they all have in common, apart from their use of possible worlds, is that they view entailment as truth preservation in the logically possible worlds. Moreover, they all appeal to truth-conditional theories of meaning. A truth-conditional theory of meaning holds that the meaning of any declarative sentence is its truth conditions. For possible worlds theorists, the truth conditions for a declarative sentence are just the set of worlds in which it is true. In evaluating positions of this kind with regard to entailment, I examine both their treatment of logical possibility and their treatment of entailment, as well as their use of the truth-conditional theory of meaning.

3.2 Kripke Semantics

I continue my discussion with some conceptual background. This conceptual back-ground was developed a decade or so after Carnap published his early work on modal logic, but it helps to set the stage for my discussion of logical possibility.

All the semantics discussed in this chapter are forms of worlds semantics. Worlds semantics was given its modern form by Kripke and others from the late 1950s and early 1960s onward.[2] In this section, I give a standard formal characterisation of a possible worlds model and then discuss the philosophical ideas behind it.

---

[2] For a history of possible world semantics see [41] and [75].

A possible worlds model is a triple $\mathfrak{M} = \langle W, R, V \rangle$, where W is non-empty set, R is a binary relation on W, and V is a function from propositional variables $(p, q, r, \ldots)$ to subsets of W. I assume for the moment that the background logic is classical propositional calculus. I discuss Kripke semantics for non-classical logics later in this chapter.

The model determines a relation, $\vDash$, between worlds (members of W) and formulas in accordance with the following inductive clauses:

- $\mathfrak{M}, w \vDash p$ if and only if $w \in V(p)$;
- $\mathfrak{M}, w \vDash A \wedge B$ if and only if $\mathfrak{M}, w \vDash A$ and $\mathfrak{M}, w \vDash B$;
- $\mathfrak{M}, w \vDash A \vee B$ if and only if $\mathfrak{M}, w \vDash A$ or $\mathfrak{M}, w \vDash B$;
- $\mathfrak{M}, w \vDash \neg A$ if and only if $\mathfrak{M}, w \nvDash A$;
- $\mathfrak{M}, w \vDash \Box A$ if and only if $\forall w' \in W(Rww' \supset \mathfrak{M}, w' \vDash A)$;
- $\mathfrak{M}, w \vDash \Diamond A$ if and only if $\exists w' \in W(Rww' \wedge \mathfrak{M}, w' \vDash A)$.

Whether formal structures of this sort adequately determine the meaning of the various connectives – in particular the meaning of logical necessity – depends in part on how one understands the various elements of a model. The members of W are called "worlds", but a world can be anything that can act as a point. And this means anything at all. In the introductory chapters of George Hughes and Max Cresswell's textbooks on modal logic, they create models out of seating arrangements of students in a classroom. In these models, each student is a possible world [82, 83]. In the mathematical theory (the model theory for modal logic), no constraint is imposed that requires a possible world to be anything like an alternative universe. I discuss the nature of worlds briefly in Section 3.5, but I try to stay away from this thorny and extremely widely discussed issue as much as I can.

The function V is a *value assignment*. It assigns to a sentence the set of those worlds that make that sentence true. In the formal language, we do not have sentences of any natural language. Instead, value assignments determine the meanings of propositional variables. Taking world talk seriously, if a set $X \subseteq W$ is assigned by V to a variable p, then p means whatever features of worlds that all and only the members of X have in common. For example, if all and only members of X have planets that have living things and cats that talk, then p in this model means "there are planets that have living things and cats that talk". In models where worlds do not have features of this sort, we can think of propositions as just meaning those sets of worlds to which they are assigned.

The relation R is called an *accessibility relation* on worlds. The idea that certain worlds are accessible to or from one another gives one some intuition of what the relation is supposed to represent. In the semantics for temporal logics, worlds are taken to be time slices of a universe. If Rww', then in the model w' comes after w (or, if time is branching, then w' might come after w). This sort of accessibility is rather straightforward. Future times, even possible futures, are accessible from this time for us. We will soon be (or might soon be) in the future. But other sorts of accessibility can be more problematic.

Consider, for example, physical necessity. All of the laws of physics are physically necessary, as are the statements that follow from them. Let R_ϕ be such that, for all worlds w, w', $R_\phi ww'$ if and only if w' is physically possible relative to w. A statement A is physically necessary at a world w if and only if for all worlds w', if $R_\phi ww'$, then $\mathfrak{M}, w' \vDash A$.

The problem here is to determine what it is for a world w' to be physically possible relative to w. One way of understanding this relation is to say that w' has all the same laws of nature as w. If we define physical accessibility in this way, then the laws of nature at w need to be determined independently of R_ϕ, on pain of circularity. There are various theories of laws that do this, but they are not my topic of interest. What is important is that the problem of determining laws independently of an accessibility relation arises with regard to laws of logic, and I discuss this in Section 3.8. For the moment, however, I put this problem aside and return to a more traditional way of understanding logical necessity.

Let R_Λ be the relation such that $R_\Lambda ww'$ if and only if w' is logically possible relative to w. Thus, where $\Box$ is logical necessity, $\mathfrak{M}, w \vDash \Box A$ if and only if, for all $w' \in W$, if $R_\Lambda ww'$ if and only if $\mathfrak{M}, w' \vDash A$. We can also define a semantic consequence relation using R_Λ:

DEFINITION 3.1 (Relative Semantic Consequence)   *In a model $\mathfrak{M}$, a formula B semantically follows from a set of formulas Γ in a world w if and only if in every world w' such that $R_\Lambda ww'$, if for all $A \in \Gamma$, $\mathfrak{M}, w' \vDash A$, then $\mathfrak{M}, w' \vDash B$. If B follows from Γ in w (in $\mathfrak{M}$), then we write '$\Gamma \vDash_w^{\mathfrak{M}} B$'.*

This form of logical consequence satisfies the three conditions set out in Chapter 1 – identity, monotonicity, and transitivity.

Where $A \dashv_3 B$ is defined as necessary material implication (i.e. as $\Box(\neg A \vee B)$), we can easily prove that

PROPOSITION 3.2 $\{A_1, \ldots, A_n\} \vDash_w^{\mathfrak{M}} B$ *if and only if* $\mathfrak{M}, w \vDash (A_1 \wedge \ldots \wedge A_n) \dashv_3 B$.

Thus, it can reasonably be claimed that in this model theory strict implication expresses a form of logical consequence and strict implication can be treated as a form of entailment.

Now I return to the problem of interpreting R_Λ. In order to represent logical necessity, this relation must relate a world w to all and only those worlds that are logically possible from the standpoint of w. The problem here (and for much of this chapter) is to determine what it could mean to say that a world w' is logically possible from the standpoint of w. One answer is that R_Λ relates every world to every world in a model. In a model, what is possible is determined by the entire set of worlds. It seems reasonable to say that what is logically possible, according to a model, is what is true in at least one of its worlds. The onus on someone who claims that logical possibilities should vary from world to world is to say what it is about different worlds that forces these changes. In Section 3.8, I discuss the claim that different worlds have different logical facts, but we can already see that accepting this commitment to logical facts is a serious metaphysical burden. Making the logical accessibility relation universal removes that burden.

Once we make logical necessity universal in this sense, we can simplify the notion of semantic logical consequence:

DEFINITION 3.3 (Universal Semantic Consequence)   *In a model $\mathfrak{M}$, a formula B semantically follows from a set of formulas Γ if and only if in every world $w' \in W$, if for all $A \in \Gamma$, $\mathfrak{M}, w' \vDash A$, then $\mathfrak{M}, w' \vDash B$. If B follows from Γ in $\mathfrak{M}$, then we write* '$\Gamma \vDash^{\mathfrak{M}} B$'.

As an instance of Proposition 3.2, we know that, if $R_\wedge$ is universal, for any world w, $\mathfrak{M}, w \vDash (A_1 \wedge \ldots \wedge A_n) \rightarrow_3 B$ if and only if $\{A_1, \ldots, A_n\} \vDash^{\mathfrak{M}} B$. This implies that exactly the same entailments are true at every world in the model. As we shall see, this fact creates some problems for this semantics of entailment.

One difficulty that we need to discuss is the problem of the *infinitary* nature of semantic consequence. In the definition of universal semantic consequence, there is no restriction placed on the size of the set of premises. It may be finite, in which case it can be represented in the language by a strict implication with a conjunction as its antecedent. If $\{A_1, \ldots, A_n\} \vDash B$, it can be represented by $(A_1 \wedge \ldots \wedge A_n) \rightarrow_3 B$. But we do not have the resources in the language to represent a semantic sequent with infinitely many premises.

This problem is avoided if the semantic consequence relation is *compact*. The usual definition of compactness may seem to have little to do with this issue. On the usual definition, a logic is compact if and only if a set of formulas Γ has a model if and only if every finite subset of Γ has a model. If the logic we start with is classical (or based on classical logic), then the proposition that Γ does not have a model can be expressed by

$$\Gamma \vDash \bot.$$

The constant $\bot$, in classical semantics, is not true in any model. And, classically,

$$\Gamma, \neg A \vDash \bot \text{ if and only if } \Gamma \vDash A.$$

Fiddling with this definition of compactness, and thinking about consequence in a classical manner, we can say that $\vDash$ is compact if and only if, for all sets of formulas Γ and all formulas A, $\Gamma \vDash A$ only if, for some finite subset Γ' of Γ, $\Gamma' \vDash A$. I take the statement that a formula follows from a set Γ if and only if it follows from a finite subset of Γ as a definition of compactness. This definition is far more useful when discussing non-classical logics, especially relevant and paraconsistent logics.

Knowing whether a semantic consequence relation is compact can be a tricky matter. In some cases, a consequence relation defined over a class of models is compact, even though the consequence relation of a particular model may not be compact. For example, consider the consequence relation defined over the class of first-order S5 Kripke models. This consequence relation is compact. But now consider a model, $\mathfrak{M}$, that has an infinite constant domain and a name for each object in the domain (names: $c_1, c_2, \ldots$). Let us also say that for each finite subset of the names, there is a world in $\mathfrak{M}$ where the predicate F holds of all and only the members of that subset. The consequence relation, $\vDash_{\mathfrak{M}}$, is not compact. The following sequent

$$F(c_1), F(c_2), \ldots, F(c_n), \ldots \vDash_{\mathfrak{M}} \forall x F(x)$$

is valid, but no finite subset of the premise set semantically implies $\forall x F(x)$. Knowing when a particular model has a compact consequence relation can be important when one is attempting to find the one true model of entailment. Recall that in Chapter 1 I point out that if 'A entails B' means that B is derivable from A, then 'A entails B' is true if and only if B is derivable from A. In order to satisfy this constraint, any reasonable theory of entailment needs to show how all and only valid derivations can be true. From a semantic point of view (and, given that I am talking about truth here, it is difficult to see what other point of view is relevant), the theory needs to indicate that there is a model (or models) in which this constraint is somehow satisfied. The model that is supposed to satisfy this condition (and satisfies the other conditions that I set out in this book) is called the "intended model".

3.3   In Search of the Intended Model

Setting out a class of Kripke models to characterise logical necessity does not fully constitute a theory of entailment. What is needed is to find a principled way to specify an *intended model*. The intended model tells us which statements are true. This is crucial for the logic of entailment. An entailment $A \rightarrow B$ is true if and only if in every world in which A is true, B is also true. Thus, we need to know what worlds there are and what sentences are true in them to determine the correct theory of entailment.

Jaakko Hintikka argued that Kripke semantics was inadequate to specify the model that we need to determine which logical necessities really hold. Hintikka said,

[T]he arbitrariness of the choice of alternatives involved in a [Kripke]-type truth definition appears completely unmotivated, especially in the case of logical modalities. When we say that something is logically necessary, we do not mean just that it is true in each member of some arbitrary set of alternatives. Rather, we mean that it is true in each *logically possible* alternative. In other words, it seems that we ought to impose much more stringent requirements on frames than are in a [Kripke]-type semantics for modal logics. We ought to require at the very least that W contains an instance of each logically possible kind of model with a fixed set of individuals I (or with a subset of I) as its domain of individuals. [80, p 284]

and

Kripke semantics is inadequate in its present form as a logic of logical modalities. And of course this was the primary application of modal logic historically. It is the purpose for which the Lewis systems were originally developed. Here there is not only a skeleton in the closet of Kripke semantics; the skeleton threatens to overturn the whole house. [81, p 2]

Hintikka demanded that an adequate model of logical necessity contained "every logically possible model". Kripke models are based on arbitrary sets of entitles that act as worlds. From the standpoint of a model, $\mathfrak{M} = \langle W, R_\wedge, V \rangle$, W is the set of all logically possible worlds. But there is nothing that requires $\mathfrak{M}$ to include all and only those worlds that are *in fact* logically possible. Thus, we can say that Kripke semantics does not give us a theory of logical necessity and possibility.

There are two ways we can understand this argument. We could see this argument, as Hintikka seemed to, as saying that there was a serious defect in Kripke semantics that could not be repaired. But we could also say the argument showed that Kripke semantics was deficient in the sense that it only gave us a formal outline of what the intended model should look like. On this reading, given some characterisation of logically possible worlds that specifies a set W and a characterisation of a related value assignment (and a universal relation R_Λ), Kripke semantics told us how the truth values of the sentences of the language were determined at all the worlds in W.

Carnap's semantics can, in some sense, lay claim to being the sort of intended model that Hintikka wants. Carnap's state-description semantics implements a strategy that David Lewis calls "linguistic ersatzism". I explain that strategy in Section 3.1, but I go over it in more depth here. Carnap's construction began with a language. Let us simplify matters for a moment and use the language of propositional logic, with conjunction, disjunction, negation, necessity, and possibility, as well as propositional variables and parentheses. Consider a truth table that gives values to every propositional variable in the language. If the language has n many propositional variables (where n is some cardinal number), then the truth table has 2^n many rows. Each of these rows is a possible world in the model that we are constructing.

Each row of a truth table is a function from propositional variables to truth values (let the truth values here be 1 and 0). We can also think of a row of a truth table as a set of propositional variables and negated propositional variables. Let r be a row of a truth table. Then if p gets the value 1 on r, we can say that $p \in r$ and if p gets the value 0 on r, we can say that $\neg p \in r$. In other words, we can either consider r to be a set of propositional variables and negated variables or as the characteristic function of that set. In the propositional setting, each row r, considered as a set in this manner, is a state description. The set of state descriptions, together with a value assignment, V, such that $r \in V(p)$ if and only if $p \in r$, is a Carnapian model for propositional modal logic.

In order to construct this sort of model for predicate logic, Carnap started with a language that also had predicates (of various arity), variables, names, and the quantifiers. A state description was then a set of atomic formulas and negated atomic formulas, such that for every atomic formula, $Pc_1 \ldots c_n$ and every state description, s, either $Pc_1 \ldots c_n$ was in s or $\neg Pc_1 \ldots c_n$ was in s, but not both. Using this, Carnap set out a truth definition for state descriptions.

$$\mathfrak{M}, s \models Pc_1 \ldots c_n \text{ if and only if } Pc_1 \ldots c_n \in s.$$

The recursive clauses for the connectives, modal operators, were the same as in Section 3.2. The treatment of the quantifiers was substitutional. That is,

$$\mathfrak{M}, s \models \forall x A \text{ if and only if for all names } c, s \models A[c/x]$$

and

$$\mathfrak{M}, s \models \exists x A \text{ if and only if for some name } c, s \models A[c/x].$$

In one sense, this model had constant domains – there was one set of names for all state descriptions, and every state description made each formula either true or its negation true. It is easy to see that the Barcan formula held in this model ($\forall x \Box A \supset \Box \forall x A$).

A formula A was said to be logically true if and only if it was true in every state description. Therefore, A was logically true if and only if $\Box A$ was logically true. If each propositional variable was given a meaning in terms of a sentence of English (or any other natural language), then a single state description will characterise the actual world (assuming that bivalence and consistency held in the actual world). On this semantics, $\Box A$ was true in the actual world if and only if A was logically true.

Or at least this was true if the language did not include identity. In adding identity to the language, Carnap specified that $c = d$ held in a state description if and only if c and d were the same name (i.e. syntactically the same) [31, 42]. His logic had the strange axiom, $c \neq d$, where c and d were distinct names. In addition, if the language contained only finitely many names, it would be logically true according to Carnap's model that there are exactly n many things in the universe, where n was the size of the domain. Moreover, if the class of names were infinite, then it would be logically true that there were more than n things in the universe for every natural number n.

If we consider the language without identity, this semantics seemed to meet Hintikka's challenge. Given a language with infinitely many names, one can specify a set of logically possible worlds, and in each and every one of those worlds, $\Box A$ is true if and only if A is a logical truth. Suppose that the formula A is in the language of non-modal first-order logic. Then it may contain the connectives, quantifiers, predicates, names, and variables, but no modal operators. Every state description describes a model for first-order logic and, given the assumption that there is a name for every element in the domain, every first-order model determines a state description. Thus, by the soundness and completeness of classical first-order logic, $\Box A$ is true in a state description if and only if A is a theorem of first-order logic. Thus, this semantics represents the first-order logical truths.

Hence we can say, with caveats concerning identity, that $\Box$ represents logical truth in Carnap's model. Perhaps what is more interesting, and more problematic, is the interpretation of the possibility operator. On this semantics, $\Diamond A$ is true (and logically true) if and only if A is satisfiable, that is, there is a model in which it is true. $\Diamond A$ is true if and only if A is true according to some consistent set of atomic formulas and negated atomic formulas. What is interesting, and problematic, about this interpretation of logical possibility is that the set of logically true formulas of the form $\Diamond A$ is not computably enumerable, since the set of satisfiable formulas of first-order logic is not computably enumerable. If it were, then the set of formulas of the form $\Diamond \neg A$ would be computably enumerable, and so the set of non-theorems of first-order logic would be computably enumerable. Since classical first-order logic is axiomatisable, its class of theorems is computably enumerable, and if both the theorems and non-theorems of a logic are computably enumerable, then it is decidable. But first-order logic is not decidable, so the class of valid formulas of the form $\Diamond A$ is not computably enumerable. This entails that the set of valid formulas is not computably enumerable (for otherwise this syntactically recognisable subset would be computably enumerable too), and so the logic is not axiomatisable. A finitely presentable axiom set gives human beings a means to understand the infinite set of theorems of a logic. One cannot understand Carnap's logic in this manner.

I am not sure that the logic of logical necessity needs to be one that we can completely comprehend. We can understand fragments of Carnap's logic. As I said, classical first-order logic does capture the modal-free fragment of Carnap's logic. And the propositional logic of Carnap's model is axiomatisable. There is, however, a more serious problem with Carnap's semantics and with the attempt to construct a single intended model.

If the class of names is infinite, then the logic of the intended model is not compact. Let P be some unary predicate. Then, $\forall x\, Px$ is true in a state description w if and only if Pi is true in w for every name i. So, Let Γ be the set $\{Pi : i \text{ is a name}\}$. Then $\Gamma \Vdash \forall x\, Pi$. Now consider an arbitrary Γ' that is a finite subset of Γ: $\Gamma' \nVdash \forall x\, Px$. Thus, $\Vdash$ is not compact. In order to provide the basis for a theory of entailment, as I argue in Chapter 1, the underlying consequence relation needs to be compact. This is a serious problem for my adopting Carnap's view in this book.

3.4   Carnap and Logical Necessity

I set aside the problem of the logic of entailment briefly, however, to consider how well this logic captures the notions of logical necessity and possibility.

Carnap amended his theory of logical truth in response to a criticism by Quine. Understanding the criticism and Carnap's reply showed something rather interesting about Carnap's theory of logical truth. In order for the theory to represent the sort of logical truths that are captured by natural language, the language he used could not be left uninterpreted. But when his language was interpreted in terms of a natural language, a problem appeared. Suppose that Bx meant 'x is a bachelor' and Mx meant 'x is married'. There was nothing in Carnap's construction that prevented Bc and Mc from being added to the same state description, and hence the model would make $\Diamond(Bc \wedge Mc)$ true, that is, it would make true, 'it is possible that a is both a bachelor and married'. Carnap did not want that. Carnap identified analytic truth with logical truth.

Carnap later developed a theory that did help to deal with this problem.[3] This was the theory of meaning postulates. Carnap considered two sorts of analytic statements: (1) statements that were true just because of the meanings of the logical connectives, such as 'Fido is black or Fido is not black', and (2) statements that relied for their truth on the meanings of non-logical vocabulary, such as 'If Jack is a bachelor, then he is not married'. The model of [31] did a good job at showing why statements of type (1) are analytic, but it did not treat statements of type (2) at all. The theory of meaning postulates was supposed to fill this gap.

---

[3] Also see Cresswell [43].

A meaning postulate was a statement that placed a constraint on the contents of state descriptions. Where A is a meaning postulate, A was true in all state descriptions. For example, if $\forall x(Bx \supset \neg Mx)$ is a meaning postulate, then it is in every state description. Thus, the construction of state descriptions in "Meaning Postulates" was more complicated than in *Meaning and Necessity*. If Bc were added to a state description s, then $\neg Mc$ would have to be added as well, and Mc could not be added. In this way, Carnap was able to bar married bachelors from his state descriptions [32, p 67].

The use of meaning postulates underlines another difficulty with Carnap's model. For Carnap, the range of possibilities was fully determined by the language – it was given by the way in which the connectives and quantifiers behaved, given the constraints introduced by meaning postulates. On this theory, any two possible worlds that made exactly the same sentences true were, in fact, the same possible world. Surely, there were properties and things for which there are no names or predicates in natural language. Thus, intuitively, there are logical possibilities that were not expressed in Carnap's model.

In Section 3.5, I examine one theory that avoids this problem by taking on a deeper metaphysical commitment.

3.5   Metaphysics of the Intended Model

One way to avoid Hintikka's criticism of Kripke semantics is to adopt David Lewis's vertebrate modal realism. On David Lewis's view, there was a set of worlds, each very much like our actual universe, and what is logically possible is just what is true in at least one of these worlds. Similarly, what is logically necessary is what is true in all these worlds. The existence and nature of these worlds was mind-independent. They were not created by us nor did their contents depend on the way that we described them. They were supposed to just exist. Thus, no worries about the expressiveness of our language affected Lewis's theory.

One problem with taking this sort of vertebrate modal realism as a theory of logical modality is the worry that there might not be enough worlds to make true everything that is, at least intuitively, logically possible. There might have been a world in which there are talking donkeys and one in which there were dogs who can dance the tango, but no world in which both occurred, even though it seemed logically possible that both these circumstances obtained together. To deal with this problem, Lewis postulated the principle of recombination.

To express the plentitude of possible worlds, I require a *principle of recombination* according to which patching together parts of different possible worlds yields another possible world. [115, p 87]

The idea here is that Lewis needed to ensure that every intuitively possible world is in his theory. If in one possible world I was in my office and the clock on my computer says that it is 10:56 am on a Thursday and in another possible world my partner Susan is in Paris and her watch said that 10:56 pm on a Wednesday, then there was a possible world in which both those things happened. The principle of recombination ensured this.

Lewis's principle of recombination went some way to ensuring that all intuitively possible worlds were in his set of worlds. But I am not sure that this principle provided enough worlds to satisfy Hintikka. Hintikka required that a model for logical necessity included a "logically possible kind of model" as a world. Lewis's theory was a theory about metaphysical necessity and possibility. Suppose that Kripke's doctrine of the necessity of one's origins was correct. This means that no one could have had different parents than one did actually have. On Lewis's theory, this would mean that my counterparts in other possible worlds all had parents who are counterparts of my actual parents. This was a thesis about metaphysical necessity. Intuitively, it was, however, not *logically* impossible for someone to have had different parents than those that he or she actually had. Despite this apparent difference between logical and metaphysical necessity, many philosophers identify them (and I think Lewis is among those that do this). So, on this view, it is logically necessary for a person to have had the parents that he or she actually had.

Thus, in adapting Lewis's modal realism as a theory of logical necessity, one conflates logical necessity and metaphysical necessity. If we are to define entailment as logically necessary implication, as Carnap did, then in some worlds some metaphysically necessary implications must fail. The doctrine of the necessity of origins is the sort of baggage that should not come with a theory of entailment. One way to satisfy Hintikka and create a language-independent theory is to treat worlds as constructs (as opposed to David Lewis's "real" worlds) out of non-linguistic elements (as opposed to Carnap's state descriptions). The theory of this sort was due to Nino Cocchiarella, whose motivation was to avoid the problems with possible world semantics that Hintikka cited.

Cocchiarella constructed a model that is very much like Carnap's, but its worlds were language-independent. He took as his starting point the notion of possibility from Wittgenstein's *Tractatus* and Wittgenstein's postulation of atomic facts (*Sachverhalten*) [219, 2-2.0141]. Like Russell and unlike Wittgenstein, however, Cocchiarella postulated both positive and negative facts.[4] He began with the set S of atomic facts and a function f from members of S to members of S. If s was an atomic fact, then $f(s)$ was the negation of s. Moreover, Cocchiarella stipulated that $f(f(s)) = s$ in order to make the principle of double negation valid: $A \equiv \neg\neg A$. A world w was a subset of S such that, for each fact s, either s was in w or $f(s)$ was in w, but it is never the case that both a fact and its negation were in the same world [40, ch 6].

Cocchiarella extended his construction to treat predicate logic. P was a set of relations of all arities and I was a set of individuals. For each relation π in P, there was its negative counterpart, $\overline{\pi}$. π and $\overline{\pi}$ were of the same arity. Where π was an n-place relation, $\overline{\pi}(i_1, \ldots, i_n)$ was the negation of $\pi(i_1, \ldots, i_n)$ [40, ch 7]. On Cocchiarella's

---

[4] For Wittgenstein, a fact was just an arrangement of objects. There was no way to construct a negative fact in the ontology of the *Tractatus*, but Russell did have negative facts in his logical atomism [187]. Cocchiarella's ontology was really much more like Russell's than like Wittgenstein's.

view, many of the now-standard essentialist propositions were not logically necessary. For example, in this model, it was not necessary for a person to have the parents that he or she actually had. Moreover, if water was an individual, it was not necessary for it to be composed of hydrogen and oxygen.

We can see that Cocchiarella's construction of worlds was a language-independent version of Carnap's state descriptions. Atomic facts took the place of atomic formulas. Hintikka's requirement that every logically possible model was present is satisfied without identifying any intuitively distinct possible worlds. This sort of language independence made it possible for Cocchiarella to avoid the problem of possibilities that were not expressed in any language and it avoided the problem of having married bachelors. If in one's language 'bachelor' was attributed in any world only to unmarried men, it would be impossible for there to have been married bachelors. The avoidance of this problem arose from the separation of the language from the frame of the model.

Cocchiarella assumed that there is a metaphysically perfect language as well. In the propositional language, this meant that every propositional variable expressed an atomic fact. Since every atomic fact was in some possible world in Cocchiarella's construction, for every propositional variable, p, p is true in some world, and hence $\Diamond p$ is true in every world. Thus, $\Diamond p$ was valid in the Cocchiarella's semantics (or, rather, what he called the "primary semantics" for logical necessity [39]).

Cocchiarella's language for first-order modal logic, however, was slightly (but importantly) different from Carnap's. In particular, in Cocchiarella's language, there were no singular constants, that is, there were no names of individuals. He very briefly discussed adding names for things, which he called "Tractarian names", but said that they are not part of the language of logical necessity [40, p 262].

Cocchiarella's theory of quantification was *objectual*, whereas Carnap's was a *substitutional* theory. As I say above, Carnap's substitutional theory of quantification forced the failure of compactness for his semantic consequence operator. Whether Cocchiarella's semantic consequence relation was compact, I do not know.

Both Carnap's logic and Cocchiarella's logic, in addition, were problematic in that they could not be given a recursive axiomatisation. Cocchiarella postulates a perfect language in which predicates each represent properties in the model. We can easily see that in his model, a formula A of first-order logic is satisfiable if and only if $\Diamond A$ is true in every world. Thus, the argument given above with regard to Carnap's logic holds with regard to Cocchiarella's.

3.6 Getting Rid of the Perfect Language

One of the problems with Carnap's and Cocchiarella's theories of modality is that they concerned "perfect languages". In both, to some extent, the language determined what is possible and what is necessary. The logical "perfection" of these languages forced their logics to be unaxiomatisable. We can do better, I think, by accepting the metaphysics of worlds in Cocchiarella's model and ditching the perfect language.

A model has two parts. The first is the frame. In Cocchiarella's model, the frame is constructed from a domain of individuals and properties. From individuals and properties, facts are constructed. Worlds are sets of facts as described above. The pair $\langle W, I \rangle$, where W is the set of worlds and I is the set of individuals, is a frame for the logic S5. Let us call this frame, $\mathfrak{F}$.

To construct a model, we need an interpretation V that assigns to each n-place predicate and world a set of n-membered sequences of individuals. For example, the sequence $\langle i, \ldots, i_n \rangle$ may be in the set determined by $V(P, w)$, where P is an n-place predicate and w is a possible world. The pair $\langle \mathfrak{F}, V \rangle$ is a model for quantified S5. Let us call this model, $\mathfrak{M}$. A value assignment, v, on $\mathfrak{M}$ is a function from variables to individuals. These value assignments allow us to give truth conditions for atomic formulas:

$$\mathfrak{M}, w \vDash_v P x_1 \ldots x_n \text{ if and only if } \langle v(x_1), \ldots, v(x_n) \rangle \in V(P, w).$$

A value assignment v' is said to be an x-variant of v if and only if v' assigns all the same individuals to variables as does v, except perhaps to x. This 'perhaps' is important since it makes v an x-variant of itself. These concepts are then used to give truth conditions for the quantifiers:

$$\mathfrak{M}, w \vDash_v \forall x A(x) \text{ if and only if for all } v' \ x-\text{variants of } v, \ \mathfrak{M}, w \vDash_{v'} A(x);$$

$$\mathfrak{M}, w \vDash_v \exists x A(x) \text{ if and only if for some } v' \text{ an } x-\text{variant of } v, \ \mathfrak{M}, w \vDash_{v'} A(x).$$

A formula is valid on $\mathfrak{M}$ if it is true in every world in $\mathfrak{M}$ for every value assignment. A formula is valid on the frame $\mathfrak{F}$ if it is valid on every model based on $\mathfrak{F}$. If the set of worlds and the domain of individuals on $\mathfrak{F}$ are both countably infinite, then it is easily proven that the formulas that are valid on $\mathfrak{F}$ are just the theorems of quantified S5 with the Barcan formula. Thus, the logic of $\mathfrak{F}$ is axiomatisable.

We can also restrict the models on $\mathfrak{F}$ to those that accord with Carnap's meaning postulates. Thus, we only allow models that make every unmarried man a bachelor, and so on. This restriction does not affect what I say in Section 3.7.

3.7   Going Two-Dimensional

It is more plausible to say that the frame $\mathfrak{F}$ represents logical truth than does any specific model on $\mathfrak{F}$. That is, the view so far constructed says that A is a logical truth if and only if A is valid on $\mathfrak{F}$. But although if A is valid on all models on A, it is true in every world in every model, there is no way in which the logical truths are distinguished or marked out in any world in any model. There is no logical operator definable in the logic that tells us that a formula is logically true. Thus, we are not given a logic of logical truth (or of entailment) by this theory.

One interesting suggestion that might be made is to incorporate all the models over $\mathfrak{F}$ into a single "two-dimensional" model. The ideas behind two-dimensional modal logic grew out of David Kaplan's semantics of indexicals [90], Robert Stalnaker's semantics for assertion [197], and Martin Davies, John Crossley, and Lloyd Humberstone's logic of actuality [47, 48]. I use Stalnaker's semantics to illustrate the idea.

In Stalnaker's theory, possible worlds have two roles. They can act as either a context of evaluation or a context of utterance. A context of evaluation is a world as I have explained it in this chapter up to this point. It is a point in a model at which a formula is given a truth value. A context of utterance, on the other hand, is a context in which the various elements of a formula are given their content. To use Stalnaker's example, consider the sentence, 'You are a fool'. This sentence cannot be evaluated until 'you' is assigned a referent. Suppose that there are two worlds w_1 and w_2. In w_1, the sentence is said to O'Leary. Suppose also that O'Leary is a fool in w_1 but is not a fool in w_2. Thus, by fixing the context of utterance at w_1, we can represent the truth values of 'You are a fool' by the following little matrix:

w_1	w_2
T	F

Suppose also that in w_2 the sentence is said to Daniels, who is a fool in w_2 but not a fool in w_1. We now can construct a slightly more complicated matrix:

	w_1	w_2
w_1	T	F
w_2	F	T

The world names across the top of the matrix indicate the context of evaluation and the world names down the side indicate the context of utterance.

I construct a two-dimensional model from Cocchiarella's frame. To make things simpler, I do so only for propositional logic, not its first-order extension. In the semantics for propositional logic, a model determines the truth or falsity of each formula at every world. Let $\mathfrak{M}_1, \mathfrak{M}_2, \ldots$ be models on $\mathfrak{F}$. For any formula A, these models determine matrices of the following form:

	w_1	w_2	$\ldots$
$\mathfrak{M}_1$	T	F	$\ldots$
$\mathfrak{M}_2$	F	T	$\ldots$
$\vdots$	$\vdots$	$\vdots$	

We can define various sorts of modal operators on these matrices. For example, we can set $\mathfrak{M}, w \vDash \blacksquare A$ if and only if for all models $\mathfrak{M}'$ on $\mathfrak{F}$, $\mathfrak{M}', w \vDash A$. And we can set $\mathfrak{M}, w \vDash \boxtimes A$ if and only if for all worlds w', $\mathfrak{M}, w' \vDash A$.

How should we understand logical truth? I say in Section 3.6 that validity on the frame $\mathfrak{F}$ seems to capture logical truth. Validity on $\mathfrak{F}$ can be represented directly in the two-dimensional model by the following defined necessity operator:

$$\Box A =_{df} \blacksquare \boxtimes A$$

The formula $\Box A$ is true in a world if and only if A is true on every world in every model on $\mathfrak{F}$, that is $\Box A$ is true in any world if and only if A is valid on $\mathfrak{F}$. Where A is a formula of classical propositional logic (and contains no modal operators), $\Box A$ is true on any world in this model if and only if A is a truth-functional tautology. For the extension of the semantics that includes the devices to treat quantificational formulas, $\Box A$ is true on a world if and only if A is valid in the class of models for first-order logic with infinite domains. There are some issues about identity that come up in this regard and are discussed in the Appendix, but I set them aside for the present.

But the same problem arises with this two-dimensional model that plagues Carnap's and Cocchiarella's models. Let possibility be defined as usual: $\Diamond A \quad =_{df} \quad \neg\Box\neg A$. Then $\Diamond A$ is true in any world in any model if and only if A is true in some model. Thus, the valid formulas of the form $\Diamond A$, where A is a formula of first-order logic, represent all and only the satisfiable formulas of first-order logic. This class is not computably enumerable and so the logic is not axiomatisable. Thus, we have added a lot of machinery to the semantics for very little gain in this regard at least.

Turning to entailment, the consequence relation of $\mathfrak{F}$ is the following:

$$\Gamma \vDash^{\mathfrak{F}} A \text{ if and only if in every model } \mathfrak{M} \text{ on } \mathfrak{F}, \ \Gamma \vDash^{\mathfrak{M}} A.$$

Where we restrict the premise set to be finite, we can represent this consequence relation by strict implication, because it is provable that

$$A_1, \ldots, A_n \vDash^{\mathfrak{F}} B \text{ if and only if } \mathfrak{M}, w \vDash \Box(A \supset B)$$

for every world model $\mathfrak{M}$ in $\mathfrak{F}$ and every world w in $\mathfrak{F}$.

I am not sure whether the consequence relation $\vDash^{\mathfrak{F}}$ is compact, but the theory of entailment to which it gives rise does have other fatal difficulties. One such difficulty is that this theory does not treat nested entailments in a reasonable manner.

The entailment connective defined in Section 3.7 is a form of strict implication. It inherits a lot of the properties of strict implication connective of S5. For example, if B is valid on the model, then

$$A \rightarrow B$$

is also true for any formula A. And, if C is valid, then so is

$$A \rightarrow (B \rightarrow C).$$

In general when a valid formula is its consequent, an entailment is valid, and even when the consequent is nested in further consequents it makes the whole entailment valid. This feature of the logic, and of any extension of S5, makes its use as a theory of proof plans severely limited.

3.8   Non-normal Worlds

One way of dealing with the problem of nested entailments is to add "non-normal" worlds to the model. A non-normal world is a world that is somehow defective in the logical sense. In particular, at non-normal worlds, not all valid formulas are true. In models with non-normal worlds, a formula is said to be valid in a model if and only if it is true at all normal worlds. Non-normal worlds were first used in models for modal logics (in particular, for Lewis's systems S2 and S3) by Kripke [75, 93].

In Kripke's semantics for S2 and S3, all necessities are false and all possibilities are true in non-normal worlds. All the other connectives behave the same as they do in normal worlds. Thus, for example, in all non-normal worlds, $A \dashv A$ is false. Since in some models, some non-normal worlds are accessible to normal worlds, $p \dashv (A \dashv A)$ is invalid. Thus, Kripke's semantics is of some help in dealing with the problem of nested entailments. The amount of help that this semantics provides, however, is very limited. If $C \dashv D$ is valid, then so is

$$(A \dashv B) \dashv (C \dashv D).$$

Thus, Kripke's semantics is inadequate to treat nested entailments.

A much more useful semantics for entailment is due to Graham Priest [164, 166]. This is his semantics for the logic N_4 (see Section 6.7). At normal worlds, an entailment $A \rightarrow B$ is true if and only if in every world in which A is true, B is also true. Thus, at normal worlds, $A \rightarrow B$ is true if and only if A entails B in the model as a whole. At non-normal worlds, however, entailments are made true randomly. $A \rightarrow B$ is true (perhaps) at some non-normal worlds and (perhaps) not at others. Entailments are not given recursive truth conditions at non-normal worlds. Rather, their truth conditions in non-normal worlds are primitive; they are not explained at all in the model theory. Priest's semantics avoids all the paradoxes of strict implication, and so it is an improvement on Kripke's semantics.

There are, however, both philosophical and formal difficulties with Priest's semantics. The philosophical difficulties have to do with the interpretation of the semantics. To explain the difficulty, I use a distinction that Priest introduces in a recent paper [167]. There he talks about the internal logic and the external logic of a world. An external logic of a world is one that describes the closure of the truths in that world. So, for example, if A is true in w but B fails to be true, then the world does not obey the entailment $A \rightarrow B$. The internal logic of a world, on the other hand, is the set of entailments that the world makes true. A world that fails to have $A \rightarrow B$ in its external logic might still have that entailment in its internal logic.[5] This is all extremely useful (in my opinion) but there is a problem with the idea of an internal logic: what is it about a world that gives it a particular internal logic?

On Priest's view, some possible worlds have different laws of logic than the actual worlds [164]. On the standard view, the truthmaker for laws of logic is the set of logically possible worlds. The laws of logic are just those that accurately describe the closure of the truths in every possible world. On Priest's view, the laws of logic need a very different sort of truthmaker.

---

[5] Priest calls such worlds "incoherent" and defines that logically impossible worlds are those that are incoherent.

I think Max Cresswell's [44] interpretation of Kripke's semantics for S2 can help understand Priest's models. Cresswell interprets Kripke's models for S2 using Kripke's models for T. He takes a standard model for modal logic with a reflexive accessibility relation and added to the language a propositional constant a. This propositional constant is true in some worlds and false in others. The worlds in which it is true were the normal worlds of the model. He then introduced a new necessity operator, N. We define NA as $a \wedge \Box A$. Thus, NA can be true only in normal worlds. In all non-normal worlds, NA is false.

Cresswell suggests understanding a as saying 'there are thinking beings'. If we understand the laws of logic to be quite literally the laws of thought, then there can be no logic without thinking beings. Let us modify Cressewell's idea a bit. We could say that a world has internal laws of logic only if it has beings that can be directed by norms, and such beings must be able to think to be directed by norms. We can go further and say that the norms differ according to the stipulations and beliefs of those beings. And so we can have different entailments true or false at various worlds. These norms can be the truthmakers for the entailments of Priest's internal logic. (I give a different interpretation of N_4 in Section 6.7.)

The problem with N_4's treatment of non-normal worlds is that there are no principles governing entailments in these worlds. This makes N_4 too weak to act as an adequate basis for a logic of entailment. Consider, for example, a proof plan of the following form: if I prove that B follows from A and that C follows from B, then I have shown that C follows from A. To represent this, as I argue in Chapter 1, the logic needs to have as a theorem some form of transitivity thesis, such as

$$((A \to B) \wedge (B \to C)) \to (A \to C)$$

or

$$(A \to B) \to ((B \to C) \to (A \to C)).$$

N_4 does not contain any form of transitivity as a theorem; it is only closed under a rule form of transitivity:

$$\frac{\vdash A \to B \quad \vdash B \to C}{\vdash A \to C}.$$

As I argue in Chapter 1, rules are not adequate to interpret proof plans. I do, however, return to the topic of N_4 in Section 6.7 and give it a different interpretation in terms of models for scientific theories.

One way to give entailments at non-normal worlds more inferential power is to use accessibility relations that give truth conditions for entailments even at non-normal worlds. Such a semantics is constructed by Alexander Sandgren and Koji Tanaka [189]. On their view, we start with a class of worlds and an accessibility relation. Two worlds are *logically different* if there are worlds accessible to one and not to the other. One world may have a weaker logic than another, or it may have logical principles that the other world does not share. If the truth conditions for the connectives other than logical necessity (and entailment) differ from world to world, then a model could be constructed in which internal logics (in Priest's sense) could differ greatly from one world to another. On the sort of normative picture of internal logics that I set out above, the logic of logical truth here could incorporate the sort of principles that one finds in deontic logics.

There is a problem, however, of how to incorporate a theory of entailment into this framework. Suppose that we treat entailment as strict implication. That is, we take the truth condition of entailment to be the following:

$$w \vDash A \to B \text{ if and only if, for all } w', Rww' \text{ and } w' \vDash A \text{ implies } w' \vDash B.$$

On this semantics, if $w \vDash A \to B$ and $w \vDash B \to C$, then $w \vDash A \to C$. In the logic of the actual world, a conjunction holds if and only if each conjunct is true. Thus, in the actual world, $((A \to B) \wedge (B \to C)) \to (A \to C)$ is true. But in the logic of other worlds, conjunction might not behave like this, and so this formula is not true in every world. A proof plan, then, such as one that has the structure of the transitivity principle given above, might be valid from the point of view of one world but not from another.

But the problem is that we get back some of the paradoxes of strict implication. In every world accessible to the actual world, the law of excluded middle holds. So, every proposition entails the law of excluded middle and this is a class of noxious cases of implosion. We could get rid of implosion by basing entailments on a second accessibility relation, say, one that relates the actual world to every world. But if we do this, we lose the proof plan given above.

In Chapter 4, I move away from the modal logical paradigm of treating entailments as telling us about the closure of truths at (possible or impossible) worlds. I look at the logic of relevant entailment and its proof-theoretic motivation.

4   Entailment and Relevance

4.1   Historical Introduction

I claim in Section 3.1 that worlds analyses of entailment do not yield a clear solution to the problem of nested entailments. In this chapter, I present a logic that I think deals well with this problem. This is the relevant logic E of entailment. My approach to E in this chapter, following Alan Anderson and Nuel Belnap [4], is proof-theoretic. I formulate and motivate the system primarily in terms of its natural deduction system. I present a semantics for E in Chapters 8 and 9.

Although relevant logics had already been set out by I. Orlov [157], Moh Shaw-Kwei [191], and Alonzo Church [37], the first logician to relate entailment and relevance in its modern sense was Wilhelm Ackermann. Orlov had given an axiomatisation of what is now known as the logic R of relevant implication. This logic, discussed in Section 4.3, is not a logic of entailment. Its implication connective does not express a necessary connection between antecedents and consequents. Even though Orlov also added a modal operator (Φ) that represented provability, the resulting logic was the system now known as NR, which is slightly different from E [49]. Moh and Church had both rediscovered R and connected it with a relevant version of the deduction theorem [4, §3]. It was Ackermann who first attempted to give an axiomatic treatment of a connective that represented a notion of relevant deducibility.

Ackermann's first attempts to produce a logic of entailment, however, were not relevant. In the early 1950s, he had been trying to formulate a "natural" set theory. The nature of the set theory itself is not my concern here. Rather I am interested in the base propositional logic of the set theory. He read '$A \rightarrow B$' as "B can be derived from A" ("B ist aus A ableitbar"). Thus, he meant the arrow to be an entailment connective. He also included in the language a provability operator, Δ. I will not give the full axiomatisation of the logic but just want to point out why the logic is, from my point of view, inadequate. Ackermann included as an axiom,

$$\Delta A \rightarrow (B \rightarrow A)$$

and as a rule,

$$\frac{\vdash A}{\vdash \Delta A}.$$

So, the following rule can be derived:

$$\frac{\vdash A}{\vdash B \to A}.$$

And this latter rule is one of the fallacies of relevance that should be avoided in a logic of entailment.

In "Foundation of Rigorous Implication", Ackermann proposed a relevant logic as a replacement for Lewis's logic. In motivating the system, he criticised Lewis's acceptance of $B \to (A \to A)$. Ackermann pointed out that "the correctness of $A \to A$ is independent of the truth of B" and so the entailment should not hold between the two [1, p 113]. Ackermann, of course, did think that $A \to A$ was derivable, and so he came to reject the rule that is derived above.

To construct his system, Ackermann used the axiomatic method much like Lewis. As such, the notion of relevance (and deducibility) that his system was supposed to capture was not clearly set out in an independent proof theory or model theory. Moh and Church, however, had designed their logics around a notion of derivability, according to which every premise has really to be used in the derivation of a conclusion. This notion was made formal in the proofs of their deduction theorems. They both designed logics (which had implication as their only connective) so as to yield such theorems. The notion of real use was then given a different formalisation, and one that was applied to a logic with all the usual connectives, by Alan Anderson and Nuel Belnap. Anderson and Belnap's formalisation was a natural deduction system.

Anderson and Belnap's idea was to constrain, in particular, the discharging of hypotheses in inferences to force hypotheses really to be used in producing their conclusions. This makes sense if the aim is to make a logic in which antecedents of valid entailments are more relevant to their consequents. This results, as I argue in this chapter, in a much more satisfying treatment of all entailments, including nested entailments.

I begin the chapter with a short introduction to Fitch-style natural deduction and then add the devices that turn the theory of natural deduction into a proof system for modal and relevant logics.

4.2   Natural Deduction: A Brief Refresher

The sort of natural deduction systems that I will be describing are Fitch systems, named after Frederick Fitch. A Fitch-style proof is a two-dimensional structure. Lines of a proof are placed under those from which they are derived. The first line in a proof is either a known theorem or a hypothesis. Subproofs of a given proof are placed to the right of the superior proofs. And vertical lines indicate the scope of subproofs. In order to make this more concrete, here is an example of a simple proof in classical logic:

1	$A \supset B$	hypothesis
2	$B \supset C$	hypothesis
3	A	hypothesis
4	$A \supset B$	1, reiteration
5	B	3,4, $\supset$E
6	$B \supset C$	2, reiteration
7	C	5,6, $\supset$E
8	$A \supset C$	3–7, $\supset$I
9	$(B \supset C) \supset (A \supset C)$	2–8, $\supset$I
10	$(A \supset B) \supset ((B \supset C) \supset (A \supset C))$	1–9, $\supset$I

The rules used in this proof, apart from the rule allowing the making of hypotheses and the rule allowing free iteration of steps from proofs into their subproofs, are the implication introduction and elimination rules. Implication elimination is just modus ponens:

$$A \supset B$$
$$A$$
$$\downarrow$$
$$B$$

Implication introduction is the rule of conditional proof:

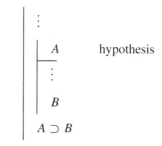

The only rules that I will discuss in this chapter other than implication rules are conjunction rules. I present the rules for disjunction and negation in Chapter 9. The classical rules for conjunction are attractively symmetrical:

$$A$$
$$B \qquad\qquad A \wedge B \qquad A \wedge B$$
$$\downarrow \qquad\qquad\quad \downarrow \qquad\qquad \downarrow$$
$$A \wedge B \qquad\quad A \qquad\qquad B$$

The introduction rule says that we can derive $A \wedge B$ from A and B and the elimination rules say that we can derive A and B from $A \wedge B$.

Here is a little proof of $(A \supset C) \supset ((A \wedge B) \supset C)$ to illustrate the use of conjunction elimination:

1	$A \supset C$	hypothesis
2	$A \wedge B$	hypothesis
3	A	2, $\wedge$E
4	$A \supset C$	reiteration
5	C	3,4, $\supset$E
6	$(A \wedge B) \supset C$	2–5, $\supset$I
7	$(A \supset C) \supset ((A \wedge B) \supset C)$	1–6, $\supset$I

And here is a proof of $((A \supset B) \wedge (A \supset C)) \supset (A \supset (B \wedge C))$ using both conjunction introduction and elimination:

1	$(A \supset B) \wedge (A \supset C)$	hypothesis
2	A	hypothesis
3	$(A \supset B) \wedge (A \supset C)$	reiteration
4	$A \supset B$	3, $\wedge$E
5	B	2,4, $\supset$E
6	$A \supset C$	3, $\wedge$E
7	C	2,6, $\supset$E
8	$B \wedge C$	5,7, $\wedge$I
9	$A \supset (B \wedge C)$	2–8, $\supset$I
10	$((A \supset B) \wedge (A \supset C)) \supset (A \supset (B \wedge C))$	1–9, $\supset$I

Each natural deduction system determines a consequence relation. In some cases, it determines more than one. In the natural deduction system for classical logic, there are two ways of defining the same relation. Perhaps the more intuitive relation is defined such that $A_1; \ldots; A_n \vdash B$ if and only if there is a derivation in which $A_1, \ldots, A_n$ are undischarged hypotheses (and the only undischarged hypotheses) and B is the conclusion. That is, there is a derivation of this form:

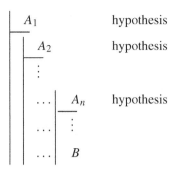

The second definition treats premises conjunctively. That is, $A_1, \ldots, A_n \Vdash B$ holds if and only if there is a valid derivation from $A_1 \wedge \ldots \wedge A_n$ to B. In terms of the natural deduction system, this can be represented as:

$$
\begin{array}{l}
A_1 \wedge \ldots \wedge A_n \qquad \text{hypothesis} \\
\vdots \\
B
\end{array}
$$

On definition 1, it is easy to show that B is a logical consequence of $A_1, \ldots, A_n$ if and only if it is a theorem of classical logic that

$$A_1 \supset (A_2 \supset \ldots (A_n \supset B) \ldots).$$

On definition 2, B is a logical consequence of $A_1, \ldots, A_n$ if and only if it is a theorem that

$$(A_1 \wedge \ldots \wedge A_n) \supset B.$$

Moreover, in classical logic, it is a theorem that

$$((A_1 \wedge \ldots \wedge A_n) \supset B) \equiv (A_1 \supset (\ldots (A_n \supset B) \ldots)).$$

So, in classical logic, $\Vdash$ and $\vdash$ are in fact the same relation and we can rewrite the semicolons in sequents as commas or vice versa.

4.3   Adding Subscripts: The Logic R

The central notion of relevant logic, as Anderson and Belnap [3, 4] present it, is of the real use of hypotheses in derivations. Their natural deduction system modifies the system for classical logic by including restriction on the rules. These restrictions require the use of subscripts on the formulas in the proof. A subscript is a set of numerals. These subscripts allow one to track the hypotheses used to produce the various lines of a proof.

I use the double arrow, $\Rightarrow$, for the implication of R to distinguish it from relevant entailment and from material implication.

To illustrate Anderson and Belnap's idea, here is a proof in their system of $A \Rightarrow ((A \Rightarrow B) \Rightarrow B)$:

1	$A_{\{1\}}$	hypothesis
2	$A \Rightarrow B_{\{2\}}$	hypothesis
3	$A_{\{1\}}$	1, reiteration
4	$B_{\{1,2\}}$	2,3, $\Rightarrow$E
5	$(A \Rightarrow B) \Rightarrow B_{\{1\}}$	2–4, $\Rightarrow$I
6	$A \Rightarrow ((A \Rightarrow B) \Rightarrow B)_{\varnothing}$	1–5, $\Rightarrow$I

When a new hypothesis is introduced, it is given a new number, and this number appears alone in the set that is the subscript of that hypothesis (as in $A_{\{1\}}$ in the first line of the proof).

The modifications to the rules for implication have to do with adding or removing numbers from subscripts. In the implication elimination rule, we add numbers to subscripts:

$$A \Rightarrow B_{\alpha}$$
$$A_{\beta}$$
$$\downarrow$$
$$B_{\alpha \cup \beta}$$

This rule says that in performing modus ponens, the conclusion is produced using all the hypotheses used to produce the major premise together will all the hypotheses used to produce the minor premise. In the introduction rule, a hypothesis number is removed from the subscript. This tells us that a hypothesis is discharged:

$$\vdots$$
$$A_{\{k\}} \qquad \text{hypothesis}$$
$$\vdots$$
$$B_{\alpha}$$
$$A \Rightarrow B_{\alpha - \{k\}}$$

where k is in α. This side condition, that mandates that a number to be in a subscript in order to discharge the corresponding hypothesis, is a *real use* requirement. The hypothesis A really has to be used to prove B for us to conclude that A implies B. This real use requirement (and its analogues concerning other connectives) is what makes the system a *relevant* logic.

The conjunction elimination rules are what one would expect:

$$A \wedge B_\alpha \qquad A \wedge B_\alpha$$
$$\downarrow \qquad\qquad \downarrow$$
$$A_\alpha \qquad\qquad B_\alpha$$

The introduction rule, however, is a little more difficult to accept:

$$A_\alpha$$
$$B_\alpha$$
$$\downarrow$$
$$A \wedge B_\alpha$$

What is difficult to understand about this rule is that it demands that the subscripts on the two premises be the same (and are the same as the subscript of the conclusion).

In order to explain why Anderson and Belnap use this introduction rule, I will contrast it with a rule that I call the "Lemmon rule", since it appears (in slightly different notation) in E. J. Lemmon's book, *Beginning Logic* [101]. Here is the Lemmon rule:

$$A_\alpha$$
$$B_\beta$$
$$\downarrow$$
$$A \wedge B_{\alpha \cup \beta}$$

I think most readers would find the Lemmon rule at first glance more intuitive than Anderson and Belnap's rule. But there are very good reasons for rejecting Lemmon's rule in favour of Anderson and Belnap's.

The addition of Lemmon's rule makes the natural deduction system for R into a system for classical logic. Here is a derivation of one of the paradoxes of material implication in the system for R together with the Lemmon rule:

1	$A_{\{1\}}$	hypothesis
2	$B_{\{2\}}$	hypothesis
3	$A_{\{1\}}$	1, reiteration
4	$A \wedge B_{\{1,2\}}$	2,3, Lemmon rule
5	$A_{\{1,2\}}$	4, $\wedge$E
6	$B \Rightarrow A$	2–5, $\Rightarrow$I
7	$A \Rightarrow (B \Rightarrow A)_\emptyset$	1–6, $\Rightarrow$I

The Lemmon rule is incompatible with the aims of R, but also with my purposes in using E as the logic of entailment. Looking slightly ahead, adding the Lemmon rule to E, allows the derivation of

$$(A \to B) \to (C \to (A \to B)).$$

Now, suppose that $A \to B$ is a theorem of the logic (E together with the Lemmon rule). So, then it is also a theorem of the logic that

$$C \to (A \to B).$$

But, as I say in Chapter 1, this is a rather poor proof plan, especially when C has nothing to do with A or B. Just because $A \to B$ is, in fact, provable, it does not mean that we can reason to it from any old proposition.

I argue in [131] that the Anderson and Belnap rule, and not the Lemmon rule, fits with the interpretation of R that I develop in *Relevant Logic*. That justification would have to be modified considerably to fit with my present project. The interpretation of entailment developed in Chapters 7 and 8, however, gives a good reason why the Lemmon rule should be rejected. The reason is straightforward given the interpretation of entailment, but without that interpretation, it makes little sense. So, I leave my justification of the acceptance of the Anderson and Belnap rule and the rejection of the Lemmon rule to Chapters 7 and 8.

4.4   Strict Subproofs and Modal Logic

In his article [66], Frederick Fitch sets out a natural deduction system for a modal logic (a deontic logic, in fact). Here I first describe a deduction system for the modal logic T. In this system, there is a distinction between standard and *strict* subproofs. A standard subproof is the sort that we have been describing so far. A strict subproof is indicated by a '□' to the left. If a formula $\Box A$ is in a proof, then we can reiterate A into a strict subproof. To illustrate, here is a derivation:

1	$\Box(A \supset B)$	hypothesis
2	$\Box A$	hypothesis
3	$\Box \quad A \supset B$	1, reiteration
4	A	2, reiteration
5	B	3,4, $\supset$E
6	$\Box B$	3–5, $\Box$I
7	$\Box A \supset \Box B$	2–6, $\supset$I
8	$\Box(A \supset B) \supset (\Box A \supset \Box B)$	1–7, $\supset$I

Schematically, the necessity introduction rule can be represented as follows:

The necessity elimination rule is quite straightforward:

$\Box A$

$\downarrow$

A

The derivation given above shows how to derive the K axiom. It is obvious that the T axiom ($\Box A \supset A$) is also derivable. To show that the rule of necessitation is also provable, we just need to think that any derivation of a theorem can be done within a strict subproof. The last line of the subproof is the theorem. We then end the subproof, and in the superior proof we can write down the necessitated theorem. Hence we have a metatheorem about the natural deduction system, that is that for any provable formula, its necessitation is also provable. Adding the K and T axioms and the rule of necessitation results in the logic T.

The system can be easily altered to be a proof system for S4. To do so, we retain the necessity introduction and elimination rules but alter the reiteration rule. For S4, the reiteration rule allows us to copy any formula of the form $\Box A$ from a proof into one of its subproofs (without deleting the outer box). This rule allows for a very easy proof of the 4 axiom:

1	$\Box A$	hypothesis
2	$\quad \Box A$	1, reiteration
3	$\Box\Box A$	2, $\Box$I
4	$\Box A \supset \Box\Box A$	1–3, $\supset$I

Once we adopt the S4 reiteration rule, we can use strict subproofs to formulate a straightforward rule of strict implication introduction. In the following proof, we can see that strict implication is introduced in the same way as material implication in classical logic derivations, except that the subproof used is strict:

1	$A \rightarrow_3 B$	hypothesis
2	$\square$ $B \rightarrow_3 C$	hypothesis
3	$\square$ A	hypothesis
4	$A \rightarrow_3 B$	1, reiteration
5	B	3,4, $\rightarrow_3$E
6	$B \rightarrow_3 C$	2, reiteration
7	C	5,6, $\rightarrow_3$E
8	$A \rightarrow_3 C$	3–7, $\rightarrow_3$I
9	$(B \rightarrow_3 C) \rightarrow_3 (A \rightarrow_3 C)$	2–8, $\rightarrow_3$I
10	$(A \rightarrow_3 B) \rightarrow_3 ((B \rightarrow_3 C) \rightarrow_3 (A \rightarrow_3 C))$	1–9, $\rightarrow_3$I

Schematically, here is the rule of strict implication introduction:

$$
\begin{array}{ll}
\square \quad A & \text{hypothesis} \\
\quad\vdots & \\
\quad B & \\
A \rightarrow_3 B & \rightarrow_3 \text{I}
\end{array}
$$

The strict implication elimination rule is the same as the material implication elimination rule:

$$
\begin{array}{c}
A \rightarrow_3 B \\
A \\
\downarrow \\
B
\end{array}
$$

The strict implication introduction and elimination rules are important to the system presented in Section 4.5.

4.5   Natural Deduction for the Logic E

Anderson and Belnap's natural deduction system for the logic E of relevant entailment combines features from the Fitch system of S4 with features from the system for the logic R: (1) it employs subscripts of the same sort that are used for R and (2) *all subproofs in this system are strict.*

The language that I am discussing now contains $\wedge$ and $\rightarrow$ (relevant entailment). Later in this chapter and in Chapter 9, I add disjunction, negation, and a propositional constant t. But for the moment, I only discuss conjunction and entailment.

A subproof is strict in the sense of the logic E if and only if the only reiterations into that subproof are formulas of the form $A \rightarrow B$. This is because entailments, if true, are necessarily true.

The introduction and elimination rules are the same as they are for R. Here are the entailment elimination and introduction rules:

$$
\begin{array}{c}
A \rightarrow B_\alpha \\[4pt]
A_\beta \\[4pt]
\downarrow \\[4pt]
B_{\alpha \cup \beta}
\end{array}
\qquad\qquad
\begin{array}{|l}
\vdots \\
A_{\{k\}} \\
\vdots \\
B_\alpha \\
A \rightarrow B_{\alpha - \{k\}}
\end{array}
$$

The sort of necessity that is incorporated into E is very S4-like. It is because of this that the introduction and elimination rules for entailment are the relevant versions of the rules in S4 for strict implication, as presented in Section 4.4.

The conjunction rules are the same as they are for R:

$$
\begin{array}{c}
A_\alpha \\[4pt]
B_\alpha \\[4pt]
\downarrow \\[4pt]
A \wedge B_\alpha
\end{array}
\qquad
\begin{array}{c}
A \wedge B_\alpha \\[4pt]
\downarrow \\[4pt]
A_\alpha
\end{array}
\qquad
\begin{array}{c}
A \wedge B_\alpha \\[4pt]
\downarrow \\[4pt]
B_\alpha
\end{array}
$$

As in R, the subscript on the conjuncts in the introduction rule must be the same.

Despite the strong similarities between E and R, the restriction on the reiteration rule is a very important difference between the two systems. Consider, for example, the following proof in R of the thesis of assertion, $A \Rightarrow ((A \Rightarrow B) \Rightarrow B)$:

1	$A_{\{1\}}$	hypothesis
2	$A \Rightarrow B_{\{2\}}$	hypothesis
3	$A_{\{1\}}$	1, reiteration
4	$B_{\{1,2\}}$	2,3, $\Rightarrow$E
5	$(A \Rightarrow B) \Rightarrow B_{\{1\}}$	2–4, $\Rightarrow$I
6	$A \Rightarrow ((A \Rightarrow B) \Rightarrow B)_\emptyset$	1–5, $\Rightarrow$I

If we replace the relevant implications with relevant entailments, then this scheme is not provable in E. The third line – the reiteration step – is not allowed in E. Instead, an instance of the assertion thesis, that is sometimes called "specialised assertion", is provable in E:

1	$A \to B_{\{1\}}$	hypothesis
2	$(A \to B) \to C_{\{2\}}$	hypothesis
3	$A \to B_{\{1\}}$	1, reiteration
4	$C_{\{1,2\}}$	2,3, $\to$E
5	$((A \to B) \to C) \to C_{\{1\}}$	2–4, $\to$I
6	$(A \to B) \to (((A \to B) \to C) \to C)_{\emptyset}$	1–5, $\to$I

The fact that E rejects the general form of assertion but accepts specialised assertion illustrates an interesting trait of that logic.

Anderson and Belnap call any formula of the form $A \to B$, where A does not contain any entailments but B is an entailment, a *fallacy of modality*. As I explain in Section 4.9, a necessity operator can be defined using the entailment connective, and, in fact we can rephrase the definition of a fallacy of entailment as any formula of the form $A \to B$, where A does not contain any instance of $\Box$ and B is of the form $\Box C$. The logic E does not contain any fallacies of modality as theorems. Most of the familiar modal logics based on classical logic do contain rallies of modality. For, as I point out in earlier chapters, most of these have theorems like $A \prec (B \prec B)$, for any A and B.

In Chapter 7, I argue that a logic of entailment should reject all fallacies of modality. That argument, however, relies on the purpose that I believe logics of entailment to have. Here, I briefly discuss the general question of whether necessitives (statements about what is necessary) do follow from statements that are only about what is.

One of the virtues of relevant logic is that it allows us to formulate modal logics that avoid all fallacies of modality. Let us consider deontic logic. In standard deontic logics, based on classical logic, where B is a theorem of the system, $A \vdash OB$, where O is the modal operator, 'it is obligatory that'. In this sort of logic, violations of this sort of the *is-ought* fallacy are forced. In them, we can infer from contingent truths to what ought to be the case. I do not want to say that there are no cases in which inferences from what is to what ought to be are valid, but they are not like this. Just because a formula is valid does not make it obligatory in any moral sense. What is needed is a logic in which the subtle and interesting cases in which oughts can be derived from contingent truths can be formulated without forcing on us these rather silly cases of pointless obligations to obey the laws of logic.

4.6   Proof Plans and E

Consider a proof plan, $(A \to B) \to ((B \to C) \to (A \to C))$. Here is a derivation of it in Anderson and Belnap's natural deduction system for E:

1	$A \to B_{\{1\}}$	hypothesis
2	$B \to C_{\{2\}}$	hypothesis
3	$A_{\{3\}}$	hypothesis
4	$A \to B_{\{1\}}$	1, reiteration
5	$B_{\{1,3\}}$	3,4, $\to$E
6	$B \to C_{\{2\}}$	2, reiteration
7	$C_{\{1,2,3\}}$	5,6, $\to$E
8	$A \to C_{\{1,2\}}$	3–7, $\to$I
9	$(B \to C) \to (A \to C)_{\{1\}}$	2–8, $\to$I
10	$(A \to B) \to ((B \to C) \to (A \to C))_{\emptyset}$	1–9, $\to$I

The hypotheses in the proof tell us the steps that we can employ to prove C. The subscripts on the hypotheses are schematic and stand for the resources from which we find the information described in the hypotheses. For example, we might be able to find in our proof of C proofs of $A \to B$ and $B \to C$ and empirical evidence that A obtains. Then, according to the plan (and the proof above), we will have deductively sufficient evidence to prove C. The derivation, moreover, describes a way in which the information extracted from these resources can be combined into a proof of C. And the subscripts show which resources were really used in this proof. In Chapters 7 and 8, I represent these resources as theories and discuss how theories can be combined to create the semantic representation of proofs.

4.7   The Axiom System for E

In this section, I set out Anderson and Belnap's axiom system for the entailment and conjunction fragment of logic E. I then show how it is related to the natural deduction system. The point of doing this is that this relationship between the two systems also determines the manner in which the entailment connective of E expresses the notion of logical consequence formalised in the natural deduction system.

The only connectives that I deal with here are conjunction and entailment. As I say in Chapter 1, the conjunction-entailment fragment of E is what I call the *core system* of entailment. The axioms and rules of the core system are as follows (where there is a standard name for a logical principle, I provide it):

Axiom Schemes

1. $A \rightarrow A$ identity
2. $(A \rightarrow B) \rightarrow (((A \rightarrow B) \rightarrow C) \rightarrow C)$ specialised assertion
3. $(A \rightarrow B) \rightarrow ((B \rightarrow C) \rightarrow (A \rightarrow C))$ suffixing
4. $(B \rightarrow C) \rightarrow ((A \rightarrow B) \rightarrow (A \rightarrow C))$ prefixing
5. $(A \rightarrow (A \rightarrow B)) \rightarrow (A \rightarrow B)$ contraction
6. $(A \wedge B) \rightarrow A$; $(A \wedge B) \rightarrow B$ simplification
7. $((A \rightarrow B) \wedge (A \rightarrow C)) \rightarrow (A \rightarrow (B \wedge C))$

Rules

$$\frac{A \rightarrow B \qquad A}{B} \text{ MP} \qquad \frac{A \qquad B}{A \wedge B} \text{ Adj}$$

In order to show that this axiomatisation is equivalent to the natural deduction system, one needs to show that all the theorems of the logic can be derived in the natural deduction system in the following form: if A is provable in the axiom system, then $A_\emptyset$ is provable in the natural deduction system. This is quite easy to prove and I leave it to the reader to do. What is more difficult to show is that everything that can be derived in the natural deduction system can be proven in the axiom system.

One way to do this is to show that every line of a natural deduction proof can be represented by a theorem in the axiom system. Let us say that $B_{\{1,...,n\}}$ is a step in a natural deduction proof where $1, \dots, n$ refer to hypotheses $A_1, \dots, A_n$, respectively. Then we can show that $A_1 \rightarrow (\dots (A_n \rightarrow B) \dots)$ is provable in the axiom system.

I use the notation $\alpha \rightarrow B$ to refer to a formula of the form $A_i \rightarrow (\dots (A_j \rightarrow B) \dots)$, where α is $\{i, \dots, j\}$. Thus, the theorem I need is the following:

THEOREM 4.1 A_α *is a line in a valid derivation in the Anderson–Belnap natural deduction system for E if and only if $\alpha \rightarrow A$ is a theorem of the axiom system for E.*

The proof of this theorem is by induction on the length of the derivation. Each hypothesis $A_{\{i\}}$ is translated into a theorem of the axiom system, $A \rightarrow A$. We can then show that all the rules of the natural deduction system transform theorems into theorems of the axiom system.

In order to show that the entailment rules are derivable in the axiom system, we need to do an induction on the length of proof. The entailment introduction rule follows immediately from the inductive hypothesis. The inductive hypothesis is that for any line in a derivation up to the one that we are examining, C_α, $\alpha \rightarrow C$ is a theorem of E. Suppose that $A_{n+1} \rightarrow B_{\{m,...,n\}}$ is a line in a derivation that results from an application of entailment introduction. Then the previous line is $B_{\{m,...,n+1\}}$. By the inductive hypothesis, $A_1 \rightarrow \dots (A_{n+1} \rightarrow B) \dots)$ is a theorem of E. But this formula is the same as the one that corresponds to the line after the application of entailment introduction.

The case of entailment elimination is much more complicated. The problem is that there are many different cases. In one case, all the numbers in the subscript of the major premise are smaller than the ones in the subscript of the minor premise. In some cases, the numbers in the minor and major are interspersed among one another. And so on. For a treatment of all of these cases, see [4]. I just treat one simple case here.

Suppose that hypothesis 1 is A and hypothesis 2 is B and that the major premise in the natural deduction proof is $C \rightarrow D_{\{1\}}$ and the minor premise is $C_{\{2\}}$. By the inductive hypothesis, $A \rightarrow (C \rightarrow D)$ and $B \rightarrow C$ are theorems. We need to prove that $A \rightarrow (B \rightarrow D)$ is also a theorem:

$$\cfrac{A \rightarrow (C \rightarrow D) \qquad \cfrac{B \rightarrow C}{(C \rightarrow D) \rightarrow (B \rightarrow D)}\ \text{suffixing}}{A \rightarrow (B \rightarrow D)}\ \text{suffixing}$$

Thus, in this case at least, the result of applying entailment elimination to theorems produces more theorems.

Using Theorem 4.1, we can then convert any natural deduction derivation to a proof in the axiom system. Here is a natural deduction proof:

1	$(A \rightarrow B) \wedge (B \rightarrow C)_{\{1\}}$	hypothesis
2	$A_{\{2\}}$	hypothesis
3	$(A \rightarrow B) \wedge (B \rightarrow C)_{\{1\}}$	1, reiteration
4	$A \rightarrow B_{\{1\}}$	3, $\wedge$E
5	$B_{\{1,2\}}$	2,4, $\rightarrow$E
6	$B \rightarrow C_{\{1\}}$	3, $\wedge$E
7	$C_{\{1,2\}}$	5,6, $\rightarrow$E
8	$A \rightarrow C_{\{1\}}$	2–7, $\rightarrow$I
9	$((A \rightarrow B) \wedge (B \rightarrow C)) \rightarrow (A \rightarrow C)_{\emptyset}$	1–8, $\rightarrow$I

And here is a proof in the axiom system that is a line-by-line translation of the natural deduction derivation. I present it linearly rather than as a tree because the tree is too wide for the page.

1	$((A \rightarrow B) \wedge (B \rightarrow C)) \rightarrow ((A \rightarrow B) \wedge (B \rightarrow C))$	Axiom 1
2	$A \rightarrow A$	Axiom 1
3	$((A \rightarrow B) \wedge (B \rightarrow C)) \rightarrow (A \rightarrow B)$	Axiom 6
4	$(A \rightarrow B) \rightarrow ((B \rightarrow C) \rightarrow (A \rightarrow C))$	suffixing
5	$((A \rightarrow B) \wedge (B \rightarrow C)) \rightarrow ((B \rightarrow C) \rightarrow (A \rightarrow C))$	3,4, suffixing
6	$((A \rightarrow B) \wedge (B \rightarrow C)) \rightarrow (B \rightarrow C)$	Axiom 6
7	$((A \rightarrow B) \wedge (B \rightarrow C)) \rightarrow (((A \rightarrow B) \wedge (B \rightarrow C)) \rightarrow (A \rightarrow C))$	suffixing, axiom 6
8	$((A \rightarrow B) \wedge (B \rightarrow C)) \rightarrow (A \rightarrow C)$	7, contraction
9	$((A \rightarrow B) \wedge (B \rightarrow C)) \rightarrow (A \rightarrow C)$	8

Clearly, we can trim this proof. It has redundant and useless lines. But the way in which we apply Theorem 4.1 is made very clear here.

The method used in this proof of the equivalence of the natural deduction and axiom systems assumes that the logic has as theorems all instances of the law of identity, $A \rightarrow A$. In Chapter 1, I claim that this law is an important principle of entailment. One reason why it is important is that it enables us to provide an intuitive axiomatisation and natural deduction system that are equivalent in a straightforward and intelligible way.

I represent an inference of the following form:

$$
\begin{array}{ll}
A^1_{\{1\}} & \text{hypothesis} \\
\ddots & \\
\quad A^n_{\{n\}} & \text{hypothesis} \\
\quad \vdots & \\
\quad B &
\end{array}
$$

in a horizontal manner as

$$A^1; \ldots ; A^n \vdash B.$$

The semicolons indicate that there is an intensional connection between premises. Strictly speaking, I should not treat sequents for E in this manner. If we were to add all the connectives to E and state all the rules of inference in a Gentzen-style sequent calculus (see Chapter 5), we would find that what we really need on the left-hand side of the single turnstile is a *structure* [178, 193]. A structure is defined as follows: Every formula is a structure. If X and Y are structures, then so is $(X; Y)$. Nothing is a structure that is not produced by means of the first two of these rules. The expression '$A^1; \ldots ; A^n \vdash B$' is an abbreviation for $(\ldots (A^1; A^2); \ldots); A^n \vdash B$. I only use the single turnstile when I want to talk about the inference to a formula from a series of hypotheses, and so I use this form of abbreviation.

One way in which the entailment connective expresses the logical consequence relation of E is captured by the following biconditional:

$$A^1; \ldots ; A^n \vdash B \text{ if and only if } \vdash_E A^1 \rightarrow (A^2 \rightarrow (\ldots (A^n \rightarrow B)\ldots)).$$

The transitivity of entailment ensures the transitivity of the consequence relation. However, the consequence relation, as understood here, is non-monotonic. Thus, it would seem that it does not satisfy the conditions that I set out in Chapter 1.

There is, however, a second way of understanding consequence in E (and in most other relevant logics). As I said, the semicolon indicates an intensional connection between premises of an inference. We can also think of premises as being connected with one another by extensional conjunction. I use this to define a second consequence relation:

$$A^1, \ldots , A^n \Vdash B =_{df} A^1 \wedge \ldots \wedge A^n \vdash B.$$

And, by the biconditional given above,

$$A^1 \wedge \ldots \wedge A^n \vdash B \text{ if and only if } \vdash_E (A^1 \wedge \ldots \wedge A^n) \rightarrow B.$$

So, $A^1, \ldots, A^n \vdash B$ is equivalent to $\vdash_E (A^1 \wedge \ldots \wedge A^n) \rightarrow B$. In classical logic, as I say in Section 4.2, $\Vdash$ and $\vdash$ are the same and the comma and semicolon represent the same structural connective. This is not true in E, but there is this limited equivalence.

The form of consequence formulated using $\Vdash$ is transitive. It satisfies the following rule:

$$\frac{\Gamma \Vdash A \qquad A \vdash B}{\Gamma \Vdash B}.$$

It also satisfies the inclusion principle:

$$\Gamma \Vdash A,$$

where A is in Γ. And it is monotonic,

$$\frac{\Gamma \Vdash B}{\Gamma, A \Vdash B}.$$

It is $\Vdash$, rather than $\vdash$, that is the principal consequence relation to be captured by the notion of entailment in the logical theory that I set out. The fact that the logic satisfies transitivity, monotonicity, and the inclusion principle shows that it satisfies three of the criteria that I set out in Chapter 1. It is also compact because it is finitary. Infinite sets of premises are not allowed.

4.8   E and R

In the 1960s it was conjectured that E was just a modalised version of R.[1] This means that, when a necessity operator is added to R, and S4-like axioms are also added, then the theorems of the resulting system – called "NR" – in the fragment that just includes negation, strict implication, conjunction, and disjunction would be just the theorems of E. Robert Meyer constructed such a system and showed that the strict implication fragment was identical to E [142]. However, Gregori Mints and Larisa Maksimova showed that the full axioms system for NR is not equivalent to E. NR contains theorems that are not present in E [16].

We can see even from the rather brief exposition I have so far given of E and R where the disconnect is between the two. It is in their different treatments of subproofs. Mints proved the following formula in NR but conjectured that it is not a theorem of E:

$$((p \rightarrow (q \rightarrow r)) \wedge (q \rightarrow (p \vee r))) \rightarrow (q \rightarrow r).$$

Maksimova proved that it is not a theorem of E [122].

The non-equivalence of NR and E is disturbing if one wants to think of entailment as strict relevant implication (in the sense of R). However, in this book, I view E as

---

[1] I am not sure who made this conjecture, but my guess is that it was Robert Meyer.

an independent system, which formalises a notion that is not represented by R, with or without modal operators. I return to the subject of NR briefly in Chapter 5. In the rest of this book, I give an interpretation of E that is largely independent of my interpretations of R and NR.

4.9   Necessity in E

A necessity operator can be defined in E. Its entailment connective incorporates a notion of necessity. We can pull out the sort of necessity in entailment by means of the following definition:

$$\Box A =_{df} (A \rightarrow A) \rightarrow A.$$

This definition, however, is extremely difficult to use. To define necessity in a more tractable way, I add a propositional constant, t. This constant can be thought of as the *weakest theorem* of the logic; t is, in effect, the conjunction of all the theorems of E. It entails every theorem. Therefore, the following rule is valid for t:

$$\frac{\vdash A}{\vdash t \rightarrow A} .$$

And t is itself an axiom. Using t, necessity can be defined as,

$$\Box A =_{df} t \rightarrow A.$$

This definition is equivalent to the previous one. Necessity, so defined, is clearly a form of logical necessity. '$\Box A$' says that A follows from the theorems of E, and so A is necessary according to E.

The natural deduction rules for t are straightforward. Its introduction rule (tI) is

$$A \rightarrow B_\alpha$$
$$\downarrow$$
$$t \rightarrow (A \rightarrow B)_\alpha$$

This rule tells us that every entailment (whether true or not) can be necessitated. The corresponding elimination tE is

$$t \rightarrow A_\alpha$$
$$\downarrow$$
$$A_\alpha$$

This elimination rule says that $\Box$ in the present sense is alethic. By tE, it immediately follows that $(t \rightarrow A) \rightarrow A$, that is $\Box A \rightarrow A$.

The t rules build in an S4-ish notion of necessity. Using tI, we can very easily prove that $(t \rightarrow A) \rightarrow (t \rightarrow (t \rightarrow A))$, that is $\Box A \rightarrow \Box\Box A$. The S3 thesis, $(A \rightarrow B) \rightarrow (\Box A \rightarrow \Box B)$, can be proven as follows:

1	$A \rightarrow B_{\{1\}}$	hypothesis
2	$t \rightarrow A_{\{2\}}$	hypothesis
3	$t_{\{3\}}$	hypothesis
4	$t \rightarrow A_{\{2\}}$	2, reiteration
5	$A_{\{2,3\}}$	3,4, $\rightarrow$E
6	$A \rightarrow B_{\{1\}}$	1, reiteration
7	$B_{\{1,2,3\}}$	5,6, $\rightarrow$E
8	$t \rightarrow B_{\{1,2\}}$	3–7, $\rightarrow$I
9	$(t \rightarrow A) \rightarrow (t \rightarrow B)_{\{1\}}$	2–8, $\rightarrow$I
10	$(A \rightarrow B) \rightarrow ((t \rightarrow A) \rightarrow (t \rightarrow B))_{\emptyset}$	1–9, $\rightarrow$I
11	$(A \rightarrow B) \rightarrow (\Box A \rightarrow \Box B)_{\emptyset}$	10, definition of $\Box$

There is a K-like theorem of the system – $\Box(A \rightarrow B) \rightarrow (\Box A \rightarrow \Box B)$ – that follows immediately from the S3 thesis, but this is not really the K-thesis of standard modal logics, since the $\rightarrow$ is not a contingent implication, but a strict implication. The disjunction forms of K, such as $\Box(A \vee B) \rightarrow (\neg\Box\neg A \vee \Box B)$ and $\neg\Box(A \vee B) \vee (\neg\Box\neg A \vee \Box B)$, are not derivable in E (see [128]).

It is easy to show that t is itself a theorem:

1	$t_{\{1\}}$	hypothesis
2	$t_{\{1\}}$	1, repetition
3	$t \rightarrow t_{\emptyset}$	1–2, $\rightarrow$I
4	$t_{\emptyset}$	3, tE

Here is a proof of the necessitation rule. Suppose that $\vdash A$. This means that there is a proof the last line of which is $A_{\emptyset}$. So

1	$t_{\{1\}}$	hypothesis
2	$A \rightarrow A_{\emptyset}$	theorem
3	$t \rightarrow (A \rightarrow A)_{\emptyset}$	2, tI
4	$A \rightarrow A_{\{1\}}$	1,3, $\rightarrow$E
5	$A_{\emptyset}$	theorem (by assumption)
6	$A_{\{1\}}$	4,5, $\rightarrow$E
7	$t \rightarrow A_{\emptyset}$	1–6, $\rightarrow$I
8	$\Box A_{\emptyset}$	7, definition of $\Box$

There is another useful rule governing t. That is, we are allowed to reiterate t across □-boundaries into subproofs. This rule is derivable in the system and we do not need to add it as a primitive rule. The constant t is equivalent to $t \to t$. One direction of the equivalence – $(t \to t) \to t$ – is just an instance of the thesis T and follows directly by the tE rule. The proof of the other direction – $t \to (t \to t)$ – is just as easy. It follows directly by the tI rule and the fact that $t \to t$ is a theorem. Once the equivalence is proven, whenever t_α occurs as a step in a proof, we can infer $t \to t_\alpha$, then this later line can be reiterated across □-boundaries, and then we can infer in the subproof that t_α. To make proofs slightly shorter, in what follows I merely reiterate t across □-boundaries.

As I say at the beginning of this chapter, another way of defining necessity in E is to set $\Box A$ to mean $(A \to A) \to A$. This definition can be shown to be equivalent to the one using t. It is very easy to show that $((A \to A) \to A) \to (t \to A)$. Here is a proof of the converse:

1	$t \to A_{\{1\}}$	hypothesis
2	$A \to A_{\{2\}}$	hypothesis
3	$t \to A_{\{1\}}$	1, reiteration
4	$A_{\{1\}}$	3, tE
5	$A_{\{1,2\}}$	2,4, $\to$E
6	$(A \to A) \to A_{\{1\}}$	2–5, $\to$I
7	$(t \to A) \to ((A \to A) \to A)_\emptyset$	1–6, $\to$I

One can distinguish between logics, such as standard modal logics, in which necessity and some form of implication are primitive and entailment is a derived connective, and those in which entailment is primitive and necessity is a derived connective. E falls into the latter group. In E, we cannot really understand the notion of necessity without first understanding entailment. This might seem untoward, since necessity is not available as an independent concept to help explain the nature of entailment. In Chapters 7 and 8, however, I show that given my interpretation of the semantics, a fairly clear notion of necessity emerges. Moreover, this notion of necessity seems appropriate as a notion of *logical* necessity and one that is incorporated into the notion of entailment.

4.10   Ackermann and Entailment

As I say at the beginning of this chapter, the logic E was created by Anderson and Belnap to capture the same notion of entailment as Wilhelm Ackermann's logic of rigorous implication. Ackermann's system, Π', includes all the axioms given in Section 4.7, together with fairly standard axioms for negation and disjunction (such as the distribution of conjunction over disjunction and contraposition). What is different about the formulation of Π' from E is that Ackermann uses four rules rather than two. The first two of these rules are the same. He calls modus ponens α and adjunction β. The rule γ is

$$\frac{A \qquad \neg A \vee B}{B},$$

and the rule δ is

$$\frac{A \to (B \to C) \qquad B}{A \to C}.$$

The rule δ can be derived in E.

In the natural deduction system, this derivation is extremely easy. We assume that $A \to (B \to C)$ and B are theorems, hence we can insert them with the empty set subscript at any point in the proof.

1	$A_{\{1\}}$	hypothesis
2	$A \to (B \to C)_{\emptyset}$	theorem
3	$B \to C_{\{1\}}$	1,2, $\to$E
4	$B_{\emptyset}$	theorem
5	$C_{\{1\}}$	3,4, $\to$E
6	$A \to C_{\emptyset}$	1–5, $\to$I

By Theorem 4.1, this derivation can be done in the axiom system as well. So, the rule δ is derivable in E.

There is, however, no derivation of γ in E. One of the reasons that Anderson and Belnap did not include γ in their formulation of E is that it could not be proven to hold in the natural deduction system [2]. Thus, the question arose as to whether the set of theorems of Π' and those of E are the same. This question was answered by Bob Meyer and Mike Dunn [144], who showed that γ is *admissible* in E. The difference between admissibility of a rule and its derivability is that if a rule is admissible, then the set of theorems conforms to it and a rule is derivable if the set of theorems of every extension of the axiom system conforms to it. So, if we add a new axiom to E, then the set of axioms will still be closed under δ, but it might not be closed under γ.[2]

The proof that γ is admissible in E has the added bonus that it shows that E is consistent in a strong sense. That is, it is characterised by the class of its consistent regular theories. A regular theory is one that includes all the theorems of the logic. This consistency result shows that E is characterised by the class of consistent Routley–Meyer models (models in which the normal worlds are all consistent). This is an aesthetically pleasing result, and it also has the consequence that E contains all the theorems of classical logic formulated using negation, conjunction, and disjunction.

---

[2] For a very nice history of the treatment of the γ rule in relevant logic, see Urquhart [208].

Ackermann gave us some hints about the nature of entailment. At the start of [1], there was a long passage about what he thought entailment was:

Rigorous implication, which we represent by $A \rightarrow B$, expresses that a logical relationship holds between A and B, that the content of B is part of the content of A, or however we want to say it. That has nothing to do with the truth or falsity of A and B. Thus one would reject the generality of the formula $A \rightarrow (B \rightarrow A)$, since the derivation [Schluss] from A to $B \rightarrow A$ has nothing to do with the truth of A, and nothing to do with whether a logical relation holds between B and A. For the same reason, the formulas $A \rightarrow (B \rightarrow (A \wedge B))$, $A \rightarrow (\neg A \rightarrow B)$, and $A \rightarrow ((A \rightarrow B) \rightarrow B)$ cannot be considered universally true. The same holds for $B \rightarrow (A \rightarrow A)$, since the correctness of $A \rightarrow A$ is independent of the truth of B. In the rejection of the last formula, the implication I employ differs from strict implication, since the notion of a proposition that contains (as in $(A \wedge \neg A) \rightarrow B$) or is contained in every proposition is not the notion of a logical connection between two statements. A formula like $(A \rightarrow B) \rightarrow ((B \rightarrow C) \rightarrow (A \rightarrow C))$, on the other hand, we can assume to be universally true because under the assumption that $A \rightarrow B$, $A \rightarrow C$ follows logically from $B \rightarrow C$. [1, p 113]

In this passage there was a rejection of the *paradoxes of material implication*, such as $A \rightarrow (B \rightarrow A)$ and $B \rightarrow (A \rightarrow A)$. First, Ackermann said that $B \rightarrow A$ "has nothing to do with the truth of A" and that the truth of $A \rightarrow A$ was independent of the truth of B. In Anderson and Belnap's natural deduction system, the idea that the consequent of a provable entailment "has to do with the truth of" the antecedent is taken up in the requirement that the antecedent really be used in the derivation of the consequent and is formalised by the rules governing the subscripted sets of hypothesis numbers. The independence of the truth of $A \rightarrow A$ from the truth of B is shown by the fact that B is not really used in the derivation of $A \rightarrow A$.

Second, Ackermann held that entailments expressed "logical relations" between antecedents and consequents. Of course everyone discussed in this book who thinks that entailment is a sensible notion thinks also that entailment expresses a relation of deducibility between antecedents and consequents, so this is nothing extraordinary. But in the first sentence of this passage, Ackermann gave us a hint of what he thinks this logical relationship is. He said that when $A \rightarrow B$ holds, the content of B was part of the content of A. Unfortunately, nowhere did Ackermann provide us with a theory of content, but much more recently Ross Brady has done so, and in Section 4.11 I move on to Brady's theory.

4.11   Content Containment

Ross Brady has given a semantical analysis of entailment in terms of the analytic containment of sentences in one another. This means that $A \rightarrow B$ is true if and only if the content of B is included in the content of A. He formalised this idea in what he calls a "content semantics" [22]. The basis of this semantics is a set of sentence contents. These contents are themselves sets of natural language sentences. For example, the content of 'John is a bachelor' includes that sentence itself, the sentences 'John is male', 'John is unmarried', and so on. In general, a content is a set of sentences closed under "analytic establishment" [25, p 263]. If sentences $S_1, \ldots, S_n$ are in a content c, and S^* follows analytically from $S_1, \ldots, S_n$, then S^* also in c.

Brady's content semantics is an algebraic semantics. The set of contents is closed under intersection and *closed union*, as well as under operators that correspond to negation and entailment. The closed union of two contents c and c' (written $c \bar{\cup} c'$) includes all the sentences that follow from the sentences in c together with the sentences in c'. Where C is the set of contents, $\langle C, \cap, \bar{\cup} \rangle$ is a lattice.

A content c entails another content c' if and only if c is a superset of c', that is $c \supseteq c'$. Thus, the idea of content inclusion is taken very literally in this semantics, with a straightforward set-theoretic reading. In addition to having a relation of entailment on contents, Brady postulates an operator that takes pairs of contents to the content that expresses that one of these contents entails the other. For this operator, I use the symbol, $\sqsupseteq$. Thus, where c and c' are contents, so is $c \sqsupseteq c'$. (I set aside the issue of negation in this semantics.)[3]

An interpretation, V, in this semantics is a function from formulas to contents. Each formula has a content. For this section only, I use '$[\![A]\!]$' for the content of the formula, A. So, where $[\![A]\!]_V = c$ and $[\![B]\!]_V = c'$, $[\![A \vee B]\!]_V = c \cap c'$, $[\![A \wedge B]\!]_V = c \bar{\cup} c'$, and $[\![A \rightarrow B]\!]_V = c \sqsupseteq c'$. This works in reverse to possible world semantics. In possible world semantics, the content (i.e. set of worlds) expressed by a conjunction is the intersection of the contents of the conjuncts, and the content of a disjunction is the union of the contents of the disjuncts. The reverse, however, makes sense in Brady's framework. Brady thinks of his contents as sets of sentences in natural language. These sentences tell us the information that is provided by the content overall. A disjunction gives us less information than either of its disjuncts, and a conjunction gives us more information than either of its conjuncts.

In addition to the set of contents, there is a set of true contents that is closed under superset, closed union, and modus ponens. This means that if c is a true content and $c \supseteq c'$, then c' is also a true content. Similarly, if c and c' are true, then so is $c \bar{\cup} c'$ and if $c \sqsupseteq c'$ and c are true, then so is c'. Moreover, if $c \cap c'$ is true, then at least one of c or c' is true and $c \sqsupseteq c'$ is true if and only if $c \supseteq c'$.

Without any additional postulates, the class of models described so far characterises an extremely weak logic. In order to obtain the usual relationships between the entailment connective and conjunction and disjunction, Brady accepts the following postulates:

1. $(c \sqsupseteq c') \bar{\cup} (c \sqsupseteq c'') \supseteq c \sqsupseteq (c' \bar{\cup} c'')$;
2. $(c \sqsupseteq c'') \bar{\cup} (c' \sqsupseteq c'') \supseteq (c \cap c') \sqsupseteq c''$.

---

[3] Brady's notion of a content is very much like C. I. Lewis's conception of an *intension* [110]. An intension of a sentence, for Lewis, was just a set of sentences closed under entailment. For Lewis, the entailment mentioned here is just the entailment that is formalised by his logic of strict implication. Brady, however, relies on a primitive notion of "analytic establishment" to describe the closure of his contents. In addition to intensions, Lewis has sense meanings and significations of sentences. A signification is a property that may or may not hold in the world. A sense meaning is a means of discovering the signification of a predicate or sentence. Brady does not, however, have anything in his theory that corresponds to Lewis's significations.

In order to make $\sqsupseteq$ seem more like $\supseteq$, Brady also adds the following postulate:

$$(c \sqsupseteq c')\overline{\cup}(c' \sqsupseteq c'') \supseteq c \sqsupseteq c''.$$

And, to ensure that provably equivalent formulas are replaceable for one another in theorems he adds an affixing rule that is an amalgam of a prefixing rule and a suffixing rule:

$$\text{If } c \supseteq c' \text{ and } c'' \supseteq c''', \text{ then } c' \sqsupseteq c'' \supseteq c \sqsupseteq c'''.$$

The logic characterised by this semantics is quite a bit weaker than E. It does not include either suffixing or prefixing, nor does it include contraction. These absences are, for Brady, not problematic, since he thinks that E is far too strong to be a viable logic of entailment.

We could, however, merely add more postulates that incorporate the various logical principles in E. We should, however, be quite wary of doing this. To use a lovely phrase I heard in a talk by Hiroakira Ono, we should not merely make our semantics a "shadow of the syntax". This is a problem generally with algebraic semantics. What we want from a semantical theory (perhaps among other features) is an independent way of justifying our logical principles. Proof theory and a semantic theory should coincide in terms of what formulas and rules they make valid, but they should lend support to one another. If they merely reflect one another in the sense that the one looks like a translation of the other into a different mathematical idiom, then this mutual support is lost. The semantics should give us a reason independent of the proof theory to understand entailment in a particular way.

There is another deeper issue that prevents me from adopting a content semantics. In Chapter 1, I argued that a logic of theory closure should not contain all the analytic truths, but clearly the intended model of the content semantics is one that does contain all analytic truths. Thus, being a logic of meaning containment is not a sufficient condition for a logic's being a theory of entailment.

There is a trivial sense, however, for which any true entailment the meaning of the antecedent must include the meaning of the consequent. I discuss this trivial sense in Section 4.12. But I don't think that meaning containment is the central notion that a logical of entailment must capture. On this topic, Richard Angell says:

[T]he concepts of deducibility and containment are two distinct concepts, and the failure to distinguish them leads to faulty attempts to merge them in formal systems. One such attempt is Anderson and Belnap's system, E, in which a Fitch-type theory of natural deduction is modified to incorporate a certain sense of "containment". Another is Parry's system, AI, of "analytic implication" which began with a more restricted sense of containment but has usually been presented as a theory of deducibility. [5, p 119]

I agree with Angell that "the concepts of deducibility and entailment are two distinct concepts", but I disagree that the natural deduction system for E is an attempt to "incorporate a certain sense of "containment". I think that Anderson and Belnap have developed a system that incorporates the notion of the real use of premises, and incorporates that notion into their understanding of the entailment connective. Real use and containment are very different notions, and the fact that it plausibly claims to formalise the notion of real use is what makes E a good candidate to be the logic of entailment.

4.12   Entailment and Analytic Implication

In the passage quoted above, Angell pointed out that Parry's logic of analytic implication (see Chapter 2) was put forward as a logic of entailment. It may seem natural, then, to combine a relevant logic of entailment with analytic implication to get a relevant analytic logic. The idea is to have a relevant logic that has the Parry property, that is, a logic in which a formula of the form $A \rightarrow B$ is provable only if all propositional variables in B are also in A. The Parry property seems a natural requirement for a relevant logic since it is a strengthening of the variable sharing property that is required for all relevant logics.

Modifying Brady's content semantics to characterise an analytic relevant entailment is a relatively straightforward manner.[4] As I say in Section 4.12, the logic of entailment should not be a logic of meaning containment. But even if we do want a logic of meaning containment, it is not clear that it needs to satisfy the Parry property. There are at least two notions of meaning containment that are salient here:

1. **Language-Relative Meaning Containment**. Given the meanings of all the expressions in the language, the meanings of certain statements contain the meanings of other statements.
2. **Agent-Relative Meaning Containment**. The meaning of a statement A contains the meaning of B if and only if knowing the meaning of A is sufficient to understand and see that the meaning of B is contained within it.

These two notions of meaning containment are rather different. Let $A \rightarrow B$ mean 'the meaning of A contains the meaning of B'. On the agent-relative notion of meaning containment, the addition axiom, $p \rightarrow (p \vee q)$, is not logically true. An agent might understand p but not know what q means. On the language-relative notion of meaning containment, however, the meanings of all the expressions in the language are presupposed and the logic merely charts their relationships. Thus, there is no reason, on the language-relative conception, to bar the addition schema.

One problem with the language-relative notion of meaning containment, however, is that almost any logic could claim to be a theory of content containment. Consider, for example, a logic with a possible world semantics. In a possible world model, every formula determines a set of worlds – those worlds in which the formula is true. This can be taken, and often is taken, to be the content of the formula. As in Chapter 3, a formula A entails B in a possible worlds model if and only if the proposition $[\![A]\!]$ is a subset of $[\![B]\!]$. (Here '$[\![A]\!]$' is the set of worlds in which A is true.) This is a very clear notion of content containment, and so any logic that has a possible worlds semantics

---

[4] I have done this with my student Shaun Hopkins in a forthcoming paper.

(and in which entailment is analysed in this way) can reasonably claim to be a logic of meaning containment in the language-relative sense.

4.13   Going Forward

My aim in this book is to give a philosophical reading and semantic grounding to the logic E. In Chapter 5, though, I look at an approach to entailment that has become widely accepted among philosophers of logic as the appropriate way to view entailment. I do not accept this view as giving us the right way to view entailment, but do wish to incorporate some of its insights into my project.

5 Reflexivity

5.1 Introduction

A reflexive logic, in the sense that I am using this term here, is a logic that interprets the rules under which the theorems of the logic itself are closed in terms of an entailment connective. This is different from some other entailment logics, in which the entailment connective represents some other, perhaps more general, notion of derivability. Here is a simple example. Suppose that a logic is formulated in terms of an axiom system that has modus ponens for material implication among its rules:

$$\frac{\vdash A \supset B \qquad \vdash A}{\vdash B}.$$

This form of modus ponens tells us that $\vdash B$ is deducible from $\vdash A \supset B$ and $\vdash A$. In order to capture the deducibility of theorems from one another, let us add a necessity connective, '$\Box$', which we read as 'it is a theorem that', as well as an entailment connective, $\rightarrow$. In a reflexive extension of this logic, modus ponens is represented as an entailment:

$$(\Box(A \supset B) \wedge \Box A) \rightarrow \Box B.$$

This scheme is an entailment form of the K axiom of classical modal logic.

I begin this chapter by briefly discussing reflexivity in the context of Peano Arithmetic (PA). As Gödel showed, PA has the resources to represent its own syntax and with it, many of its own proof-theoretic concepts. In particular, the notion of provability can be represented in PA. An entailment $A \rightarrow B$ can be represented in PA by a formula that says that $A \supset B$ is provable. This definition, however, yields a rather strange theory of entailment and I discuss its inadequacies. I then examine Dana Scott's reflexive approach to entailment. Scott presented a quite general method for turning a logic, which was presented as a proof theory, into a logic of entailment. Although I come at the end of the chapter to criticise this view of entailment, it does have some rather attractive features. In particular, it gives a precise meaning to the relationship between deducibility and entailment.

In subsequent chapters, I adopt one of these attractive features in my own theory of entailment. I require that any theory of entailment be a theory of the logic of entailment. This means that the theorems of the logic must be closed under the principles of entailment of that logic. I call this as the *weak* form of reflexivity. A logic is *strongly* reflexive if and only if its set of theorems can be generated by a finite set of axiom schemes and rules, which rules are represented as entailments within the logic itself. This chapter is mostly about this strong form of reflexivity.

There are at least two philosophical motivations for the project of creating a logic that is strongly reflexive. The first of these is an epistemological motivation. If every rule of inference used to construct a logic is also a logical entailment, then the rule could be said to be underpinned by a logical truth. If we have a theory of how we come to know logical truths or how these truths are justified, then we know how this rule is justified. In other words, the reflexivity project is supposed to lighten the epistemological load – one sort of justification is all that is needed for logical rules and the axioms of the logic.

The second project, which I think was Scott's project, is the *interpretational project* of creating a reflexive logic. The idea is to understand entailment just as the representation of a logic's rules within itself. This meaning of the entailment connective is to be understood just as representing in the logic what the logic already tells us is a valid inference.

In this chapter, I argue against both the epistemological and interpretational projects.

5.2   Gödel's Provability Predicate and Entailment

The best-known attempt to produce a reflexive logic is Gödel's construction of a provability predicate in PA. In that construction, Gödel defined an encoding that allowed him to represent as natural numbers each expression of the language, and structures of these expressions, such as the sequences of formulas that constituted proofs. Using these encodings, Gödel then constructed a predicate expression that was usually abbreviated as '*Bew*' (for 'beweisbar') or '*Prov*' (for 'provable'). I use '*Prov*'. I follow standard practice and use corner quotes to indicate the Gödel encoding of expressions. Gödel showed that *Prov* was a provability predicate in the sense that $Prov(n)$ was *provable* in PA if and only if n was the code of a formula that is provable in PA.[1]

Provability can be used in PA to define a strict implication:

$$A \to B =_{df} Prov(\ulcorner A \supset B \urcorner).$$

The question for us is whether this strict implication is really a form of entailment. *Prov* is a provability predicate in the sense that

$$\vdash_{PA} Prov(\ulcorner C \urcorner) \text{ if and only if } \vdash_{PA} C$$

---

[1] I should be distinguishing the use of 'n' in a logical scheme, in which case it is a metavariable representing expressions in the object language that refer to numbers, from its use in the metatheory, in which case it is a parameter that represents natural numbers. However, introducing more notation at this point will only make a complicated topic more complicated and add nothing of real value.

and so, in particular,

$$\vdash_{PA} Prov(\ulcorner A \supset B \urcorner) \text{ if and only if } \vdash_{PA} A \supset B,$$

that is,

$$\vdash_{PA} A \to B \text{ if and only if } \vdash_{PA} A \supset B.$$

In order to see whether this arrow is really a form of entailment, we need to select some consequence relation that strict implication is supposed to express. I use a fairly standard definition of proof in PA to define a consequence relation.

Let Γ be a finite set of formulas and A be some formula. A proof from Γ to A is a sequence of formulas such that A is the last member of the sequence, and every member of the sequence either is a member of Γ, a theorem of PA, or follows from previous members of the sequence by modus ponens for the material conditional. Where A is provable from Γ in PA, let us write '$\Gamma \Vdash_{PA} A$'. The deduction theorem for classical first-order logic can be applied to show that

$$A \Vdash_{PA} B \text{ if and only if } \vdash_{PA} A \supset B.$$

Moreover, *Prov* is a provability predicate, and so

$$\vdash_{PA} A \supset B \text{ if and only if } \vdash_{PA} Prov(\ulcorner A \supset B \urcorner).$$

$Prov(\ulcorner A \supset B \urcorner)$ is a form of strict implication. This strict implication can be said to express the consequence relation in the sense that what strict implications are provable correspond to the proofs of PA. But is it really an entailment connective?

If the connective $\to$ in PA is to be considered an entailment, then for any formulas A and B, $A \to B$ must be true if and only if $A \Vdash_{PA} B$. To see whether this is so, we need to discuss models for PA. There are some models for PA in which strict implication clearly does not represent deducibility in that model. To show this, I appeal to a famous theorem of Martin Löb. Löb showed that, for any formula A, if it is provable in PA that $Prov(\ulcorner A \urcorner) \supset A$, then A is provable (see [21]). If PA is consistent, however, there are some formulas that are not provable in PA, so there are some formulas A for which $Prov(\ulcorner A \urcorner) \supset A$ is not provable. This means that there are some models in which $Prov(\ulcorner A \urcorner)$ is true but A is false. This is a failure of reflexivity of a sort, but not really the sort that I am interested in; it does, however, give rise to a failure of the salient sort of reflexivity.

In classical logic, for any formula A, it is a theorem that

$$1. \ A \equiv ((A \supset A) \supset A).$$

Let $\mathfrak{M}$ be a model for PA and A be a formula such that $\mathfrak{M} \models Prov(\ulcorner A \urcorner)$ and $\mathfrak{M} \not\models A$. Then we know that

$$2. \ \mathfrak{M} \not\models (A \supset A) \supset A.$$

From 1, we can derive in classical logic (and hence in PA) that

$$3. \ A \supset ((A \supset A) \supset A).$$

Now I appeal to the fact that the provability predicate acts like a regular modal operator, that is to say, it obeys the following rule:

$$\frac{\vdash B \supset C}{\vdash Prov(\ulcorner B \urcorner) \supset Prov(\ulcorner C \urcorner)}.$$

From 3 and regularity, we can prove that

4. $Prov(\ulcorner A \urcorner) \supset Prov(\ulcorner (A \supset A) \supset A \urcorner)$.

Thus, we have from 4 and the assumption that $\mathfrak{M} \vDash Prov(\ulcorner A \urcorner)$,

5. $\mathfrak{M} \vDash Prov(\ulcorner (A \supset A) \supset A \urcorner)$.

From 5 and the definition of $\rightarrow$, we obtain

6. $\mathfrak{M} \vDash (A \supset A) \rightarrow A$.

But $A \supset A$ is true in all models of PA and $\mathfrak{M} \nvDash A$. Thus, in $\mathfrak{M}$, there are true entailments that do not obey modus ponens. This means that $\rightarrow$ in PA is not really an entailment. If it is true that A is derivable from $A \supset A$ and it is true that $A \supset A$, then A. So, $\rightarrow$ cannot really be said to represent derivability in PA and so it is not really an entailment connective.

One might reply that, whereas the arrow does not represent entailment in all models, it does do so in the standard model of arithmetic. The standard model of arithmetic contains all and only those arithmetical statements that are actually true. For the standard model, $\mathfrak{S}$, if $\mathfrak{S} \vDash A \rightarrow B$ and $\mathfrak{S} \vDash A$, then $\mathfrak{S} \vDash B$. The reasoning behind this is rather simple. If $\mathfrak{S} \vDash Prov(\ulcorner A \supset B \urcorner)$, then it is true that $A \supset B$ is provable in PA. Since $\mathfrak{S}$ is a model of PA, then $\mathfrak{S} \vDash A \supset B$. If $\mathfrak{S} \vDash A \supset B$ and $\mathfrak{S} \vDash A$, then $\mathfrak{S} \vDash B$.

This appeal to the standard model is, from the perspective of a theory of entailment, problematic. There is still a failure of reflexivity. The provability predicate is for PA, not for the standard model. Since there are some models for PA in which strict implication does not represent provability, we have to say that the axioms and rules of PA do not capture an adequate notion of entailment. In the standard model strict implication may represent entailment for PA, but why should that be considered to be *entailment* simpliciter? PA would seem to lose its special status if it is merely considered to be a subsystem of the one true system (i.e. the set of formulas true in the standard model) and that this latter system needs to be considered in order to describe entailment in PA.

Jc Beall and Julian Murzi [12] give a rather different proof that PA does not contain its own entailment. Their argument is more general than mine. It is supposed to show that PA *cannot* contain its own entailment predicate. This means that the argument is supposed to show that no predicate definable in PA can be interpreted as representing entailment. I follow these authors in using '$Val(\ulcorner A \urcorner, \ulcorner B \urcorner)$' to mean that B is validly derivable from A. Thus, $Val(x, y)$ is supposed to be an entailment predicate. They place two constraints on Val, that it must satisfy in order to be considered an entailment predicate:

VS1 If $A \Vdash_{PA} B$, then $\vdash_{PA} Val(\ulcorner A \urcorner, \ulcorner B \urcorner)$;
VS2 $\vdash_{PA} Val(\ulcorner A \urcorner, \ulcorner B \urcorner) \supset (A \supset B)$.

VS1 says simply that provable Val-statements must represent at least all of the valid derivations in PA. Clearly, this is the minimal constraint needed on an entailment predicate. VS2 says that Val-statements must preserve truth. I discuss the problem with thinking this with regard to provability statements, and so we should be wary about VS2. But with both provability and validity statements, VS2 is intuitive.

I now give a simple model-theoretic version of the argument. By Gödel's fixed-point theorem, we know that there is a formula Π for which it is provable that

$$\Pi \equiv Val(\ulcorner \Pi \urcorner, \ulcorner \bot \urcorner),$$

where $\bot$ is an arbitrary contradiction. Let us assume for a moment that there is an arbitrary model, $\mathfrak{M}$ such that

(1) $\mathfrak{M} \vDash \Pi$.

By (1) and the fixed-point theorem,

(2) $\mathfrak{M} \vDash Val(\ulcorner \Pi \urcorner, \ulcorner \bot \urcorner)$.

By (2) and VS2,

(3) $\mathfrak{M} \vDash \Pi \supset \bot$.

Thus, by (1) and (3),

(4) $\mathfrak{M} \vDash \bot$,

which is impossible. So, no model satisfies Π and, hence, no model satisfies $Val(\ulcorner \Pi \urcorner, \ulcorner \bot \urcorner)$.

Thus, by soundness,

(5) $\nvdash_{PA} Val(\ulcorner \Pi \urcorner, \ulcorner \bot \urcorner)$.

By (5) and VS1,

(6) $\Pi \nVdash_{PA} \bot$.

Note that PA is a theory of first-order classical logic. As such, it follows from the completeness theorem for classical first-order logic that PA is complete over the class of its first-order models. (This does not conflict with Gödel's first incompleteness theorem that states, in effect, that PA is incomplete over its standard model.) By this form of semantic completeness, however, from (6) we obtain that there is some model $\mathfrak{M}'$ such that

(7) $\mathfrak{M}' \vDash \Pi$

and

(8) $\mathfrak{M}' \nvDash \bot$.

But, as steps (1) to (5) show, (7) cannot be true. So, it seems that there cannot be a predicate definable in PA that satisfies VS1 and VS2.

One problem that underlies the difficulties with both the provability and valid deducibility interpretations of entailment in PA is that PA contains enough mathematics to prove the fixed-point theorem. In that sense, there is *too much reflexivity in* PA. The reply that is put forward to the validity-Curry paradox is to keep out the mathematical mechanism that allows the proof of the fixed-point theorem, Löb's theorem, and similar theorems that produce these sorts of paradoxes of reflexivity. In the remainder of the chapter, I examine reflexive theories of entailment based on logics that do not contain arithmetic or any other sort of mathematics that allow one to prove these sorts of theorems.

5.3   Reflexivity and Axiom Systems

As a gentle lead-in to Scott's theory of entailment, I introduce his concepts in a somewhat simpler framework. Consider the following standard axiomatisation of classical propositional logic:

- $A \supset (B \supset A)$
- $(A \supset (B \supset C)) \supset ((A \supset B) \supset (A \supset C))$
- $(A \supset \neg B) \supset (B \supset \neg A)$
- $\neg\neg A \supset A$

$$\frac{\vdash A \supset B \qquad \vdash A}{\vdash B}.$$

To the language, I add the entailment connective, $\rightarrow$, and a propositional constant, $\top$ (called 'top'). This constant is similar to the constant t that I use in Chapter 4 to formulate necessity in E, in that both constants represent all the theorems of their respective logics. There are, however, some important differences. In classical logic, every theorem is equivalent to every other theorem, but this is not true of relevant logics. We can take $\top$ to be an arbitrary theorem of classical logic. (Recall that we understand t to be the conjunction of all the theorems of E.) Moreover, $\neg\top$ is equivalent to $\bot$ (see Section 5.2).

In order to respect the meaning of entailment and top, we need to add $\top$ itself as an axiom and add the rules:

$$\frac{\vdash A}{\vdash \top \rightarrow A} \qquad \frac{\vdash A \rightarrow B \qquad \vdash A}{\vdash B}$$

And, following Scott, I formulate a necessity operator, $\Box$, that means 'is a theorem':

$$\Box A =_{df} \top \rightarrow A.$$

This definition says that a formula is a theorem if and only if it follows from the theorem, $\top$. This makes sense for classical logic, since all its theorems are equivalent to one another. Given this definition, we can rewrite the left-hand rule as

$$\frac{\vdash A}{\vdash \Box A}.$$

This, of course, is the rule of necessitation.

The rule of necessitation itself needs to be represented in the logic. This is what the reflexivity project is all about. This representation is done by the procedure of "horizontalization" – the rewriting of the rule as a formula. The horizontalization of the necessitation rule is

$$\Box A \to \Box\Box A.$$

This tells us that A's being a theorem entails that $\Box A$ is also a theorem, and that would seem to be what the rule says too. This formula is the 4 principle, which will be familiar to many readers from modal logic.

In order to represent modus ponens for the material conditional, we need to use conjunction to depict there being two premises. Conjunction is here defined as usual as

$$A \wedge B =_{df} \neg(A \supset \neg B).$$

Then the horizontalization of modus ponens is

$$(\Box(A \supset B) \wedge \Box A) \to \Box B.$$

Let us assume a deduction theorem (see Section 2.10) for the final logic (with entailment and top), of the form *if B is derivable from $A_1, \ldots, A_n$, then $(A_1 \wedge \ldots \wedge A_n) \supset B$ is a theorem of the logic*. This assumption allows us to claim that $(A_1 \wedge \ldots \wedge A_n) \to B$ represents not only the derivability of B from $A_1, \ldots, A_n$ but also the theoremhood of $(A_1 \wedge \ldots \wedge A_n) \supset B$. Since, $\Box C$ represents $\vdash C$, then we can say that $(A_1 \wedge \ldots A_n) \to B$ means the same thing as $\Box((A_1 \wedge \ldots A_n) \supset B)$. In particular, we take $A \to B$ to be equivalent to $\Box(A \supset B)$.

Because of the simplicity of the current logic (it does not contain arithmetic of any sort), we can add a version of the principle VS2 from Section 5.2:

$$(A \to B) \to (A \supset B).$$

This principle says that entailment preserves truth. An instance of this is

$$(\top \to A) \to (\top \supset A).$$

By the definition of $\Box$ and the fact that $\top \supset A$ is equivalent in classical logic to A, we obtain

$$\Box A \to A$$

and this is a necessitated version of the T axiom for modal logic. Adding the salient version of VS2 then allows us to prove all of the theorems of the modal logic S4. In particular, the system includes all the axioms, that is, it includes all the theorems of classical propositional calculus, together with

K $\Box(A \supset B) \supset (\Box A \supset \Box B)$,
T $\Box A \supset A$,
4 $\Box A \supset \Box\Box A$,

and the rules of necessitation and modus ponens for the material conditional. Taking $A \to B$ to be defined as $\Box(A \supset B)$, we can prove all the principles that I have cited here in S4. Thus, the system (as understood thus far) is just S4.

Turning briefly to quantification, it is common in axiom systems to include a rule form of universal generalisation, such as,

$$\frac{\vdash A(x)}{\vdash \forall x A(x)}.$$

Clearly, we do not want to include $\Box A(x) \to \Box \forall x A(x)$ in our system. Instead, one might take '$\vdash A(x)$' to be telling us that $A(x)$ is proven to hold of any object in the domain of any model for the logic. In Chapters 10 and 11, I argue that this reading of '$\vdash A(x)$' is inadequate, but I set that aside here. As such, we may interpret $\vdash A(x)$ in the object language as

$$\forall x \Box A(x)$$

and so the entailment corresponding to the rule of universal generalisation is

$$\forall x \Box A(x) \to \Box \forall x A(x).$$

This is the famous *Barcan formula*. In Chapter 10, I reject the Barcan formula, because it does not fit neatly within my semantic framework. My rejection of the reading of the Barcan formula as an adequate interpretation of the rule of universal generalisation allows me to abandon the Barcan formula as well.

5.4   Reflexivity and Sequent Systems

In "Engendering an Illusion of Understanding" [190], Scott constructed his method of determining a logic of entailment around a Gentzen-style sequent calculus for classical logic. The outcome of this method was the same version of S4 that is constructed in Section 5.3, but Scott's use of a more robust proof theory is illuminating and helps with my discussion of relevant logic and reflexivity later in this chapter.

I do not go into length about the nature of Gentzen systems here, since they do not play a role in any other section in this book, but I do need to say a few words about their form in order to explain what Scott was doing. A sequent in Gentzen's system for classical logic is a structure of the form

$$\Gamma \Vdash \Delta,$$

where both Γ and Δ are sets of formulas. In the case where both Γ and Δ are finite, we can read '$\Gamma \Vdash \Delta$' as saying that the disjunction of formulas in Δ is derivable from the conjunction of formulas in Γ. Where Γ contains only a finite set of formulas, say, $A_1, \ldots, A_n$, I follow the standard practice and write '$A_1, \ldots, A_n \Vdash \Delta$' instead of '$\{A_1, \ldots, A_n\} \Vdash \Delta$', and similarly for Δ.

There are three types of rule in a Gentzen system. First, there are axioms. In the standard Gentzen system for classical logic, the sequent

$$A \Vdash A$$

is an axiom, for every atomic formula A. For each connective, there are rules that allow the introduction of formulas in which that connective is the main operator on the left-hand side of the turnstile and rules that allow for the introduction of such formulas on the right-hand side of the turnstile. For example, the introduction on the right rule for the material conditional is

$$\frac{\Gamma, A \Vdash B, \Delta}{\Gamma \Vdash A \supset B, \Delta} \, .$$

This rule says that if the formulas in Γ, conjoined with A, logically imply B disjoined with the formulas in Δ, then the conjunction of the formulas in Γ logically implies the disjunction of $A \supset B$ along with the formulas in Δ. The horizontalisation of this rule is

$$((G \wedge A) \to (B \vee D)) \to (G \to ((A \supset B) \vee D)).$$

The introduction rule for the material conditional on the left is

$$\frac{\Gamma, B \Vdash \Delta \qquad \Gamma' \Vdash A, \Delta'}{\Gamma, \Gamma', A \supset B \Vdash \Delta, \Delta'} \, .$$

Its horizontalisation is

$$(((G \wedge B) \to D) \wedge (G' \to (A \vee D'))) \to ((G \wedge G' \wedge (A \supset B)) \to (D \vee D')).$$

The material conditional introduction on the left rule takes a while to get used to, as does its horizontalisation, but both are very useful.

The third type of rule is a structural rule. Structural rules can be represented without any of the connectives and concern with the nature of deducibility itself, rather than the meanings of any particular connectives. For example, the rule of weakening on the left says that we can add any formula to the premise set of a sequent and still have a provable sequent:

$$\frac{\Gamma \Vdash \Delta}{\Gamma, A \Vdash \Delta} \, .$$

The horizontalisation of weakening is straightforward:

$$(G \to D) \to ((G \wedge A) \to D).$$

A more interesting structural rule is the cut rule:

$$\frac{\Gamma, A \Vdash \Delta \qquad \Gamma' \Vdash A, \Delta'}{\Gamma, \Gamma' \Vdash \Delta, \Delta'} \, .$$

An instance of this rule is the following transitivity rule:

$$\frac{B \Vdash C \qquad A \Vdash B}{A \Vdash C} \, .$$

The horizontalization of transitivity is the conjunctive syllogism scheme, which plays an important role in the logic of entailment discussed in Chapter 7:

$$((B \rightarrow C) \wedge (A \rightarrow B)) \rightarrow (A \rightarrow C).$$

This example illustrates how Scott's method, applied to a robust proof theory, can lead us to understand nested entailments. I apply this in Section 5.5 to the natural deduction system for the relevant logic R.

5.5   Reflexivity and Relevance

In order to provide more relevant content in this section, I use the logic R as a base. Some of the ideas I employ here come from Kosta Došen [50], but the logic that results here is somewhat different from his. Like Došen, I start from a natural deduction system, but unlike him I use Anderson and Belnap's system (see Section 4.3).

I begin with the logic R, as presented in Section 4.3. I add to the language three pieces of vocabulary. One is the entailment connective, $\rightarrow$, and another is t (from the language of logic E), which I will define a logical necessity connective. The third piece is a binary connective called "fusion" ($\circ$). Fusion is introduced to help translate some natural deduction rules into the object language. Fusion could be dispensed with, but the logic R together with entailment is more elegant with it.

Fusion is an intensional form of conjunction. Its similarity to extensional conjunction can be seen in its introduction rule ($\circ$I):

$$A_\alpha$$
$$B_\beta$$
$$\downarrow$$
$$A \circ B_{\alpha \cup \beta}$$

This rule generalises the extensional conjunction introduction rule ($\wedge$I). It does not require that the subscripts on the two conjuncts be the same. This rule gives fusion a close connection to implication, which is illustrated in the following derivation:

1	$(A \circ B) \Rightarrow C_{\{1\}}$	hypothesis
2	$A_{\{2\}}$	hypothesis
3	$B_{\{3\}}$	hypothesis
4	$A_{\{2\}}$	2, reiteration
5	$A \circ B_{\{2,3\}}$	3,4, $\circ$I
6	$(A \circ B) \Rightarrow C_{\{1\}}$	1, reiteration
7	$C_{\{1,2,3\}}$	5,6, $\Rightarrow$E
8	$B \Rightarrow C_{\{1,2\}}$	3–7, $\Rightarrow$I
9	$A \Rightarrow (B \Rightarrow C)_{\{1\}}$	2–8, $\Rightarrow$I
10	$((A \circ B) \Rightarrow C) \Rightarrow (A \Rightarrow (B \Rightarrow C))_\emptyset$	1–9, $\Rightarrow$I

The scheme $((A \circ B) \Rightarrow C) \Rightarrow (A \Rightarrow (B \Rightarrow C))$ is the principle of exportation. It is not valid in R if we replace fusion with extensional conjunction. This connection between fusion and implication is made even closer by the fusion elimination rule. The rule $\circ$E is a complex form of implication elimination:

$$A \circ B_\alpha$$
$$A \Rightarrow (B \Rightarrow C)_\beta$$
$$\downarrow$$
$$C_{\alpha \cup \beta}$$

This elimination rule allows for the very easy derivation of the converse of exportation, that is, the principle of importation: $(A \Rightarrow (B \Rightarrow C)) \Rightarrow ((A \circ B) \Rightarrow C)$. Note that fusion in R is commutative and associative.[2]

The entailment connective is used to represent derivations in the object language. Entailment and fusion together can be used to represent lines in a natural deduction proof. For example, suppose that the step $B_{\{1,2,3\}}$ is provable in a valid deduction, where the hypotheses 1, 2, and 3 are A_1, A_2, and A_3, respectively. Then we represent this in the extended logic by making

$$(A_1 \circ A_2 \circ A_3) \to B,$$

a theorem. Where $A_\emptyset$ is provable in a derivation, then $t \to A$ is a theorem of the logic. It is not enough, however, just to have these rules. They do not tell us what formulas with nested entailments are valid. In order to do so, I add mechanisms and rules to the natural deduction system to treat t and entailment.

In order to present nested entailments, we need to write derivations within derivations. In Gentzen systems this is done by having sequents derivable from sequents. In natural deduction systems, nested entailments are represented by subproofs within proofs. The sort of subproofs that give rise to entailments are not ordinary subproofs in a derivation in the system for R, but strict subproofs that only allow information about entailments to be reiterated from superior proofs into subproofs.

In effect, what I do now is justify the rules for entailment and t from Section 4.9. The introduction rule for entailment is:

$$\square \quad A_{\{k\}} \qquad\qquad \text{hypothesis}$$
$$\vdots$$
$$B_\alpha$$
$$A \to B_{\alpha-k} \qquad \to\text{I}$$

where k is in α. Only entailments are allowed to be iterated across the strict scope indicator, and so what is concluded from the discharging of this scope indicator concerns only entailments. The elimination rule is the relevant version of modus ponens.

---

[2] Fusion is neither commutative nor completely associative if added to E or to weaker relevant logics.

$$A \rightarrow B_\alpha$$
$$A_\beta$$
$$\downarrow$$
$$B_{\alpha\cup\beta}$$

The entailment elimination rules tell us that entailment preserves truth, and so is similar to condition VS2 from Section 5.2.

I adopt rules for t from Section 4.9. Recall that the t-elimination rule is:

$$t \rightarrow A_\alpha$$
$$\downarrow$$
$$A_\alpha$$

This rule is easy to justify. It is an enthymatic version of modus ponens. It suppresses the assumption that t is a theorem. The t-introduction rule says that any entailment, if true, is a theorem:

$$A \rightarrow B_\alpha$$
$$\downarrow$$
$$t \rightarrow (A \rightarrow B)_\alpha$$

$A \rightarrow B$ says that there is a logically valid deduction from A to B. If $A \rightarrow B$ is logically valid, and the arrow is really an entailment connective, then $A \rightarrow B$ must be a theorem.

I also need to add two rules that allow t to interact with contingent relevant implication ($\Rightarrow$):

$$
\begin{array}{cc}
t \Rightarrow A_\alpha & A_\alpha \\
\downarrow & \downarrow \\
A_\alpha & t \Rightarrow A_\alpha
\end{array}
$$

These two rules are not as easy to justify. I try to do so in Section 5.6.

The resulting logic is the system that Robert Meyer constructed to capture entailment in a modalised version of R, that is, it is the logic NR (see Section 4.8). The modal axioms of NR are the following:

EK $(\Box A \wedge \Box B) \Rightarrow \Box(A \wedge B)$;
IK $\Box(A \Rightarrow B) \Rightarrow (\Box A \Rightarrow \Box B)$;
T$_\Rightarrow$ $\Box A \Rightarrow A$;
4$_\Rightarrow$ $\Box A \Rightarrow \Box\Box A$.

EK is the version of the modal logic thesis K concerning extensional conjunction. IK, on the other hand, concerns entailment, which is intensional. EK is easily proven in the natural deduction system. T$_\Rightarrow$ follows directly from t-elimination, and 4$_\Rightarrow$ follows from t-introduction. Here is a derivation of IK:

1	$t \rightarrow (A \Rightarrow B)_{\{1\}}$	hypothesis
2	$t \rightarrow A_{\{2\}}$	hypothesis
3	$\square$ $t_{\{3\}}$	hypothesis
4	$t \rightarrow (A \Rightarrow B)_{\{1\}}$	1, reiteration
5	$A \Rightarrow B_{\{1,3\}}$	3,4, $\rightarrow$E
6	$t \rightarrow A_{\{2\}}$	2, reiteration
7	$A_{\{2,3\}}$	3,6, $\rightarrow$E
8	$B_{\{1,2,3\}}$	5,7, $\Rightarrow$E
9	$t \rightarrow B_{\{1,2\}}$	3–8, $\rightarrow$I
10	$(t \rightarrow A) \Rightarrow (t \rightarrow B)_{\{1\}}$	2–9, $\Rightarrow$I
11	$(t \rightarrow (A \Rightarrow B)) \Rightarrow ((t \rightarrow A) \Rightarrow (t \rightarrow B))_\emptyset$	1–10, $\Rightarrow$I

Therefore, by the definition of $\square$, this proof shows that $\square(A \Rightarrow B) \Rightarrow (\square A \Rightarrow \square B)$. The proof of the necessitation rule follows directly from t-introduction. Hence, all of the theorems of NR are provable in this logic.

Conversely, all the theorems of this logic can be proven in NR. To show this, we can adopt the natural deduction system for NR that is given in Anderson, Belnap, and Dunn's *Entailment* [3, pp xxv–xxvii], and prove that all the rules of the present logic are derivable in it. This is easily done, and I do not do this here. However, because these logics coincide in their theorems (and, in fact, in their natural deduction rules), I refer to the present logic as NR.

Unfortunately, the reflexive approach does not give a complete interpretation of the logic. In order for it to do so, as in Scott's classical project, the non-modal rules would have to be understood prior to the introduction of the modal rules.

5.6   Entailment and Information

In order to give a full interpretation of the proof theory for NR, I turn to the theory of situated inference that I present in *Relevant Logic* [129]. In order to explain this interpretation, I first briefly explain the Routley–Meyer semantics.

A Routley–Meyer frame for a relevant logic consists of a set of indices, which I think of as situations in the sense of Barwise and Perry [8]. A situation is a perhaps partial and perhaps impossible representation of a world. The set of situations is divided into normal situations and non-normal situations. And, there is a *three place* accessibility relation, R, on the set of situations. This accessibility relation is used to give a satisfaction condition for implication:

$$a \vDash A \Rightarrow B \text{ if and only if for all } b \vDash A, \ Rabc \text{ implies } c \vDash B.$$

My way of understanding the relation R is to think about what one can deduce in a situation. If one is in a world in which a and b both obtain, then he or she is deductively warranted in holding that some c also obtains in that world such that $Rabc$. The information in a situation may warrant that the presence of an A situation indicates the presence of a B situation. For example, physicists armed with the laws of gravity infer from the nature of the Earth's tidal bulge that the moon is being forced away from the Earth. We can analyse in terms of situations: one that contains the physicists and the information concerning the laws of gravity, one that contains the features of the Earth's tidal bulge, and many situations that contain the moon's moving away from the Earth. Given the first two situations, scientists can infer that the universe contains one of the third sort of situations.

The laws of nature used by scientists are examples of information that links other pieces of information together. Another sort of linking information is logical information. A piece of logical information tells us about the general closure conditions on the information in situations. On this view, the information that A entails B tells us that every situation that contains the information that A also contains the information B. Some situations give us accurate information about the world (or the domain of situations) and some do not. Some situations even have inaccurate logical information – what they say about the closure of situations is not always true.

In order to model entailment, I add a three-place relation E on situations such that $Eabc$ says that the logical information in a tells us that if b obtains in the same world as a, then a situation like c also obtains there. By 'obtains' in this context, I mean that the information contained in a situation accurately describes a world. Then I set

$$a \vDash A \rightarrow B \text{ if and only if for all } b \vDash A, \; Eabc \text{ implies } c \vDash B.$$

Clearly, logical information is a form of linking information. Thus, E is really an abstraction from R in the sense that

$$\text{If } Rabc \text{ then } Eabc, \text{ for all situations } a, b, \text{ and } c.$$

$Eabc$ holds less commonly than $Rabc$ does. This gives us the semantic consequence,

$$A \rightarrow B \vDash A \Rightarrow B.$$

In every situation in which A entails B, A also implies B. By virtue of this connection, entailment picks up certain properties from implication. For example, for all situations, $Raaa$. If a tells us that A implies B and it also tells us that A, it tells us that B as well. Thus, $A \Rightarrow B, A \vDash B$. Now suppose that $A \rightarrow B$ holds in a and so does A. Then $A \Rightarrow B$ and a are both true at a, but then so is B. Thus, we get $A \rightarrow B, A \vDash B$.

Turning to the issue of reflexivity, we can easily get if $A \vDash B$ holds in a model, then $A \rightarrow B$ is valid in that model. To do so we need a class of situations O (for 'origin')[3] that contain all the correct logical information. $A \rightarrow B$ obtains at every situation in O if and only if all situations that contain A also contain B. To get this, we postulate that there is a situation o in O such that $Eoab$ if and only if $a \leq b$, where '$a \leq b$'

---

[3] O is the set of normal situations.

says that b contains all the information that a contains. We then say all and only those formulas satisfied at every member of O are valid in the model.

This manoeuvre by itself, however, does not allow us to prove that the semantic correlate of horizontalisation holds. We cannot prove, for example, that $((A \to B) \wedge (B \to C)) \to (A \to C)$, even though it is valid on the class of models that if $A \to B$ and $B \to C$ are valid, then so is $A \to C$. But we can do much better.

Let us set $\Box A =_{df} t \to A$ and have t satisfied by all and only situations in O. We can then construct a binary modal accessibility relation, M. Mab if and only if for some o in O, $Eaob$. From this definition, we can easily derive that

$$a \vDash \Box A \text{ if and only if for all } b, Mab, b \vDash A.$$

Now, we need to postulate that

$$Eabc \text{ if and only if there is some situation } x, Max \text{ and } Rxbc.$$

This says that semantically $\to$ is the strict form of $\Rightarrow$. If we add that M is reflexive, then we can derive that $Eabc$ implies $Rabc$.

It is easy to prove that $(\Box A \wedge \Box B) \to \Box (A \wedge B)$ is valid in the class of these models. So it is also easy to prove that

$$(\Box (A \Rightarrow B) \wedge \Box (B \to C)) \to \Box ((A \Rightarrow B) \wedge (B \Rightarrow C))$$

is also valid, since it is merely an instance of the previous formula. But

$$(A \Rightarrow B) \wedge (B \Rightarrow C) \vDash A \Rightarrow C$$

is valid on the class of R models. So we get

$$(\Box (A \Rightarrow B) \wedge \Box (B \to C)) \to \Box (A \Rightarrow C),$$

which is equivalent to

$$((A \to B) \wedge (B \to C)) \to (A \to C).$$

We have proven the horizontalisation of the transitivity of entailment.

To get $\Box A \to \Box \Box A$, one only needs to add that M is transitive. In order to derive all of NR, one needs to add

$$\Box (A \Rightarrow B) \to (\Box A \Rightarrow \Box B).$$

This is an intensional form of the standard K postulate of modal logic. As I show in Section 5.5, this is derivable given the addition of the rules for entailment in the natural deduction system for R. Unfortunately, what we have to do to the model theory to make this scheme valid is not pretty. The only way is to add the following postulate (or something similar) [183]:

$$\exists x (Rabx \wedge Mxc) \text{ implies } \exists x (Mbx \wedge Eaxc).$$

I am not sure how to give this postulate an interpretation in the present framework that makes it plausible.

Setting aside the specifics of the mathematical mechanism used to model the logic R and its modal extension, there is a philosophical point to be made here. This project of extending R to a theory of entailment makes the notion of entailment dependent on there being in the logic a notion of contingent implication. There is nothing untoward about adding modal operators to a relevant logic. I spent a good deal of my early logical career doing so. The tasks that I have in mind for a logic of entailment, however, do not seem to require an underlying theory of contingent implication. The project of the construction of a logic of theory closure and the representation of proof plans in particular do not appear to demand the presence of a contingent form of implication. Of course, this does not mean that there is no such operator lurking in the background, but there needs to be an argument for this view if it is the case. It is my purpose in this book to motivate E independently of my views about R, and if we start with the logic R, we end up with a logic (NR) that is different from E.

5.7   Reflexivity and the Logic E

The logic E has its own entailment connective. That, after all, is the point of E. We can think of E as reflexive in at least three different ways. First, Anderson and Belnap's formulation of E captures its own primitive rules as theorems. These rules are modus ponens and adjunction:

$$\frac{\vdash A \to B \quad \vdash A}{\vdash B} \qquad \frac{\vdash A \quad \vdash B}{\vdash A \wedge B}$$

Defining logical necessity as $\Box A =_{df} t \to A$ allows for easy proofs of

$$(\Box(A \to B) \wedge \Box A) \to \Box B$$

and

$$(\Box A \wedge \Box B) \to \Box(A \wedge B),$$

which are the horizontalisations of modus ponens and adjunction.

Second, the relationship between the entailment connective and natural deduction proofs is also rather straightforward. If there is a valid deduction of the form,

$$A^1_{\{1\}} \qquad \text{hypothesis}$$

$$A^n_{\{n\}} \qquad \text{hypothesis}$$

$$B_{\{1,\dots,n\}}$$

then

$$A^1 \to (\dots (A^n \to B)\dots)$$

is a theorem of E. This follows from repeated applications of the entailment introduction rule. The nested entailments represent the various subproofs of the derivation. This is not surprising since the notion of a subproof is introduced primarily to formulate the rule of entailment introduction. Subproofs play similar roles in the treatment of classical, modal, and intuitionist logics. The important difference between E and these other logics is that, because of the use of subscripts, provable entailments track proofs in which all the premises are really used in the proofs of their conclusions.

What is more interesting is the relationship between E and its *derivable* rules. A rule,

$$\frac{A_1 \quad \cdots \quad A_n}{B}$$

is derivable in a logical system if and only if when all the premises $A_1, \ldots, A_n$ (together with their substitution instances) are added to the axioms of the logic, then B is derivable from the new axiom set and the rules of the system. For example, in E we can derive the rule

$$\frac{A \to (B \to C) \quad B}{A \to C} \ .$$

Here is a sketch of a proof of this rule:

$$\frac{\dfrac{\vdash B}{\vdash t \to B} \quad \dfrac{\vdash (t \to B) \to (((t \to B) \to C) \to C)}{\vdash ((t \to B) \to C) \to C} \quad \dfrac{\vdash A \to (B \to C)}{\vdash A \to ((t \to B) \to C)}}{\vdash A \to C} \ .$$

In this proof, it is assumed that the premises of the rule to be derived are theorems of the logic, and some of the actual theorems of E are used as well.

The notion of a derivable rule can be captured in E using the device of *enthymatic entailment*. A formula A enthymatically entails B if and only if it is true that $(A \land t) \to B$. This means that A enthymatically entails B if and only if A, together with the theorems of E, entails B. It is called 'enthymatic' entailment because the use of t indicates that the deduction from A to B might be an enthymeme and to be made explicitly valid it might require the use of one or more theorems. So, the theorems are suppressed premises of a sort.

It can be easily proven, as a corollary of the completeness theorem, that if the rule

$$\frac{A_1 \quad \cdots \quad A_n}{B}$$

is derivable in E, then it is a theorem that

$$(A_1 \land \ldots \land A_n \land t) \to B$$

see [4]. For a relevant logician, it is quite useful to be able to distinguish between cases in which a formula entails another *simpliciter* and those in which a formula entails another when in the presence of the theorems of the logic. Hence the distinction between valid derivations and derivable rules and the parallel distinction between entailments *simpliciter* and enthymatic entailments.

Ackermann, however, formulated his version of E with a disjunctive syllogism rule, and this rule is not represented in the logic. The disjunctive syllogism rule is *admissible* in E, in the sense that the set of theorems of E is closed under it, but it is not a derivable rule. Similarly, the rule of explosion, viz.,

$$\frac{A \qquad \neg A}{B},$$

is also admissible in the system. Since no substitution makes both premises into theorems of the logic, the logic could be said to be closed under this rule. But, thankfully, the rule is not derivable. There are some admissible rules, such as these two, that are not represented as entailment theorems of any sort.

The question arises, then, whether this sort of failure of reflexivity is an important problem. By itself, reflexivity is not a virtue in any strong sense. There are rules in a proof system that are meant to capture the rules of good reasoning. For relevant logicians, explosion and disjunctive syllogism – perhaps except in special circumstances – do not represent good reasoning. Logical systems are supposed to characterise the very general principles of good deductive reasoning and if very local or specialised principles are used to characterise the class of general principles, so be it.

5.8 Internal and External Rules

Let us look in more depth at the notion of specialised rules of reasoning. Consider again Ackermann's use of disjunctive syllogism as a rule of his logical system. He must have known, because of his familiarity with Lewis and Langford's book, that he could not add the principle,

$$((A \vee B) \wedge \neg A) \rightarrow B,$$

to the logic without thereby obtaining some of the paradoxes of strict implication. So, Ackermann must have thought of the rule of disjunctive syllogism as a rule of inference that is appropriate to apply to theorems of a logic but not to all other subject matters. In "Relevant Logic and the Philosophy of Mathematics" [132], I make a very similar distinction. There, I distinguish between the inferential and generative role of logical principles. A logical rule or entailment has an inferential role if it is warranted for it to be applied to any subject matter. Principles that have an inferential role (or "inferential principles") are perfectly general. On the other hand, we sometimes state a rule because it helps to generate the theorems we want in a theory. This is the principle's generative role. A purely generative principle is one that has (in a particular theory) only a generative role.[4]

---

[4] As a referee pointed out to me, Timothy Smiley made the distinction in 1963, distinguishing between "rules of proof" and "rules of inference" see [195] and [85], scooping Paoli and me by a wide margin.

In "Logical Consequence and the Paradoxes" [135], Francesco Paoli and I make a similar distinction between internal and external rules of inference. An internal rule is one that can be horizontalised and represented within the logic. It is internal to the logic. Purely external rules are ones that we can use, for example, to derive theorems from theorems or sequents from sequents, but are not represented within the logic itself. The logic that I present in Chapter 7, called 'GTL', has some purely external rules, one of which is modus ponens:

$$\frac{\vdash A \to B \qquad \vdash A}{\vdash B}$$

Neither of the following is a theorem of the logic:

$$((A \to B) \land A) \to B$$

$$(\Box(A \to B) \land \Box A) \to \Box B$$

According to this logic, the logical laws are closed under modus ponens, but not every acceptable theory is closed under modus ponens.

When faced with the issue of whether GTL (or any other system) is an adequate logic of entailment, I think the question is not so much about whether it is reflexive but whether the theories that it allows are all and only the sorts of things that we would deem to be theories. I say a lot more about the nature of theories in Chapters 6–8 and I give an argument in Chapter 8 that GTL is inadequate because it does not license certain intuitive and common inferences about theories.

This does not mean that the issue of reflexivity is completely without interest. If one has purely genetic or external rules that determine our logical system, then the onus is on him to explain why his logic should be closed under these rules that are not generally acceptable rules of inference. One way of providing such a justification is to look at the logic determined by these rules (and the other rules and axioms of the system) to see what sort of universally valid inferences it licences.

5.9   S4 or S5?

In Chapter 3, I discuss various approaches to logical necessity based on the possible worlds semantics for modal logic. There, an entailment is defined as a logically necessary implication. In that chapter, I discuss Hintikka's idea that a model for logical possibility should include as worlds every logically possible model. Logical necessity on this model is determined by the entire set of logical possibilities and the logic that is characterised by this model is S5 (or rather, an extension of S5). In Chapter 4 and this chapter, I examine proof- theoretic approaches to entailment. The forms of necessity that emerge from these approaches are classical or relevant variants on the necessity of the logic S4.

It looks like we have a stand-off here. On one hand, if we take a model-theoretic perspective, we should prefer S5. On the other hand, if we take a proof-theoretic perspective, S4 is the better system. I think, however, that this apparent opposition is misleading.

There are (at least) two ways of understanding logical necessity: truth in all logically possible worlds and what follows from the axioms of logic. The logic S5 (or some variant of it) seems to represent the first notion quite well. Its universal semantics, in which every world is accessible from every world, seems to represent the idea that necessity is truth in all worlds very well. But the classical version of S5 represents the notion of entailment very poorly. On it, every necessary truth follows from every proposition. This is quite a useless notion of entailment.

One might think that we can capture both notions of logical necessity with a relevant version of S5. Such logics have been constructed [70, 126]. The most important problem with these logics is that their proof theories are not as elegant or intuitive as that of E. The semantics of these systems is quite fiddly as well. I think it is better to think of these logics as characterising different sorts of necessity, but allow that E captures the notion of following from the axioms.

5.10   Logic and Other Theories

One thing that is good about the project to interpret entailment through a reflexive analysis of a logical system is that it yields a way of understanding nested entailment. For example, it tells us that we can understand a transitivity axiom for entailment as representing a transitivity rule for deducibility. When combined with relevance, the reflexivity project gives us a non-trivial manner of treating nested conditionals. There is, however, something wrong about the interpretive project.

The interpretive project treats a logic of entailment as being largely about itself. It turns the logic of entailment into a formal theory of navel gazing. A logic of entailment, however, is supposed to be a very general theory. It is supposed to tell us what follows from what regardless of subject matter.

In order to comprehend the importance of this issue, let us think about the two main tasks that a logic of entailment should be able to perform. First, it should be able to formalise proof plans in an intuitive manner and, second, it should be able to formalise theories in a reasonable manner. As I say in Chapter 1, these two tasks are closely linked. When we are talking about proof plans, we often are talking about proving things about or within theories.

My approach to entailment is to concentrate first on the logic of theory closure. There has been a lot of mathematical and philosophical work done in this subject since it was introduced by Tarski in 1930. In Chapter 6, I examine Tarski's project and his reasons for abandoning and, in Chapter 7, I generalise some of the concepts in Tarski's theory to produce a semantics for a weak relevant logic. Then, in Chapter 8, I bring in the problem of proof plans and make connections between this semantics and the natural deduction system to construct and motive a semantics for the logic E of relevant entailment.

The acceptability of a logic as a logic of entailment, in the present project, is to be judged in accordance with whether it provides an appropriate logic for the closure of theories and the analysis of proof plans. One thing that it must do is treat theories without adding to (or subtracting from) the subject matters of theories. The logical view that I construct separates theories and the logic under which those theories are closed. This is not traditionally done. Typically, when logicians have constructed theories, they have included the theorems of their logic in each theory. This is especially clear in formalised theories in mathematics.

When working in classical logic, classically based modal logic, or intuitionist logic there is no choice but to include the logic in every theory closed under that logic. Each of those logical systems is closed under the rule of implosion:

$$\frac{\vdash B}{\vdash A \to B}.$$

For any non-empty theory T, the closure of T under one of these logics also contains all the theorems of that logic. For any of these logics, in fact, there is no difference between the logical closure of T and the logical closure of the union of T and the set of theorems of that logic. These logics, one could say, are *self-obsessed* in the sense that they make every theory about them.

I propose a philosophy of logic that is neither self-obsessed nor navel-gazing. It judges the logical system in terms of how it treats other theories (as well as itself) and it does not force every theory to contain all of its own theorems.

Part II

Theories and Entailment

6   Theories and Closure

A Guide to Part II

In Part I, I adopt Anderson and Belnap's logic E as the correct logic of entailment. However, this logic needs a philosophical interpretation. In Part II, I develop a general semantic theory that is to act as a basis for this interpretation. In Chapter 1, I claim that one of the key tasks of a logic of entailment is to give general principles of theory closure. The semantic theory that I adopt makes the theory closing aspect of a logic into the way in which the logic is to be understood.

The model that I construct in Chapter 7 is based on a set of theories. The satisfaction of formulas is relativised to theories in much the same way as the satisfaction of formulas is relativised to worlds in Kripke semantics for modal logic. Some of these theories are theories of entailment, that is, they satisfy formulas of the form $A \rightarrow B$. Each of these theories determines a consequence operator. The satisfaction conditions for entailment formulas are given in terms of these consequence operators, and nested conditionals are understood in terms of how theories treat other theories that satisfy entailment formulas. I adopt some simple postulates for the model theory that characterise a logic that I call 'GTL' (for 'Generalised Tarski Logic'). The job of adding the postulates that characterise the logic E is left to Part III.

In Chapter 6, I introduce closure operators. I present Tarski's axioms for closure operators and his motivations for adopting them. After briefly discussing an argument Tarski later gave for abandoning his syntactic-operator approach to theories, I discuss how to integrate the operator view into a more modern, semantic view of scientific theories. In Chapter 7, I use this semantic view as the metaphysical basis for my model.

6.1   Introduction

At the end of Chapter 5, I say that there is a need for a logic of theory closure. In this chapter, I start to build a semantical theory for that logic through an examination of Alfred Tarski's postulates for theory closure operators. By 'operator' here, I mean a function that takes objects of a type as arguments and returns an object of the same type. In this case, closure operators take sets of formulas and return sets of formulas. Building on Casimir Kuratowski's axiomatised theory of closure operators in point-set topology, Tarski developed a theory of functions that turn sets of formulas into sets that he thought of as having the structure of scientific theories.

Before I begin to examine Tarski's view, however, I need to make clearer how I intend for the relationship between entailment and theories to be understood. This view is to be understood normatively. I think that if a set of statements is put forward as belonging to one theory, then the commitments of that set follow the principles of the logic of entailment.

Of course, there may be cases in which a theory should be dissected into sub-theories. Bryson Brown and Graham Priest [26] put forward such a view in order to deal with cases in which a collection of postulates is used which is in tension with one another or contradicts one another. For example, Nils Bohr's early quantum theory was combined with a classical theory of electrodynamics which predicted that electrons would eventually lose energy and collapse into nuclei. On Brown and Priest's view, the theory should be dismembered into sub-theories, called "chunks". This idea seems right to me, but it does not affect the claim that a theory should be treated as closed under entailment. Rather, in cases like this, some scientific theories should be treated as networks of other theories, and each of these other theories should be understood as closed under the logic of entailment.

In this chapter, I look at Tarski's notion of a closure operator and construct a simple logic (Tarski Logic – TL) that captures this concept. I also discuss criticisms of Tarski. His theory was attacked from a variety of points of view. Tarski himself came to reject the syntactic treatment of theories and logical consequence that was at the heart of his 1930 analysis of closure operators [201]. He developed instead a model-theoretic understanding of theory closure (known as his theory of logical consequence) [202], which posed a very serious problem for any theory of entailment. The logical consequence relation that Tarski proposed was not compact, and as I say in Chapter 3, no finitary logic of entailment can adequately represent a non-compact consequence relation. I argue, however, the example that Tarski uses to motivate his approach does not show that we should abandon the idea that a reasonable consequence relation should be compact.

Before I go on, I want briefly to discuss the word 'theory'. In logic, the word 'theory' usually is taken to mean a set of formulas closed under the principles of the logic. Sometimes logicians talk of "L-theories" for some logic L. In this book, I sometimes use 'theory' in this way, but I also use it to denote scientific theories. I think that the context largely disambiguates what I am talking about, but sometimes when necessary, I explicitly state which notion I am using.

6.2   Tarski's Consequence Operators

In his 1930 paper, "Fundamental Concepts of the Methodology of Deductive Sciences" [201], Tarski set out his famous conditions for consequence operators on sets of sentences. His aim was to develop a general theory of scientific theories that tells us to what statements a theory is committed.

Tarski assumed a language S. He identified the language with its set of grammatically well-formed statements. His first axiom was concerned with the nature of S itself. This axiom was

Axiom 1: $|S| \leq \aleph_0$.

This axiom said that the set of formulas of S is countable. This implied that each category of grammatical elements (propositional variables, predicates, names, variables) is at most countably infinite. For if there were uncountably many individual names and even one predicate, for instance, uncountably many formulas could have been constructed. Tarski said that "the contrary hypothesis would be unnatural and would lead to unnecessary complications in proofs" [201, p 63]. I follow Tarski in using a countable language, although I briefly discuss uncountable languages in Section 6.5.

The second axiom said that consequence operators were conservative, in the sense that no statements were lost by closing under an operator:

Axiom 2: If $\Gamma \subseteq S$, then $\Gamma \subseteq \mathcal{C}(\Gamma) \subseteq S$.   [1]

Axiom 2 captured the intuition that entailments of a theory included the axioms. The axioms of a theory were stated as parts of the theory and were not just used to generate the theorems of the theory and so should appear in the theory that results from the closure of the axiom set under the consequence operator.

The third axiom asserted that one application of the consequence operator is enough:

Axiom 3: If $\Gamma \subseteq S$, then $\mathcal{C}(\mathcal{C}(\Gamma)) = \mathcal{C}(\Gamma)$.

We can break this axiom down into two parts: (i) $\mathcal{C}(\Gamma) \subseteq \mathcal{C}(\mathcal{C}(\Gamma))$ and (ii) $\mathcal{C}(\mathcal{C}(\Gamma)) \subseteq \mathcal{C}(\Gamma)$. (i) is just an instance of axiom 2 and is redundant. (ii), however, is very important. A *theory* of a logic is a set of formulas closed under the entailments of the logic. In the sections that follow this one, a close relationship between closure operators and entailments is constructed. Given that relationship, we can define a theory to be a set of formulas Γ such that $\mathcal{C}(\Gamma) \subseteq \Gamma$. This tells us that all the consequences of the formulas in Γ according to $\mathcal{C}$ are also in Γ. $\mathcal{C}(\mathcal{C}(\Gamma)) \subseteq \mathcal{C}(\Gamma)$ says that $\mathcal{C}$ is a theory-forming operator. It takes a set of formulas and returns a theory. Any consequence of the formulas in $\mathcal{C}(\Gamma)$ is in $\mathcal{C}(\Gamma)$. No further application of $\mathcal{C}$ is necessary.

Axiom 4 is a bit more technical-looking than the others:

Axiom 4: If $A \subseteq S$, then $\mathcal{C}(A) = \bigcup_{Finite\ X \subseteq A} \mathcal{C}(X)$.

The fourth axiom, in more modern terms, says that the logic of the consequence operator is compact. Any consequence of a set is a consequence of a finite number of formulas in that set. Many of the logics that are most familiar to students, such as classical propositional logic and classical first-order logic, are compact, but many

---

[1] When Tarski's axioms are repeated now, the condition '$\mathcal{C}(\Gamma) \subseteq S$' is usually omitted. This condition merely says that Γ is a set of formulas.

logics are not. Full second-order logic, for example, is not compact, nor are many modal logics and infinitary logics.

Tarski used the fourth axiom to derive certain properties of consequence operators that were extremely useful. One such property is the monotonicity of C:

(Mono) If $A \subseteq B$, then $C(A) \subseteq C(B)$.

Mono is an immediate consequence of axiom 4. If A is a subset of B, then every finite subset of A is also a finite subset of B. Since $C(A)$ and $C(B)$ are just the unions of the consequence sets of their finite subsets, everything that is in A is also in B.

In more recent accounts, Mono is taken to be a separate axiom. In those presentations it replaces axiom 4. Although Mono can be derived from compactness, it does not assume compactness. Thus, dropping axiom 4 in favour of Mono leads to a treatment of consequence that applies to a wider range of logical systems. As I argue in Section 6.5, any logic that can be considered a logic of entailment must be compact, so I accept Tarski's axiom 4.

Tarski was likely influenced in his choice of axioms by Casimir Kuratowski's axiomatisation of closure operators in topology.[2] In his 1922 paper, "Sur l'operation $\overline{A}$ de l'Analysis Situs" [96], Kuratowski gave four axioms for the closure of a set of points in a topological space. The closure of a set, as he characterised it, was "composed of a set together with its limit points" [96, p 183]. Here are Kuratowski's axioms translated into more modern notation:[3]

1. $C(\Gamma \cup \Delta) = C(\Gamma) \cup C(\Delta)$
2. $\Gamma \subseteq C(\Gamma)$
3. $C(\emptyset) = \emptyset$
4. $C(C(\Gamma)) = C(\Gamma)$

Kuratowski's axioms 2 and 4 were the same as Tarski's axiom 2 and 3, respectively. Axiom 1 was not provable in Tarski's theory. A similar but slightly weaker postulate, however, was derivable in Tarski's theory:

LEMMA 6.1   *For any sets of formulas Γ and Δ, $C(C(\Gamma) \cup C(\Delta)) = C(\Gamma \cup \Delta)$.*

Proof   First, I prove that $C(\Gamma \cup \Delta)$ is a subset of $C(C(\Gamma) \cup C(\Delta))$.

$$\frac{\dfrac{\Gamma \subseteq C(\Gamma) \qquad \Delta \subseteq C(\Delta)}{\Gamma \cup \Delta \subseteq C(\Gamma) \cup C(\Delta)}}{C(\Gamma \cup \Delta) \subseteq C(C(\Gamma) \cup C(\Delta))} \text{Mono}$$

Now, I prove that $C(C(\Gamma) \cup C(\Delta)) \subseteq C(\Gamma \cup \Delta)$.

$$\frac{\dfrac{\dfrac{C(\Gamma) \subseteq C(\Gamma \cup \Delta) \qquad C(\Delta) \subseteq C(\Gamma \cup \Delta)}{C(\Gamma) \cup C(\Delta) \subseteq C(\Gamma \cup \Delta)}}{C(C(\Gamma) \cup C(\Delta)) \subseteq C(C(\Gamma \cup \Delta))} \text{Mono}}{C(C(\Gamma) \cup C(\Delta)) \subseteq C(\Gamma \cup \Delta)} \text{Axiom 3}$$

Putting these two conclusions together, we get $C(C(\Gamma) \cup C(\Delta)) = C(\Gamma \cup \Delta)$.   ■

The monotonicity property was derivable from Kuratowski's first axiom. For suppose that $\Gamma \subseteq \Delta$. Then $\Delta = \Gamma \cup (\Delta - \Gamma)$. By axiom 1, $\mathcal{C}(\Delta) = \mathcal{C}(\Gamma \cup (\Delta - \Gamma)) = \mathcal{C}(\Gamma) \cup \mathcal{C}(\Delta - \Gamma)$. Hence, $\mathcal{C}(\Gamma) \subseteq \mathcal{C}(\Delta)$.

Tarski did not accept Kuratowski's third axiom. This is because Tarski had classical logic (and perhaps intuitionist logic and Łukaseiwicz many-valued logic) in mind. For these logics, it was reasonable to think that $\mathcal{C}(\emptyset)$ was the set of theorems of the logic. I do accept Kuratowski's axiom, however, and take it to be a requirement of an acceptable closure operator that when applied to the empty set it returns the empty set. I say more about this in Chapter 7.

Before I leave the topic of Tarski on closure operators, it is interesting that Tarski, together with J. C. C. McKinsey, showed that S4 could be represented on a topological space in which possibility is a Kuratowski closure operator [141]. To put this in more modern (and more familiar) terms, consider a Kripke model for S4. The sets that we deal with here are the sets of worlds determined by formulas. Thus, for example, $[\![A]\!]$ is the set of worlds in the model in which A is true. Let us take $\Diamond$ to be the operator on a set such that $\Diamond[\![A]\!]$ is the set of worlds that can see some world in which A is true. In other words, $\Diamond[\![A]\!] = [\![\Diamond A]\!]$. It is easy to see that $\Diamond$ is a closure operator in Kuratowski's sense. Together with Bjiarni Jónsson, Tarski went on to represent operators like $\Diamond$ using binary relations [89]. Thus, Tarski and Jónsson invented a precursor to Kripke semantics (see [75]).

6.3   From Operators to Consequence Relations

In earlier Chapters 4 and 5, I discuss consequence relations for various logics. In this section, I explore some correlations between the properties of closure operators and the properties of consequence relations.[4]

First, I define the concept of a consequence relation that is determined by a closure operator. Given an operator $\mathcal{C}$, there is a consequence relation, $\Vdash_\mathcal{C}$, such that the following biconditional holds:

$$\Gamma \Vdash_\mathcal{C} B \text{ if and only if } B \in \mathcal{C}(\Gamma).$$

I use the symbol '$\Vdash_\mathcal{C}$' to represent the consequence relation correlated with a closure operator instead of '$\vdash_\mathcal{C}$', because here collections of premises are treated as sets or as if they are conjunctions of the premises, like the relation $\Vdash$ in Chapter 4. Note that $\Vdash_\mathcal{C}$ is infinitary; $\Gamma \Vdash_\mathcal{C} A$ may hold even when Γ is infinite.

The operator $\mathcal{C}$ operates on sets. The elements of a set are not ordered in the set. Thus, if

$$\Gamma, A, B \Vdash_\mathcal{C} D,$$

then

$$\Gamma, B, A \Vdash_\mathcal{C} D.$$

---

[4] For a general discussion of consequence relations and their relationships to consequence operators, see Humberstone [86, ch 1].

In Gentzen systems (see Section 5.4), this is called the rule of *exchange*. In addition, $\{A\} \cup \{A\} = \{A\}$, so the rule of *contraction* also holds:

$$\frac{\Gamma, A, A \Vdash_C B}{\Gamma, A \Vdash_C B} .$$

And from Mono, we can derive that the consequence relation obeys the rule of *weakening*:

$$\frac{\Gamma \Vdash_C B}{\Gamma, A \Vdash_C B} .$$

The weakening rule looks like a violation of relevance; it allows the addition of irrelevant premises to a deduction. I explain below why it is not problematic.

Axiom 3 together with Mono implies that the consequence relation is transitive, viz.,

$$\frac{\Gamma \Vdash_C A \qquad A \Vdash_C B}{\Gamma \Vdash_C B} .$$

Here is a derivation of the transitivity principle.

$$\begin{array}{c} \text{Mono} \\ \text{Axiom 3} \end{array} \frac{\dfrac{\text{def.} \dfrac{\Gamma \Vdash_C A}{A \in \mathcal{C}(\Gamma)}}{\dfrac{\{A\} \subseteq \mathcal{C}(\Gamma)}{\dfrac{\mathcal{C}(\{A\}) \subseteq \mathcal{C}(\mathcal{C}(\Gamma))}{\mathcal{C}(\{A\}) \subseteq \mathcal{C}(\Gamma)}}} \qquad \dfrac{A \Vdash_C B}{B \in \mathcal{C}(\{A\})} \text{def.}}{\dfrac{\dfrac{B \in \mathcal{C}(\Gamma)}{\Gamma \Vdash_C B} \text{def.}}{}}$$

'def.' stands for 'definition' and it refers to the definition of the consequence relation from the closure operator.

The consequence relation also obeys a generalisation of the transitivity rule. This more general rule is the *cut rule*:

$$\frac{\Gamma \Vdash_C A \qquad \Delta, A \Vdash_C B}{\Gamma, \Delta \Vdash_C B} .$$

The cut rule has an important place in the history of Gentzen systems, but I do not need to discuss it further.

The rules that govern $\Vdash_C$ come from the Gentzen systems for classical and intuitionist logics. The weakening rule in its usual form is banned from relevant logic. The reason that the rules given above are benign is that they are not combined with the usual rules governing the connectives rules. In particular, as I say in Chapter 4, the following rule of entailment introduction is invalid:

$$\frac{\Gamma, A \Vdash_C B}{\Gamma \Vdash_C A \to B} .$$

This rule, together with weakening, would allow us to prove

$$B \Vdash_C A \to B$$

and other fallacies of relevance. Here the difference between $\Vdash$ and $\vdash$ is important. In the logic E, the only form of entailment introduction that is valid is

$$\frac{A_1; \ldots; A_n; B \vdash C}{A_1; \ldots; A_n \vdash B \to C}.$$

As I set out in Chapter 4, $\vdash$ relates structures to formulas as opposed to sets and formulas. The full logic of $\vdash$ is quite complicated but is not important here.

6.4 Closure Operators, Consequence Relations, and Entailment

In Section 6.3, I say that the standard entailment introduction rule is unacceptable. This raises the question of the relationship between the entailment connective and the turnstile, $\Vdash_C$. In order to explain the connection between them, I also need to use conjunction. I begin by explaining the relationship between $\Vdash_C$ and conjunction.

I add to Tarski's axioms the following condition regarding conjunction:

$$A \wedge B \in \mathcal{C}(\Gamma) \quad \text{if and only if} \quad A \in \mathcal{C}(\Gamma) \text{ and } B \in \mathcal{C}(\Gamma).$$

Adding this condition enables us to prove that

$$\Gamma, A, B \Vdash_C C \quad \text{if and only if} \quad \Gamma, A \wedge B \Vdash_C C.$$

It also makes valid the following rules:

$$\frac{\Gamma, A \Vdash_C C}{\Gamma, A \wedge B \Vdash_C C} \qquad \frac{\Gamma, B \Vdash_C C}{\Gamma, A \wedge B \Vdash_C C}$$

$$\frac{\Gamma \Vdash_C A \quad \Gamma \Vdash_C B}{\Gamma \Vdash_C A \wedge B} \qquad \frac{\Gamma \Vdash_C A \wedge B}{\Gamma \Vdash_C A} \qquad \frac{\Gamma \Vdash_C A \wedge B}{\Gamma \Vdash_C B}$$

The goal here is to formulate a logic in which entailment represents these rules. I call this logic 'Tarski Logic' or 'TL'. I characterise TL axiomatically. It has one axiom scheme:

$$A \to A$$

and it is closed under the following rules:

$$\frac{A \to B \quad A \to C}{A \to (B \wedge C)}$$

$$\frac{A \to C}{(A \wedge B) \to C} \qquad \frac{B \to C}{(A \wedge B) \to C}$$

$$\frac{A \to B \quad B \to C}{A \to C}$$

Note that none of the axioms or rules are formulated using any scheme that has nested entailments. This fact is important for what I say about TL later in this chapter and in later chapters.

Despite the fact that this logic does not deal very well with nested entailments, it does have many of the properties one needs from a logic. For example, one can prove in it that conjunction is both commutative and associative. Here is a proof of the commutativity of conjunction:

$$\frac{\dfrac{B \to B}{(A \wedge B) \to B} \quad \dfrac{A \to A}{(A \wedge B) \to A}}{(A \wedge B) \to (B \wedge A)}$$

And here is a proof of one direction of the associativity of conjunction:

$$\frac{\dfrac{A \to A}{(A \wedge (B \wedge C)) \to A} \quad \dfrac{\dfrac{B \to B}{(B \wedge C) \to B}}{(A \wedge (B \wedge C)) \to B}}{(A \wedge (B \wedge C)) \to (A \wedge B)} \quad \frac{\dfrac{C \to C}{(B \wedge C) \to C}}{(A \wedge (B \wedge C)) \to C}$$
$$(A \wedge (B \wedge C)) \to ((A \wedge B) \wedge C)$$

The proof of the other direction is similar.

I define a consequence relation, $\Vdash_{TL}$, for the logic TL. For any set Γ of sentences, we say that $\Gamma \Vdash_{TL} B$ if and only if there are some $A_1, \ldots, A_n$ in Γ (for some $n > 0$) such that $(A_1 \wedge \ldots \wedge A_n) \to B$ is a theorem of TL. (Note that I have made tacit use of the associativity of conjunction in the definition.) Using this consequence relation, I define a consequence operator, C_{TL}. For any set of formulas Γ,

$$B \text{ is in } C_{TL}(\Gamma) \text{ if and only if } \Gamma \Vdash_{TL} B.$$

I set $C_{TL}(\emptyset)$ to be empty. This is just Kuratowski's third axiom. Its importance is made clear in Chapter 7.

The operator C_{TL} obeys Tarski's four axioms. The definition of the language ensures that the set of formulas is countable. Hence C_{TL} satisfies Tarski's first axiom. The definitions of $\Vdash_{TL}$ and C_{TL} immediately imply Tarski's second and fourth axiom. In order to prove that Tarski's second axiom holds, it suffices to show that $C_{TL}(C_{TL}(\Gamma))$ is a subset of $C_{TL}(\Gamma)$.

Here is a proof of this fact. Take an arbitrary formula C in $C_{TL}(C_{TL}(\Gamma))$. By the definition of C_{TL}, there are $B_1, \ldots, B_m$ in Γ such that $(B_1 \wedge \ldots \wedge B_n) \to C$ is a theorem of TL. Moreover, by the same definition, for each of these B_i, there are $A_1^i, \ldots, A_{n_i}^i$ in Γ, such that $(A_1^i \wedge \ldots \wedge A_{n_i}^i) \to B_i$ is a theorem of TL. From this we can infer that

$$(A_1^1 \wedge \ldots \wedge A_{n_1}^1 \wedge \ldots \wedge A_1^m \wedge \ldots \wedge A_{n_m}^m) \to (B_1 \wedge \ldots \wedge B_m)$$

is also a theorem of TL. Together with $(B_1 \wedge \ldots \wedge B_n) \to C$ and the transitivity rule, we then obtain that

$$(A_1^1 \wedge \ldots \wedge A_{n_1}^1 \wedge \ldots \wedge A_1^m \wedge \ldots \wedge A_{n_m}^m) \to C$$

is a theorem of TL. Since all of $A_1^1, \ldots, A_{n_1}^1, \ldots, A_1^m, \ldots, A_{n_m}^m$ are in Γ, C is in $C_{TL}(\Gamma)$. Generalising, therefore, we have shown that $C_{TL}(C_{TL}(\Gamma))$ is a subset of $C_{TL}(\Gamma)$, which is what I set out to show.

So, I have shown that the Tarski axioms determine the logic TL, which in turn characterises a consequence relation, $\Vdash_{TL}$, which determines a consequence operator C_{TL}, which satisfies Tarski's axioms. This is a very satisfying circle.

6.5 Tarski's Rejection of His Earlier Theory of Consequence

Perhaps the most important criticism of Tarski's theory of consequence operators came from Tarski himself. His famous 1935 paper "On the Concept of Logical Consequence" contained an argument that had two important conclusions that told against his earlier view: (1) no adequate notion of logical consequence was compact and (2) the real notion of logical consequence was not syntactic but model-theoretic.

I quote this argument at length:

Some years ago I gave a quite elementary example of a theory which shows the following peculiarity: among its theorems there are such sentences as:

A_0. *0 has the given property P,*
A_1. *1 has the given property P,*

and, in general, all particular sentences of the form

A_n. *n has the given property P,*

where 'n' stands for any symbol which denotes a natural number in a given (e.g. decimal) number system. On the other hand, the universal sentence:

A. *Every natural number possesses the given property P,*

cannot be proven on the basis of the theory in question by means of the normal rules of inference. This fact seems to me to speak for itself. It shows that the formalised concept of consequence, as it is generally used by mathematical logicians, by no means coincides with the common concept. Yet intuitively it seems certain that the universal sentence A follows in the usual sense from the totality of particular sentences $A_0, A_1, \ldots, A_n, \ldots$. Provided all these sentences are true, the sentence A must also be true. [202, pp 410–411]

Tarski's argument was quite straightforward. The rule he stated was often called the *omega rule*, since it appealed to the omega-sequence of natural numbers. There are non-standard models for first-order Peano arithmetic in which all of the natural numbers (as we understand them) have a particular property, but not all the numbers according to the model have that property. These non-standard models have non-standard numbers in addition to the standard ones. The Gödel's incompleteness theorems show that there is no finitely axiomatic system, with finitely stable rules in which this rule is derivable. This rule is derivable in full second-order Peano arithmetic, but that system cannot be axiomatised.

The "formalised concept of consequence, as it is generally used by mathematical logicians" to which Tarski referred, was one that could be axiomatised using only rules that had finitely many premises. This concept could not capture the "common concept", according to which if each individual standard natural number had a property, then every natural number had that property. Thus, the traditional formalised concept is not the common concept.

If sound, this argument shows that there is a problem with the project of constructing a logic of entailment. The logic of entailment, as I understand it, contains finitary rules and allows only for formulas of finite length. The omega rule cannot be captured by a formula of finite length.

There is, however, a serious problem with Tarski's argument. He is right that most people, and most mathematicians, would think that the omega rule is valid. But there are, I think, two senses of validity here that are being conflated. The rule is valid in the sense that our common notion of the natural numbers is closed under it. It is, in the terminology that I introduced in Chapter 5, an external rule. This does not mean, however, that it needs to be made an internal rule of the logical system. The fact that the omega rule is infinitary shows that it is not a rule that can be used to make inferences, at least not in a direct way.

Consider a common example of the way in which the omega rule is supposedly used in reasoning about arithmetic. Peano arithmetic is a finitely axiomatised theory and is not closed under the omega rule. Consider the theory of *true arithmetic* (TA). TA contains all and only the sentences of arithmetic that are in fact true. TA is not finitely (or recursively) axiomatisable but is closed under the omega rule. Now let us take a two-place predicate Pr such that '$Pr(x, y)$' means that x encodes a proof of y in Peano arithmetic, and let g be the encoding of the Gödel sentence G, that is, the sentence that Gödel constructed to show that Peano arithmetic is incomplete.

We know that for each ordinary natural number n,

$$TA \Vdash \neg Pr(n, g),$$

since, otherwise, Peano arithmetic would be inconsistent. So, we infer from this that

$$TA \Vdash \forall x \neg Pr(x, g),$$

because we know that TA is closed under the omega rule. The qualification 'ordinary' before 'natural number' is important. Peano arithmetic has non-standard models that contain *non-standard numbers*. It is for this reason that the omega rule cannot be proven from the induction axiom of first-order Peano arithmetic. Just because every ordinary natural number $(0, 1, 2, \ldots)$ has a property P, does not mean that $\forall x(P(x) \supset P(x'))$, where x' is the successor of x, for in a given model x might be a non-standard number.

We cannot represent the above reasoning as a finitary *logical* rule. If we could do so, it would undermine Gödel's incompleteness theorems. In particular, we cannot give a complete, but finite, formal characterisation of the notion of an ordinary natural number. At some point, we have to *appeal* to the fact that we somehow know what we are talking about when we talk about the standard natural numbers. This reasoning, therefore, is not completely formal. It cannot be codified without a lot of ellipses $(1, 2, 3, \ldots)$ and necessarily tacit assumptions. These are assumptions that we cannot fully state. When constructing models of TA, say, we start with the natural numbers and assume that everyone knows what we are talking about. But as finite beings, we cannot formally specify this model fully.

Tarski's argument, however, raises the very interesting issue of entailment connectives that express *non-compact* consequence relations. If the language we are dealing with only contains connectives that join together finite numbers of formulas, then none of these connectives can fully represent a non-compact consequence relation, for there will be some infinite sets Γ and formulas A such that $\Gamma \Vdash A$ that cannot be represented in the finitary language.

One apparent solution is to move to infinitary languages. Where χ and λ are infinite cardinal numbers, an infinitary language, $\mathcal{L}_{(\chi, \lambda)}$, allows conjunctions and disjunctions of length less than χ and formulas with fewer than λ many bound variables (see [14]). If we have an infinitary language $\mathcal{L}_{(\aleph_1, \aleph_1)}$, then we can represent the omega rule in the language as

$$\bigwedge \{ P(\overline{n}) : \ n \in \mathbb{N} \} \supset \forall x \, P(x).$$

The move to infinitary languages, however, does not completely solve the problem. The project of producing a logic of entailment that corresponds to a given consequence relation requires that the consequence relation be compact in an extended sense. A consequence relation $\Vdash$ is said to be χ-*compact* if and only if, for every set of formulas Γ and formula A, if $\Gamma \Vdash A$, then there is a subset Γ' of Γ such that Γ' has fewer than χ many formulas and $\Gamma' \Vdash A$. Then the consequence relation that is χ-compact can be expressed by a language that allows conjunctions of length up to (but not necessarily including) χ. Unfortunately, infinitary logics are largely not compact, even in this extended sense, except when χ is extremely large [14]. Instead of turning to infinitary logics, I think that I can capture the notion of entailment adequately in a finitary logic with a model theory that has a consequence relation that is compact in the standard sense.[5]

Tarski, however, rejected compactness and rejected his earlier syntactic notion of logical consequence in favour of a model-theoretic conception:

The sentence A follows logically from the class Γ if and only if every model for the class Γ is a model for A. [202, p 417].

There has been an extremely interesting debate about the range of structures that should count as models in a theory of logical consequence [59, 192]. I set aside this debate here because it has little to do with my own view. Instead, I go on to develop my positive view.

I do accept a model-theoretic notion of consequence, but one that is somewhat different from Tarski's. My view utilises his notion of a consequence operator. I give the formal details of this view in Chapters 7 and 8. The model theory, like Kripke's semantics for modal logics, is indexical. A model contains a collection of points at which formulas may be satisfied. The points in these models are theories, but not sets of formulas in the sense of Tarski's early view of logical consequence. Rather, they

---

[5] Also, see Chapter 1 for a discussion of finitary languages that contain names for infinite sets of formulas.

are abstract objects of a different sort. In order to give a philosophical underpinning for this view, I now turn to the semantic notion of scientific theories.

6.6   The Semantic View of Theories

In order to give an answer to the question of what a theory is, I turn to the literature on theories in the philosophy of science. A standard narrative has developed in that literature, according to which an older "received view" about theories has been replaced with a more modern semantic view of theories.

The received view, roughly, was the view held by the logical positivists. On this view, a theory was a set of sentences of some appropriate language, which was axiomatised by a finite subset of those sentences (along with some rules of inference). Tarski's early theory was a way of generalising the received view of theories.

By itself, a set of sentences could not act as a scientific theory. The sentences in the theory had to be given meaning in some way. The logical positivists held that the meaning of a theory was given by the conditions under which it could be verified. The positivists, at least according to Frederick Suppe and the other proponents of the semantic view, made a hard distinction between observable and theoretical scientific terms [199, pp 45–53].[6] An observable term was one that could be directly and empirically confirmed to hold of a thing. Colour and shape terms were clear examples of observable language. The important question for the positivists was how meaning could be conferred on theoretical terms.

Perhaps the best-known method of attributing empirical meaning to theoretical terms was due to Ramsey [173, p 131] and Carnap [28]. On Ramsey's method, we replace the theoretical terms with free variables in the axioms of the theory. Then, we bind the free variables with existential quantifiers and produce what is now known as a "Ramsey sentence". The meaning of theoretical terms, according to Ramsey's method, is the role that they play in an empirical theory, and whatever fills this role in contexts in which the theory is verified is what the corresponding theoretical terms denote.

The received view, in this form, required a precise boundary between empirical and theoretical terms. But this was problematic. In the early 1960s, Grover Maxwell gave a slippery-slope style argument to the effect that there can be no non-arbitrary way of making this distinction [139]. Carnap himself agreed with Maxwell that there was a continuum of degrees of observability and the point on this continuum at which we decide to draw the line between what we count as observable and what is theoretical was somewhat arbitrary [35, p 266]. Moreover, as Suppe argued, the notion of observability that scientists used was very different from the one that was being imposed on science by the positivists [199, p 47]. Scientists would often talk about observable properties that are detectable in experiments but were not directly observed in the positivists' sense.

---

[6] Carnap, for one, explicitly said that this distinction was not hard, but somewhat arbitrary and vague. See below.

Thus, it seemed to some philosophers of science that the received view had to be abandoned. The prevailing alternative to the received view is now the semantic view of scientific theories. On the semantic view, the content of a theory is given by a class of models. Every term in a theory can be given meaning in terms of what it denotes in the models of the theory. Some semantic theorists, like Bas van Fraassen, identify a theory with the class of its models itself. This identification has the advantage of making a theory language independent [209, p 44]. This means that, unlike the received view, this version of the semantic view does not require that a theory be represented in any particular language.

By 'model' here, it was meant something as simple as a set of entities in some arrangement, or something rather more complicated like a system evolving over time according to some set of laws. Van Fraassen considered the following three simple axioms for a geometrical theory [210, p 219]:

1. For any two lines, at most one point lies on both.
2. For any two points, exactly one line lies on both.
3. On every line there are exactly two points.

He gave a model with seven points, but one could give a model of these axioms that has only three points and two lines that are perpendicular to each other. Then we could have one point at the intersection of the two lines and one point on each line not at the intersection. In this case the domain of the model was made up of abstract objects (points and lines) rather than physical objects.

Ronald Giere gave a very different example of a model [72]. His example was a harmonic oscillator. This oscillator consists of a body b, the motion of which depends only on a single force F. F in turn depends only on the position x of b and a constant k. In classical mechanics, this is expressed as $F = -kx$. An example of a harmonic oscillator is a body attached to a spring anchored from above, pulled, and then released (with no other forces acting on it). A graph of the motion of the oscillator forms a wave pattern. This is a model of Newtonian mechanics. Giere's model, in this case, was an idealisation of a physical system.

Other sorts of models of Newtonian mechanics could be the motion of an object along a frictionless plane (a favourite of my high-school physics teachers) and a simple spinning top isolated from all external forces. These like the simple harmonic oscillator are highly idealised models. Similarly, models used in economics often abstract away from difficulties found in actual situations and attribute to individuals or firms, for example, perfect information.

To accommodate these very idealised models, Giere thought of models as abstract objects:

I propose that we regard the simple harmonic oscillator and the like as *abstract entities* having all and only the properties ascribed to them in the standard texts. [72, p 78]

A model qua abstract object was useful for the purposes of science, in particular, for explanation and prediction, to the degree to which it approximated real physical systems.

Although I agree with the semantic theorists in taking models to be abstract objects, I do have some criticisms of their view. I defer these criticism until after I briefly explore a simple semantics for a logic of entailment based on the semantic view of theories.

6.7   Making a Model out of Models

In this section, I construct a model, in the logical sense, out of the models of the semantic view of theories. I begin with a set M of scientific models. In order to make this into a model for a language, I need a treatment of the connectives. Conjunction and disjunction are straightforward in this framework, but negation is a bit more complicated.

Scientific models such as Giere's harmonic oscillator, for the most part, are incomplete models of reality. They do not tell us what else, if anything, exists. Thus, the treatment of negation as failure – where $\neg A$ holds when A fails – is inappropriate. We need to allow for truth gaps with regard to truth in a scientific model. I also allow for truth gluts – where a model could be inconsistent. There has been some use of inconsistent models in the philosophy of mathematics. Chris Mortensen, Robert Meyer, Graham Priest, and Zach Weber have constructed a variety of inconsistent models for mathematical theories [146, 148, 165, 215]. Surely, it would be just as constructive to build inconsistent models of certain problematic scientific theories, for example, scientific theories that make use of inconsistent mathematical theories, such as Newton's use of his infinitesimal calculus in his physics (see, e.g., [211, 212]).

In order to allow both gaps and gluts, I adopt in this section Dunn's four-valued semantics, although more or less in the form presented by Priest [166].[7] A value assignment in this semantics is a *relation* ρ between models, propositional variables, and the truth values T and F. ρ determines a satisfaction relation $\vDash^+$ and anti-satisfaction $\vDash^-$ such that the following clauses hold for conjunction, disjunction, and negation:

- $m \vDash^+ p$ if and only if $p\rho_m T$;
- $m \vDash^- p$ if and only if $p\rho_m F$;
- $m \vDash^+ A \wedge B$ if and only if $m \vDash^+ A$ and $m \vDash^+ B$;
- $m \vDash^- A \wedge B$ if and only if $m \vDash^- A$ or $m \vDash^- B$;
- $m \vDash^+ A \vee B$ if and only if $m \vDash^+ A$ or $m \vDash^+ B$;
- $m \vDash^- A \vee B$ if and only if $m \vDash^- A$ and $m \vDash^- B$;
- $m \vDash^+ \neg A$ if and only if $m \vDash^- A$;
- $m \vDash^- \neg A$ if and only if $m \vDash^+ A$.

What determines the falsity of a propositional variable in a scientific model is a topic that I do not discuss here. I leave it as an intuitive notion.

---

[7] Jc Beall suggests that the logic of first-degree entailments (the $\rightarrow$-free fragment on the present logic) is the universal logic of the closure of theories [10].

What the semantics needs is a treatment of entailment. I take this from Priest [166]. I add to the class M an origin model O that satisfies all the true entailments. It does this by means of the following clause:

$$O \vDash^+ A \to B \text{ if and only if for all } m \in M, \text{ if } m \vDash^+ A, \text{ then } m \vDash^+ B$$

And, the anti-satisfaction clause at O is:

$$O \vDash^- A \to B \text{ if and only if for some } m \in M, m \vDash^+ A \text{ and } m \nvDash^- B$$

For points in the model other than O, ρ makes entailments satisfy or anti-satisfy entailments without further explanation.

A formula is said to be valid on a model if and only if it is satisfied by O. The class of these models characterises the logic N_4 [166]. To axiomatise N_4, I add the following axiom schemes to the axiomatic basis for TL:

$$(A \lor B) \leftrightarrow \neg(\neg A \land \neg B)$$

$$A \leftrightarrow \neg\neg A$$

$$(A \land (B \lor C)) \to ((A \land B) \lor (A \land C))$$

And the rules:

$$\frac{\vdash A \to C \qquad \vdash B \to C}{\vdash (A \lor B) \to C} \qquad \frac{\vdash A \leftrightarrow B}{\vdash C \leftrightarrow C'}$$

where C' results from C by the replacement of B for A in any context in C in which A does not occur in the scope of an entailment. To make the replacement rule more general, in [164] Priest adds a set of propositions to the model theory (see Chapter 10) and states the satisfaction condition for entailment in non-O theories in terms of propositions rather than formulas. The biconditional is defined as usual: $A \leftrightarrow B =_{df} (A \to B) \land (B \to A)$.

One might object that this semantics is not a general semantics for the analysis of scientific models. In models of quantum physics, for example, the rule of the distribution of conjunction over disjunction often fails. This is a serious issue about the current theory. In Chapter 8, I claim that the logic that I put forward is the logic of *standard theories*. When faced with theories like quantum physics, we might have to retreat to what I call the *core logic* of entailment, which includes only principles governing conjunction and entailment. The same thing can be said with regard to TL and N_4. When faced with non-standard theories, we might have to retreat to a more basic logical theory – TL.

This notion of retreating or retrenching to a more basic logic is made more plausible by the fact that N_4 is a conservative extension of TL. This means that if we isolate the theorems of N_4 that can be stated using only conjunction and entailment (and propositional variables and parentheses), we find that this is exactly TL. This can be shown rather easily using the model theory.

A TL model is a triple $\langle M, O, v \rangle$ such that M is a non-empty set, O is in M and v is a function from propositional variables to subsets of M. O and all the other entities in

M obey the satisfaction (not anti-satisfaction) clauses for conjunction and entailment. It is easy to show that every theorem of TL is valid on the class of TL models. And it is easy to show that TL is complete over this class of models.

Let us take a TL model $\langle M, O, v \rangle$ and extend v to a relation ρ such that for all m in M and all propositional variables p, $p\rho_m T$ if and only if $m \in v(p)$. The resulting model is an N_4 model. Similarly, every N_4 model is a TL model. By the completeness of N_4 and TL, all the theorems of N_4 that can be formulated only with propositional variables, parentheses, conjunction, and entailment are also theorems of TL. Thus, N_4 is a conservative extension of TL.

A theory T is a subset of M and $T \vDash A$ if and only if $m \vDash^+ A$ for all models m in T. I also define a consequence operator, C_O. Where Γ is a set of formulas, $C_O(\Gamma)$ is the set of formulas satisfied by all m in M such that every member of Γ is satisfied by m. It is easy to prove that C_O satisfies Tarski's axioms.

6.8   What Is a Theory?

I say in Chapter 1 that one of the problems in constructing a logic of entailment is the difficulty in understanding nested entailments. In the semantics given above, this problem is addressed by having entailments satisfied (or not satisfied) by points in the model. But this semantics does not really help to understand the meaning of nested entailments – the satisfaction condition is not at all illuminating. Entailments are satisfied or fail to be satisfied at points other than O in a random fashion.

In Chapter 7 and the remainder of the book, I postulate that each theory determines a consequence operator, and the entailments satisfied by a theory are correlated in a formal way with its consequence operator. The way in which I make these operators formal is by setting out a model theory based on a set of theories. The relationships that these theories bear to one another are determined by these consequence operators.

What is a theory in this semantics? It is an abstract object. That is not to say very much. According to the received view of scientific theories, a theory is a set of sentences, which is an abstract object. According to the semantic view, a theory is a set of models, also an abstract object. A theory, on my view, is an abstract object that somehow determines a set of models and a consequence operator. I think the semantic view is preferable to the received view in that the cognitive content of a scientific theory certain extends beyond a set of sentences and confirmation (or disconfirmation) conditions. We do need to know what sort of physical systems would make a theory true in order to understand it fully.

I do think, however, that reducing a theory to a set of models is to ignore what constitutes the unity of a theory. Tim Maudlin has claimed, for example, that quantum physics is not a theory [138]. There are plenty of models of quantum physics but what is needed for a theory is some sort of conceptual unity that is lacking if Maudlin is right. For a very simple example, I return for a minute to van Fraassen's geometrical example. He uses a small list of axioms and then illustrates this list by his seven-point model. Both the axioms and the model play an important role in understanding what is going on. If one constructs another model (as I did) of the axioms, then it is the axioms that unify the models as models of a single theory. This is not to say that all theories must be finitely axiomatisable. Rather, I only claim that theories need some sort of conceptual unity in order to be considered theories.

In the following chapters, I set out a model theory in which theories are the basic elements. They are the primitives of the semantics. I could spend more time on the nature of these theories, but I want to remain at least slightly neutral on the sorts of questions that would arise if I were to do so. My main concern is the role that theories can play in the semantics rather than the nature of theories themselves.

7 Theories of Entailment

7.1 A Strong Programme in Philosophy of Logic?

In this chapter, I build on the idea that a logic of entailment is primarily a logic of theory closure. I start with the fact that a logical system is itself a theory. In particular, a logic of entailment is a theory about entailment. And, as we see in earlier chapters of this book, there are many alternative theories of entailment. As a theory of theory closure, the one true logic should tell us about the closure of these other theories of entailment. Moreover, it should determine conditions under which it is itself closed. This latter point can be understood as saying that the logic of entailment should be reflexive in the weak sense of Chapter 5 – it should be closed under its own principles of entailment. In order to satisfy reflexivity in this sense, a logic of entailment must be closed under modus ponens: if $A \rightarrow B$ and A are in theorems, then so is B.

Returning to the first point that a logic of entailment is a theory about the closure of theories, in order for it to be useful as a tool for reasoning about theories, it needs to be applicable to both true and false theories, and to theories that seem bad as well as those that seem good. These remarks should be taken to be about theories of entailment and other sorts of theories. I interpret nested entailments in terms of a theory of entailment stating the conditions under which it and other theories of entailment are closed. For example, the formula $(A \rightarrow B) \rightarrow (C \rightarrow D)$ says that if a theory says that A entails B, then it is also committed to saying that C entails D.

I see the current project of constructing a theory of entailment in rather similar terms to the way in which David Blood, Barry Barnes, et al. saw their *strong programme in the sociology of science*. The strong programme was characterised by the following four principles [18, p 7]:

1. Causality: A theory in the sociology of science must give causal explanations of what scientists do, of what theories they propose and support, the experiments they take to be important, and so on.
2. Impartiality: A theory in the sociology of science must study both true and false theories alike.
3. Symmetry: A theory in the sociology of science must give similar explanations for the discovery of true and false theories alike.
4. Reflexivity: A theory in the sociology of science must give the same sort of explanation for its own discovery or invention as it does for other theories.

I think that a theory of entailment should satisfy analogous principles to those of impartiality, symmetry, and reflexivity.

The principle of causality has a clear role in the sociology of science. Bloor, Barnes, and the other members of their school attempt to give social explanations of why scientists do what they do. I have no interest in this book, however, in presenting or evaluating causal explanations of how scientists, mathematicians, or anyone else discovers or invents theories. Giving a causal explanation of why logicians do what they do is an interesting project, but it is not my project. So, I set aside the principle of causality.

The principles of impartiality and symmetry, on the other hand, are of central importance in the study of entailment. Both true and false theories, and even "crazy theories", need to be studied in terms of their logical closure. Scientists often deduce predictions from theories in order to determine whether they accord with the available data. In the simplest cases, if the data does not fit with the theory, then the theory (as it stands) is false. With regard to empirical theories, of course, empirical data is of paramount importance. With regard to mathematical or logical theories, on the other hand, we appeal to other sorts of data, of the sort discussed in Chapter 1. In more complicated cases, more than one theory may be involved. I discuss these more complicated situations in Chapters 8 and 11. The point is that often to determine whether a theory is true one must use deductive inferences to determine whether that theory fits with the data. Thus, both true and false theories need to be studied logically. Hence, the principle of impartiality should be accepted in the study of entailment.

Moreover, both true and false theories can be said to be closed under the same logical laws. When we reason about a false theory, we usually want to know what it would be like for that theory to be true. We do not, typically, use counterfactual reasoning to determine the contents of theories, because we do not want to add anything contingent to them. We just want to determine what can be deductively determined about them from what we already know about them. Therefore, there is no special logic for true theories. Logic is perfectly general. It applies to any theory at all. Thus, the principle of symmetry is a principle of the study of entailment.

As I say above, every theory of entailment should satisfy a weak form of reflexivity. It should be closed under the principle of modus ponens. However, as is discussed in Chapter 5, there is another view that claims that a logic should be closed under all and only its own principles. I argue in Chapter 5 that theories, including theories of entailment, may be closed under non-logical principles that should not be represented in logic as entailments. Thus, I only accept a weak form of reflexivity.

7.2 Models and the Intended Model

In this chapter, I construct a class of models for a logic of entailment that I call 'Generalised Tarski Logic' (GTL). Each GTL model is based on a set of points. Each point is a theory. I represent arbitrary theories by lower case roman letters – $a, b, c, \ldots$. In each model, there is a *master theory*, Σ, that represents the closure conditions for all the theories in its model. Thus, for all formulas A and B, $\Sigma \vDash A \rightarrow B$ if and only if, for all theories a in the model of Σ, $a \vDash A$ implies $a \vDash B$.

I claim in Chapters 1 and 3 that having a class of models is not enough to characterise a logic of entailment. An entailment $A \rightarrow B$ is *true* if and only if B is derivable from A. Thus, we need to have an intended model. This is a model that can plausibly be claimed to represent the actual world and so show us which entailments are true and show that these entailments correspond to logically valid derivations. Thus, I need to designate one model as the intended model, in which all and only valid derivations correspond to true entailments. This model is the canonical model, constructed in detail in the Appendix of this book.

In specifying the class of models, I use the double turnstile symbol '$\vDash$' to represent both the satisfaction relation between theories and formulas, as in '$a \vDash A$', and as a semantic consequence relation between sets of formulas and individual formulas. Where Γ is a non-empty set of formulas,

$$\Gamma \vDash A$$

means that in any theory a in any model, if all the members of Γ are satisfied by a, then B is also satisfied by a. This is the same as the consequence relations discussed in Chapter 3. I also use a consequence relation for each individual model. The expression,

$$\Gamma \vDash_{\mathfrak{M}} A$$

means that in every theory a in $\mathfrak{M}$, if every member of Γ is satisfied by a, then A is also satisfied by a.

Where $\mathfrak{M}$ is the intended model, as I argue in Chapter 3, the consequence relation $\vDash_{\mathfrak{M}}$ must be compact. The consequence relations of the intended models of this chapter and Chapter 8 are compact. This means that

$\Gamma \vDash_{\mathfrak{M}} A$ if and only if there are $G_1, \ldots, G_n$ in Γ (for some finite $n \geq 1$) such that
$$G_1, \ldots, G_n \vDash_{\mathfrak{M}} A.$$

This biconditional, in turn, implies that

$\Gamma \vDash_{\mathfrak{M}} A$ if and only if there are $G_1, \ldots, G_n$ in Γ (for some finite $n \geq 1$) such that
$$\Sigma \vDash (G_1 \wedge \ldots \wedge G_n) \rightarrow A,$$

where Σ is the master theory of the intended model. Thus, the intended model represents every valid derivation (in the sense of $\Vdash$) as an entailment that is true in the master theory. In the Appendix, I prove that the consequence relation of the canonical model is compact, and this lends credence to my claim that the one true model of the logic (the intended model) has a compact consequence relation.

7.3   Applying Theories to One Another

In Section 7.2, I say that models contain theories and a designated theory, Σ. In order to treat nested entailments, I add additional structure to models. An entailment, in my view, is a closure principle for theories. Let us again think of theories syntactically, that is, as sets of formulas. Suppose that T is a theory of entailment, that is, it contains some entailments. We can construct an operator C_T such that, for any set of formulas Γ, $C_T(\Gamma)$ is the set of formulas B where there are some A in Γ such that $A \to B$ is in T. Let us also stipulate that Γ is closed under conjunction – if C and D are in Γ, then so is $C \wedge D$.

I call $C_T(\Gamma)$ the 'application of T to Γ'. I represent application of theories in the model by the operation $\circ$. In a model, $a \circ b$ is the result of applying a to b. Another way of writing '$a \circ b$' is '$C_a(b)$'. In some circumstances, it is more intuitive to think in terms of the application operator and sometimes it is more intuitive to think in terms of consequence operators. The idea here is that nested entailments are to be understood in terms of what happens when theories are applied to other theories, which behave in certain ways when applied to other theories, and so on. Here is a satisfaction condition for entailments in these models:

$$a \vDash A \to B \text{ if and only if, for all } b \vDash A, a \circ b \vDash B.$$

This is called a *satisfaction condition* rather than a truth condition because it is about the content of theories, and only a relatively small number of theories are actually true.

The satisfaction condition for conjunction is the standard one:

$$a \vDash A \wedge B \text{ if and only if } a \vDash A \text{ and } a \vDash B.$$

Thus, the set of formulas satisfied by a theory a is closed under conjunction.

Alasdair Urquhart [205] was the first to give satisfaction conditions for entailment using an application operator, and Kit Fine [61] was the first to combine this sort of operator with a relational semantics to give a semantics for all of E. Similar methods have been used by Yale Weiss [217] and Lloyd Humberstone [84] to give semantics for negation-free relevant logics. My theory is based on Fine's.

A model is a pair $\langle \mathfrak{F}, \vDash \rangle$, where $\mathfrak{F}$ is a theory-frame and $\vDash$ is a satisfaction relation on $\mathfrak{F}$. A theory-frame is, in turn, a quadruple, $\langle \Upsilon, \Sigma, \circ, \leq \rangle$, where Υ is a non-empty set (of theories), Σ is a member of Υ, $\circ$ is a binary operator on Υ, and $\leq$ is a binary relation on Υ, and the following all hold:

- $\leq$ is a partial order, that is, it is reflexive, transitive, and anti-symmetrical;
- $\Sigma \circ a = a$ for all $a \in \Upsilon$;
- If $a \leq a'$ and $b \leq b'$, then $a \circ b \leq a' \circ b'$.

The satisfaction relation has one constraint imposed on it:

$$\text{If } a \leq b \text{ and } a \vDash p, \text{ then } b \vDash p.$$

This constraint is used together with the satisfaction conditions for the connectives and the semantic postulates given above to prove the following *hereditariness lemma*:

LEMMA 7.1 (Hereditariness) *For all formulas A and all theories a and b, if $a \leq b$ and $a \vDash A$, then $b \vDash A$.*

From the hereditariness lemma, the semantic entailment theorem follows:

THEOREM 7.2 (Semantic Entailment)   *In a model $\mathfrak{M}$, $\Sigma \vDash A \rightarrow B$ if and only if, for every theory a, $a \vDash A$ only if $a \vDash B$.*

This is the relationship between the master theory and the set of theories that I stated to hold in Section 7.2.

In these models, all the points are theories of the master theory, Σ. In addition, the application of one theory to another is always defined. This means that for any theories a and b, $a \circ b$ is also a theory of Σ. This fact makes valid a strong transitivity rule called the "affixing rule":

$$\frac{A \rightarrow B \qquad C \rightarrow D}{(B \rightarrow C) \rightarrow (A \rightarrow D)}.$$

This rule can be analysed into the following more standard transitivity rules:

$$\text{Prefixing } \frac{B \rightarrow C}{(A \rightarrow B) \rightarrow (A \rightarrow C)} \qquad \frac{A \rightarrow B}{(B \rightarrow C) \rightarrow (A \rightarrow C)} \text{ Suffixing}$$

Let us consider suffixing to illustrate why these rules are valid. Suppose that $A \rightarrow B$ holds in Σ. And suppose that there is a theory a that satisfies $B \rightarrow C$. By the satisfaction condition for entailment, for any $b \vDash B$, $a \circ b \vDash C$. Assume that $c \vDash A$. Then, by the semantic entailment theorem, $c \vDash B$. So, $a \circ c \vDash C$. Generalising, for any theory b that satisfies A, $a \circ b \vDash C$. Therefore, by the satisfaction condition for entailment, $a \vDash A \rightarrow C$. And so, by the semantic entailment theorem and generalising, $\Sigma \vDash (B \rightarrow C) \rightarrow (A \rightarrow C)$.

7.4   Generalising Tarski

In Section 7.3, I outline a model theory for a very simple logic. However, this logic is very weak. In this section, I begin the search for additional semantic postulates.

The first thought that may occur to one is that models should generalise all of Tarski's conditions for consequence operators. Here again are the modern versions of Tarski's conditions:

1. $\Gamma \subseteq C(\Gamma)$ (Inclusion);
2. If $\Gamma \subseteq \Delta$ then $C(\Gamma) \subseteq C(\Delta)$ (Monotonicity);
3. $C(C(\Gamma)) \subseteq C(\Gamma)$ (Transitivity).

If we translate C as C_Σ and $\subseteq$ as $\leq$, all of these are provable in the models I define in Section 7.2. The inclusion condition, translated into the language of models, is $a \leq C_\Sigma(a)$, or $a \leq \Sigma \circ a$. But $a = \Sigma \circ a$, and so the condition holds. The other two conditions follow similarly from the fact that $a = \Sigma \circ a$ for all theories a.

The important question here is what happens when we generalise these conditions to be about C_a for every theory a. We then get

1. $b \leq C_a(b)$ or $b \leq a \circ b$;
2. if $b \leq c$, then $a \circ b \leq a \circ c$;
3. $C_a(C_a(b)) \leq C_a(b)$ or $a \circ (a \circ b) \leq a \circ b$.

Condition 2 is provable in the models defined in Section 7.2. It is included because it is needed to prove Lemma 7.1. Neither conditions 1 nor 3, however, are provable in this class of models. I accept condition 3 but reject condition 1.

The fact that I do not accept all three of these conditions raises an interesting problem. All three conditions hold in models when we replace 'a' throughout with 'Σ'. What is it that makes Σ special so that it justifies condition 1? The answer is that Σ is, according to the model, the true theory. This means that everything that the model counts as a theory is closed under the consequence operator determined by Σ, that is C_Σ. Thus, $b = C_\Sigma(b)$ for every theory b, and this implies $b \leq C_\Sigma(b)$.

If it were that $b \leq a \circ b$ for arbitrary theories a and b, then we would have to claim that every theory contains all the entailments in Σ. Suppose that $\Sigma \models A \rightarrow B$ and $b \models A$. By the satisfaction condition for entailment and $b = \Sigma \circ b$, $b \models B$. If $b \leq a \circ b$, by the hereditariness lemma, $a \circ b \models B$. Thus, for any theory $b \models A$, $a \circ b \models B$ and hence $a \models A \rightarrow B$. This shows that a contains all the entailments in Σ. Thus, a theory can be wrong about entailment only by containing some false entailments, not by failing to contain some true entailments. This is far too restrictive as a notion of a theory of entailment.

I claim that conditions 2 and 3 are contained in the concept of a consequence operator, but for the reason just given, condition 1 is not.

The justification for condition 2 (the monotonicity condition) is given formal justification in Section 7.2 – it enables the proof of the hereditariness lemma. Its philosophical justification is quite straightforward and is given already in Chapter 6. To recap, suppose that one is working with a theory a of entailment and that one derives a statement B from a theory b using a – formally $a \circ b \models B$. One might for nonlogical reasons extend b to b'. One should still be able to accept that B is derivable from b', that is, $a \circ b' \models B$. This justifies the monotonicity postulate, $b \leq b'$ implies $a \circ b \leq a \circ b'$.

I call condition 3, 'reverse contraction' (RC) for reasons that become clear in Section 8.3 . The warrant for RC is a bit more involved. It requires the notion of a closure operator (see Chapter 6). A set of formulas Γ is closed under an operator C if the result of applying C to Γ does not produce any formulas that are not already in Γ. Formally, Γ is C-closed if and only if

$$C(\Gamma) \subseteq \Gamma.$$

Now, if C is a closure operator, then the result of applying it to any set of formulas Γ produces a C-closed set, that is,

$$C(C(\Gamma)) \subseteq C(\Gamma).$$

This much I have already discussed in Chapter 6. In that chapter, I agree with Tarski that the correct consequence operator should be a closure operator. Here, I make the stronger claim that it is contained in the very idea of a consequence operator that it is a closure operator. In terms of the model theory, this means that if, say, C_a is a consequence operator for some theory a, then for any theory b,

$$C_a(C_a(b)) \le C_a(b)$$

or, to say the same thing in different notation,

$$a \circ (a \circ b) \le a \circ b.$$

And this is the postulate that I am trying to justify.

While I am discussing the topic of closure operators, I note that a syntactic closure operator (one that operates on sets of formulas) based on a relevant logic is closer in form to Kuratowski's definition of a closure operator than one based on classical logic. Consider the empty set, $\emptyset$. If C is based on, say, a classical modal logic, then $C(\emptyset)$ will include all the theorems of that logic. If, on the other hand, C is based on a relevant logic, then $C(\emptyset) = \emptyset$, which is postulate 3 of Kuratowski's definition of a topological closure operator (see Section 6.2).

Condition 2 is derivable in the definition of a model given in Section 7.2. Condition 3, however, is new. The addition of condition 3 makes valid the *conjunctive syllogism* (CS) scheme:

$$(CS)\ ((A \to B) \wedge (B \to C)) \to (A \to C).$$

Conjunctive syllogism tells us that entailment is transitive, but in a different way than the prefixing and suffixing rules. It is about the closure of theories under chains of entailments. If $A \to B$ and $B \to C$ are both true according to a theory a, then so is $A \to C$. The assumption that theories are closed under the transitivity of entailment is a very useful tool in doing proofs. I claim in Chapter 8 that we require stronger transitivity axioms than conjunctive syllogism, but CS is the transitivity scheme in the logic of this chapter.

Here is a little proof that CS holds in models that satisfy RC. It illustrates how one uses the semantic entailment theorem in proving soundness of relevant logics:

Proof Let a be a theory in a model and suppose that $a \vDash (A \to B) \wedge (B \to C)$. Now take an arbitrary theory b in that model such that $b \vDash A$. I show that $a \circ b \vDash C$. $a \vDash A \to B$, so $a \circ b \vDash B$ and $a \vDash B \to C$, so $a \circ (a \circ b) \vDash C$. By the postulate that $a \circ (a \circ b) \le a \circ b$ and the hereditariness lemma (Lemma 7.1), $a \circ b \vDash C$. Generalising, by the satisfaction condition for entailment, $a \vDash A \to C$. Thus, by the semantic entailment theorem, $\Sigma \vDash ((A \to B) \wedge (B \to C)) \to (A \to C)$. ∎

7.5 The Logic GTL

A formula is valid on a model if and only if it holds in the master theory of that model. A formula is valid on a class of models if and only if it is valid on every member of that class. A logic is *characterised* by a class of models if and only if the set of formulas valid on that class is exactly the set of theorems of that logic. In other words, a logic is characterised by a class of models if the logic is complete over that class. This form of completeness is weak completeness. If the consequence relation of the class of models is also compact, then the logic is strongly complete over that class of models.

I call the logic that is characterised by the class of models that satisfy RC as 'Generalised Tarski Logic' or 'GTL'. It has the axioms and rules given below. I prove a strong completeness result for GTL in the Appendix.

Axiom Schemes

1. $A \to A$
2. $(A \wedge B) \to A; (A \wedge B) \to B$
3. $((A \to B) \wedge (A \to C)) \to (A \to (B \wedge C))$
4. $((A \to B) \wedge (B \to C)) \to (A \to C)$

Rules

$$\frac{\vdash A \to B \qquad \vdash A}{\vdash B} \qquad \frac{\vdash A \qquad \vdash B}{\vdash A \wedge B}$$

$$\frac{\vdash A \to B \qquad \vdash C \to D}{\vdash (B \to C) \to (A \to D)}$$

In terms of the syntactic consequence operator determined by GTL, C_{GTL}, axiom 1 states that for all sets of formulas Γ,

$$\Gamma \subseteq C_{GTL}(\Gamma).$$

This, of course, is the inclusion postulate. Axioms 2 and 3 give us the syntactic version of the satisfaction conditions for conjunction. That is, for any GTL-theory Γ, $A \wedge B$ is in Γ if and only if both A and B are in Γ. Axiom 4, of course, is conjunctive syllogism.

Note that GTL is not strongly reflexive in the sense of Chapter 5. It contains primitive rules for which there is no formula counterpart that is a theorem of the logic. One such rule is modus ponens. The logic does not contain as a theorem the scheme $((A \to B) \wedge A) \to B$. However, we can see here modus ponens can be justified without claiming that it holds universally of theories. Together with the adjunction rule, it tells us that the master theory is a theory of itself, and so states the weak reflexivity condition on theories. The case for affixing is similar. It is justified by the desire (or need) to have a model that consists solely of Σ-theories. This justifies having affixing without accepting its horizontalisation in the sense of Chapter 5.

In my definition of application in Section 7.3, I assume that conjunction is transitive. To prove one direction of associativity, we can first proof $(A \wedge (B \wedge C)) \to (A \wedge B)$, as follows:

$$\frac{(A \wedge (B \wedge C)) \to A \qquad \dfrac{(A \wedge (B \wedge C)) \to (B \wedge C) \qquad (B \wedge C) \to B}{(A \wedge (B \wedge C)) \to A}}{(A \wedge (B \wedge C)) \to (A \wedge B)}$$

Then, we can prove $(A \wedge (B \wedge C)) \rightarrow C$:

$$\frac{(A \wedge (B \wedge C)) \rightarrow (B \wedge C) \qquad (B \wedge C) \rightarrow C}{(A \wedge (B \wedge C)) \rightarrow C}$$

Using axiom 3 and modus ponens, we obtain

$$(A \wedge (B \wedge C)) \rightarrow ((A \wedge B) \wedge C).$$

The other direction can be proven first by showing that $((A \wedge B) \wedge C) \rightarrow A$:

$$\frac{((A \wedge B) \wedge C) \rightarrow (A \wedge B) \qquad (A \wedge B) \rightarrow A}{((A \wedge B) \wedge C) \rightarrow A}$$

And then proving that $((A \wedge B) \wedge C) \rightarrow (B \wedge C)$:

$$\frac{\dfrac{((A \wedge B) \wedge C) \rightarrow (A \wedge B) \qquad (A \wedge B) \rightarrow B}{((A \wedge B) \wedge C) \rightarrow B} \qquad ((A \wedge B) \wedge C) \rightarrow C}{((A \wedge B) \wedge C) \rightarrow (B \wedge C)}$$

Then, by using axiom 3 and modus ponens again, we obtain

$$((A \wedge B) \wedge C) \rightarrow (A \wedge (B \wedge C)).$$

Putting all of this together, we obtain the principle of the associativity of conjunction.

7.6   The Denecessitation Operator

I extend the language by adding the logical truth constant t – discussed in Chapter 4. There is one axiom governing t (that t itself is an axiom) and a necessitation rule:

$$\frac{\vdash A \rightarrow B}{\vdash t \rightarrow (A \rightarrow B)}.$$

In the semantics t has the following satisfaction condition:

$$a \vDash t \text{ if and only if } \Sigma \leq a.$$

As in Chapter 4, I here define logical necessity in terms of t and entailment:

$$\Box A =_{df} t \rightarrow A.$$

And I define an operator $\Box^-$ in the semantic to treat necessity, viz.,

$$\Box^- a =_{df} a \circ \Sigma.$$

The theory $\Box^- a$ is called the *denecessitation* of a. Recall that $\Box A =_{df} t \rightarrow A$. The following proposition, then, tells us about the relationship between $\Box$ in the object language and the semantic operator, $\Box^-$.

PROPOSITION 7.3   *For all theories a and all formulas A, $\Box^- a \vDash_v A$ if and only if $a \vDash_v t \rightarrow A$.*

To prove Proposition 7.3, first suppose that $a \vDash_v t \to A$. By the satisfaction condition for t, $\Sigma \vDash_v t$, hence $a \circ \Sigma \vDash_v A$. By the definition of the semantic operator $\Box^-$, $\Box a \vDash_v A$. Now, suppose that $\Box^- a \vDash_v A$. Then suppose that $b \vDash_v t$, hence $\Sigma \leq b$. By semantic postulate two, $a \circ \Sigma \leq a \circ b$. Since $\Box^- a = a \circ \Sigma$, $\Box^- a \leq a \circ b$. Thus, by the hereditariness theorem, $a \circ b \vDash_v A$. Generalising, by the semantic entailment theorem, $a \vDash_v t \to A$.

Proposition 7.3 shows that, with regard to logical necessity, in the present semantics an operator takes the place of the binary accessibility relation in Kripke semantics (see Chapter 3). A necessity, $\Box A$, obtains in a theory a if and only if A obtains in a theory $\Box^- a$. This sort of semantics is often called an "operational semantics" because operators take the place of the relations in Kripke semantics (see Chapter 3).[1] The notation '$\Box^-$' is taken from Hughes and Cresswell [83, ch 6].[2] In [83], where Γ is a set of formulas, $\Box^- \Gamma$ is the set of formulas A such that $\Box A$ is in Γ.

It is easy to show that the valid formulas in a model are closed under the rule of necessitation. Suppose that $\Sigma \vDash_v A$. And suppose that, for an arbitrary theory b, $b \vDash_v t$. Then, $\Sigma \leq b$, and by the hereditariness theorem, $b \vDash_v A$. By the semantic entailment theorem, then, $\Sigma \vDash_v t \to A$, that is, $\Sigma \vDash_v \Box A$.

The master theory is identical to its own denecessitation. That is, $\Sigma = \Box^- \Sigma$. For, $\Sigma \circ a = a$ for all theories a, so $\Sigma \circ \Sigma = \Sigma$. Therefore, $\Box^- \Sigma = \Sigma$. I call any theory that is identical to its own denecessitation, a *pure theory of entailment*. In Chapter 8, pure theories of entailment are shown to be very interesting. Here, I only make the philosophical point that a pure theory of entailment looks, at least to some extent, like a logic since it tells us only about those statements that are logically necessary.

There is also a formal point about the fact that the master theory is its own necessitation. This fact makes valid the rule of necessitation. For suppose that $\Sigma \vDash A$. Since $\Sigma = \Box^- \Sigma$, $\Box^- \Sigma \vDash A$. But then, by Proposition 7.3, $\Sigma \vDash \Box A$. Thus, in all my models for entailment, necessitation holds.

Just as in the relational semantics for modal logic, we can place conditions on the semantic box operator to make valid further schemes. Here are some standard schemes and their semantic postulates:

Name	Postulate	Scheme
T	$\Box^- a \leq a$	$\Box A \to A$
4	$\Box^- a \leq \Box^- \Box^- a$	$\Box A \to \Box\Box A$
K$_\to$	$\Box^- a \circ \Box^- b \leq \Box^-(a \circ b)$	$\Box(A \to B) \to (\Box A \to \Box B)$

All of these schemes are derivable in the logic E (see Chapter 4), and their corresponding postulates are deducible in the frame theory for E (see Chapter 8).

The operational semantics provides an elegant treatment of necessity. Unfortunately, it is far less clear how to give a reasonable treatment of possibility in this

---

[1] Katalin Bimbo, J. M. Dunn, and Nicholas Ferenz [17] call this semantics a "hybrid" semantics because it uses both operators, like $\circ$ and $\Box^-$, and relations, like $\leq$.
[2] In Chellas [36], the same operator is written '$\Box^{-1}$'.

framework. Of course, if we define the possibility operator as $\Diamond A =_{df} \neg\Box\neg A$, then, given a semantics for negation (as in Chapter 9), we can derive a satisfaction condition for possibility. However, this derived satisfaction condition lacks any intuitive grounding. Clearly, this area needs more work. In this book, however, possibility plays almost no role, and so I set this topic aside.

7.7   The Nature of Necessity

The functions of entailment, in the present semantics, are to determine closure operators and to state the closure conditions on theories. Entailment connectives, however, were introduced into logic to represent the notion of a derivation in the language of logic. As I say in Chapter 1, the concept of a derivation incorporates some notion of logical necessity. The premises of a valid derivation do not only imply the conclusion, the conclusion follows necessarily from the premises. In order to make a case that this semantics is a good semantics for a logic of entailment, I need to show how its entailment incorporates some notion of necessity.

The obvious way to show an important connection between necessity and entailment in this semantics is to find a strong relationship between entailment and the denecessitation operator. However, in the GTL semantics, this relationship seems rather weak. In Chapter 8, I show that in the semantics of E, there is a clear and important connection between entailment and necessity. Here, I indicate what is missing in the GTL semantics in this regard and preview the relationship between denecessitation and entailment in E.

Consider the denecessitation operator in the semantics for GTL. Suppose that $A \to B$ is satisfied by a GTL-theory a. In $\Box^- a$, there may not be any trace of that entailment. The entailment operator is a primitive of the system; it is not defined in terms of another conditional and a necessity operator. The semantics for E, on the other hand, obeys the following principle:

$$a \circ b = \Box^- a \circ b.$$

I call this the "Application Postulate" or "App". What App tells us is that the entailments of a theory a are all contained in $\Box^- a$. In order for this to occur, given the definition of $\Box^-$, if $a \vDash A \to B$, then $a \vDash \Box(A \to B)$. Since $\Box A$ is defined using entailment, as $t \to A$, then adding App implies adding the 4 principle given above. Here is a little derivation of 4 from App in the model theory:

$$\begin{array}{llll} 1. & a \circ \Sigma & = & \Box^- a \circ \Sigma \quad \text{App} \\ 2. & \Box^- a & = & \Box^-\Box^- a \quad \text{1, def } \Box^- \end{array}$$

As we can see, App is slightly stronger than 4 in the sense that it entails the identity of $\Box^- a$ and $\Box^-\Box^- a$ and not just the inclusion of the first in the second. App corresponds to the scheme $(A \to B) \leftrightarrow \Box(A \to B)$.

Adding App also gives the denecessitation operator a very close connection to the notion of consequence operators. For a theory a, the consequence operator C_a is defined such that $C_a(b) = a \circ b$ for all theories b. Suppose that for two theories a and a', $a \circ b = a' \circ b$ for all b. Then, in particular, $a \circ \Sigma = a' \circ \Sigma$, hence $\Box^- a = \Box^- a'$. Therefore, if applying a to a theory always results in the same theory as applying a', then the denecessitations of a and a' are the same, but so are the consequence operators determined by a and a'. The adding of App, then, will enable us to represent consequence operators even more directly in the semantical theory. But I wait to do so until Chapter 8.

7.8   Fallacies of Modality

Sometimes necessary truths follow from contingent truths. For example, $p \rightarrow (p \vee \neg p)$ is true (and necessarily so), but $p \vee \neg p$ is a necessary truth and p might be (depending on its interpretation) a contingent truth. This does not seem problematic. In $p \rightarrow (p \vee \neg p)$, the consequent is a necessary truth but it is not stated to be a necessary truth. What is difficult to accept is:

$$p \rightarrow \Box(p \vee \neg p).$$

In this formula, the consequent is not just necessarily true, but the consequent tells us that $p \vee \neg p$ is necessarily true. A formula that states that a formula is necessarily true is a *necessitative*. The fallacies of modality (that I am interested in) tell us that a contingent proposition entails a necessitive.

Of course some contingent propositions do entail necessitives. The formula $(p \wedge \Box q) \rightarrow \Box q$ is a theorem of GTL and, depending on the interpretation of p, it can be a contingency. But the antecedent does contain a necessitive, $\Box q$. Building on this, Anderson, Belnap, and Coffa formulate the notion of a fallacy of modality in terms of formulas that contain boxes and those that do not. Their definition of a modal fallacy is: A formula $A \rightarrow B$ is modally fallacious if and only if B is of the form $\Box C$ and A contains no occurrences of $\Box$ [4, §21.1.2].[3] In other words, no modal formula is entailed by a non-modal formula. GTL and E satisfy the requirement that they do not contain any fallacies of modality.

One issue that arises here and elsewhere in discussing the logic of entailment is the role of logical form. $p \rightarrow \Box q$ is surely true for some interpretations of p. For example, if p is used to represent the same proposition as $\Box q$, then $p \rightarrow \Box q$ is true. In Chapter 1, however, I claimed that the theorems of a logic of entailment must be closed under the rule of uniform substitution. It is this that makes the logic a *formal logic* of entailment. The principles of the logic should be understood schematically. They tell us which inference forms are valid.

---

[3] Larisa Maksimova has a more sophisticated definition of a fallacy of modality, but the simple one will do for my purposes here.

A criticism one might make of both the fallacies of modality story and the treatment of the logic of entailment as schematic is that there are specific apparently contingent sentences that entail specific necessities. For example, identity statements are said to entail their own necessitation, that is, all formulas of the form $i = j \rightarrow \Box i = j$ are said to be theorems. One simple way of treating this sort of example so that it does not conflict with the formulation of the fallacies of modality is to claim that the identity predicate has logical necessity built into it – to say that i and j are identical is just to say that they are necessarily the same thing. I do not, however, take this route to avoiding the problem. The doctrine of the necessity of identities is a metaphysical doctrine, and not one that I think we should commit theories in general. Let us consider the nature of theories, as I have described it in Chapter 6. An individual or type of thing in a theory is really a role, which things can play in models. Just because in the models for a theory two roles always coincide in the same thing does not seem to imply that the theory holds that these roles coincide of logical necessity. It might be that we would want to say of, for example, a theory in biology, that if two roles coincide in this way in all its models then that theory holds that it is biologically necessary that a particular identity holds, but with regard to logical necessity it seems better to treat identity as contingent. A biological theory, and its models, do not tell us what would happen in worlds that obey different laws of biology.

Thus, I reject the principle $i = j \rightarrow \Box i = j$. This is not to say that there is no weaker link between a true identity and its necessitation. It may be that $i = j \Rightarrow \blacksquare i = j$, where '$\Rightarrow$' is the relevant implication of the logic R and '$\blacksquare$' is metaphysical necessity. This means that I do not have to give an argument to show where the arguments for the doctrine of the necessity of identity go wrong. I merely have to reinterpret these arguments. This sort of reinterpretation, moreover, is quite easy. If sound, the arguments show that in all metaphysically possible worlds if an identity holds in one, then it holds in all of them (in which the salient objects exist). This claim is clearly different from the claim that if an identity holds in a theory, then that theory tells us that this identity is logically necessary.

7.9   The Intended Model

On the definition of entailment, $A \rightarrow B$ is *true* if and only if B is derivable from A. The intended model, therefore, must make all and only entailments true that correspond to valid deductions. My intended model is based on the *canonical model* used in completeness proofs. I define the canonical model in the Appendix. Here we only have to take note of certain features of the canonical model.

The master theory, Σ, of the canonical model is just the set of theorems of the logic. Let '$\vDash_C$' be the consequence relation for the canonical model, such that for any set of formulas, Γ, $\Gamma \vDash_C A$ if and only if for every theory a in the canonical model, if $c \vDash G$ for all G in Γ, then $a \vDash A$. By the semantical entailment theorem (Theorem 7.2),

$$\Sigma \vDash (G_1 \wedge \ldots \wedge G_n) \rightarrow A \text{ if and only if } \{G_1, \ldots, G_n\} \vDash_C A,$$

for any finite and non-empty set of formulas, $\{G_1, \ldots, G_n\}$. In the Appendix, I also show that $\vDash_C$ is compact. This means that for any set of formulas Γ (finite or infinite), all of its consequences are represented in a finite way by the entailments that hold in Σ.

These results bring GTL close to satisfying the requirements for a logic of entailment that I set out in Chapters 3 and 6, but more is needed. It is necessary to show that it is plausible that Σ contains all and only true entailments. In the Appendix, I show that for the logic E, there are world-like theories that have Σ as their sets of entailments, and this result, I suggest, completes the argument that E is a reasonable logic of entailment. This proof can be adapted with no difficulties to treat GTL when it is embedded in a logic that contains the other standard connectives – disjunction and negation. In Section 7.10, I look briefly at conservative extensions of GTL with these other connectives.

7.10 GTL and Dialectical Logics

Among relevant logics, those that are conservative extensions of GTL are the *Dialectical Logics*. The dialectical logics were introduced by Richard Routley and Robert Meyer in the mid-1970s in a paper in the journal *Studies in Soviet Thought* [184]. Despite the title and the venue of the article, the only significant property these logical systems had in common with the logics of Hegel and Marx is the ability to tolerate contradictions without thereby becoming trivial. The logic at the centre of Routley and Meyer's study was DL. This logic had as its entailment and conjunction fragment the logic GTL.

The primitive connectives of the language of DL are conjunction, entailment, and negation. Disjunction is defined in the usual De Morgan manner:

$$A \vee B =_{df} \neg(\neg A \wedge \neg B).$$

Here is an axiomatisation of DL.

Axiom Schemes

1. $A \to A$
2. $((A \to B) \wedge (B \to C)) \to (A \to C)$
3. $(A \wedge B) \to A; (A \wedge B) \to B$
4. $((A \to B) \wedge (A \to C)) \to (A \to (B \wedge C))$
5. $(A \wedge (B \vee C)) \to ((A \wedge B) \vee (A \wedge C))$
6. $\neg\neg A \to A$
7. $(A \to \neg B) \to (B \to \neg A)$
8. $(A \to \neg A) \to \neg A$

The rules are modus ponens, adjunction, and affixing. In the original formulation of DL in [184, p 7], there is an additional axiom, $p_0 \wedge \neg p_0$, where 'p_0' is a propositional constant. This says that there is a true contradiction. This is an interesting axiom because its inclusion makes the logic fail to be closed under the rule of uniform substitution.

Despite having a contradiction as an axiom, DL has the law of non-contradiction as a theorem: $\neg(A \wedge \neg A)$. Here is a sketch of a proof of it.

$$\frac{\dfrac{\dfrac{(A \wedge \neg A) \to \neg A}{(A \wedge \neg A) \to (\neg A \vee \neg\neg A)}}{(A \wedge \neg A) \to \neg(A \wedge \neg A)} \quad ((A \wedge \neg A) \to \neg(A \wedge \neg A)) \to \neg(A \wedge \neg A)}{\neg(A \wedge \neg A)}$$

The law of excluded middle, $A \vee \neg A$, is equivalent to the law of non-contradiction, by De Morgan principles. Axiom 8 plays an important role in this proof, and non-contradiction is not provable in this system without the use of that axiom.

Later, Routley adopted a weaker logic, DK, which replaced axiom 8 with the law of excluded middle and did not include the contradictory axiom. Ross Brady adopted a logic act that was still weaker than DK, called 'DJ', which also dropped axiom 8 and the contradiction without replacing it with anything. Brady then added "metarules" to DJ to create MC (his logic of meaning containment). The 'K' and 'J' in the names of these logics merely indicate successive weakenings of DL.

The logics DL, DK, and DJ seem, given what I have said so far, like reasonable candidates for the logic of entailment. The fact that, as formulated here, DK contains an axiom that is not an entailment seems somewhat inappropriate. It is supposed to be a logic of entailment. If so, what does this axiom – excluded middle – tell us about entailment? But DL and E both have excluded middle as a theorem, and hence as a side-effect of their principles of entailment. It may be that we can add an entailment as an axiom to DJ and obtain DK. I do not know whether this is possible.

There is another more important fact that we do need to take into account if we accept a logic that contains excluded middle. Recall that in our intended model, an entailment must be true if and only if the corresponding deduction holds. If the law of excluded middle is true, then the truths of the actual world make up a maximal consistent set of sentences. I show in the Appendix that there are maximal consistent sets of sentences which contain all and only the entailments provable in E. In the intended model of E, one of these worlds is supposed to represent the actual world.

7.11   The Functional Theory of Meaning

As I say at various points in this book, I do not adopt a purely truth-functional theory of meaning for entailment. Instead, I adopt what I call a *functional* theory of meaning. There are two "functional" meanings that I attribute to entailment. The first is that the function of an entailment of the form $A \to B$ is to express a universal closure condition on theories. The second is that the function of such an entailment is to help express the closure operator of a theory. That is, if $a \vDash A \to B$, and if $b \vDash A$, then $C_a(b) \vDash B$.

In terms of the universal closure function of an entailment, there is an interesting dichotomy between the master theory and the other theories in the model. All the entailments in the master theory are true according to the model. They all accurately state closure properties of all the theories in the model (including the master theory itself). But the non-master theories may lack entailments that are true in the model or contain ones that are false. These non-master theories are defective in this sense but are still worthy objects of study and as such belong in the intended model.

We can think of the intended model and its master theory in two ways. From the *bottom up*, we can say that the choice of a master theory is vindicated by the range of theories that it describes. If these theories include all and only those that we intuitively think of as theories, then this is good justification for accepting Σ as a general description of the closure properties of theories. From the *top down*, we can say that the range of theories described is correct if the general conditions that are satisfied by closure operators seem to give these operators the right properties.

The functional theory of meaning should not be seen as supplanting the truth-conditional theory of meaning for the other connectives. Some connectives, such as entailment and conjunction, can be understood in part in terms of their functions with regard to theories. But conjunction also has a truth-conditional meaning – $A \wedge B$ is true if and only if A is true and B is true. Understanding this in part constitutes an understanding of conjunction. Entailment has a truth condition, that is, $A \rightarrow B$ is true if and only if every theory that contains A also contains B, but clearly this truth condition is parasitic on entailment's functional meaning.

7.12   Proofs and Proof Processes

Proof plans are used to make inferences about the contents of theories. A theory that we may be most interested in is our theory that is supposed to contain all and only the truths about the world. But, of course, we are interested in other theories, some of which contain only some truths, and some of which contain at least some falsehoods. I suggest in this book that the logic of entailment is a theory about theories – a theory about the principles under which the statements of theories are closed. In this section, I outline a proof theory that uses the central features of the model theory to enable one to use the model theory and the logic together to make inferences about what theories contain.

This proof theory is a natural deduction system. It is a form of labelled deduction system, which was developed by Dov Gabbay and others (see, e.g., [9]). Labelled deduction systems have elements that refer to items in the semantics, like possible worlds and relations, so that the deduction system can inherit some of the mathematical power of the semantical theory. I use a labelled natural deduction system to make the relationship between the proof theory of the logic and the semantics very explicit. In Section 8.9, I make this relationship more precise. Here, I use the labelled system to show how the proof theory and the semantics can be employed together to demonstrate how a proof plan is to be implemented in what I call a "proof process".

The labelled natured deduction system is not a Fitch-style system like the derivation system of Chapter 4. Some of the features of the Fitch system are not needed here, such as the strict scope lines. The steps of the system are *judgements*. A judgement is an expression of the form '$\alpha \vDash A$', where 'α' is a term that represents theories and A is a formula. I use variables, $x, y, z, \ldots$, and the binary operator $\circ$ together with parentheses to make up theory terms. I also sometimes use $T, T_1, T_2, \ldots$ as parameters to represent specific theories. To illustrate how all this works, here is a little annotated derivation of a theorem of GTL:

$$
\cfrac{
  \cfrac{
    \cfrac{
      \cfrac{[x \vDash (A \to B) \land (B \to C)]^2}{x \vDash A \to B} \land E \qquad [y \vDash A]^1
    }{x \circ y \vDash B} \to E \qquad
    \cfrac{[x \vDash (A \to B) \land (B \to C)]^2}{x \vDash B \to C} \land E
  }{
    \cfrac{
      \cfrac{
        \cfrac{x \circ (x \circ y) \vDash C}{x \circ y \vDash C} \text{CS} *
      }{x \vDash A \to C} \, 1 \to I
    }{\Sigma \circ x \vDash A \to C} \Sigma I
  } \to E
}{\Sigma \vDash ((A \to B) \land (B \to C)) \to (A \to C)} \, 2 \to I
$$

To derive $x \circ y \vDash C$ from $x \circ (x \circ y) \vDash C$, I use a *structural rule*, rather like the structural rules in the natural deduction systems of John Slaney [193] and Greg Restall [178]. The asterisk marks the step at which the structural rule is used. The idea is that if $a \leq b$ is provable in the semantics, then given the hereditariness lemma (Lemma 7.1), we can infer $b \vDash A$ from $a \vDash A$. In this case, the structural rule uses the semantic postulate CS, which says that $a \circ (a \circ b) \leq a \circ b$.

This proof can be used as the basis of a proof process. Suppose that one knows that the theory T contains the entailments $A \to B$ and $B \to C$, then the process by which we prove that $A \to C$ is in T can be represented as

$$
\cfrac{
  \Sigma \vDash ((A \to B) \land (B \to C)) \to (A \to C) \qquad
  \cfrac{T \vDash A \to B \qquad T \vDash B \to C}{T \vDash (A \to B) \land (B \to C)}
}{
  \cfrac{\Sigma \circ T \vDash A \to C}{T \vDash A \to C}
}
$$

This representation of a proof process is the application of the logic, and this natural deduction system, to tell us about specific theories, rather than just to derive theorems.

Now that the proof method has been illustrated with some examples, let us move on to the formal details of the system. The square brackets around $x \vDash (A \to B) \land (B \to C)$ in the derivation of conjunctive syllogism indicate that this formula is a discharged hypothesis. The superscripted 2 and the 2 to the right of the last consequence line indicate where the discharge takes place. One can use a hypothesis as many times as needed. In this proof, the hypothesis $x \vDash (A \to B) \land (B \to C)$ is used twice. As in the Fitch-style rule, the entailment introduction rule requires the use of a hypothesis:

$$
[x \vDash A]
$$

$$
\vdots
$$

$$
\cfrac{a \circ x \vDash A}{a \vDash A \to B} \to I
$$

where x does not occur in the theory term a. This rule says that if $a \circ x$ for any theory $x \vDash A$, then $a \vDash A \rightarrow B$. This is just one direction of the satisfaction condition for entailment. The entailment elimination rule recapitulates the other direction:

$$\frac{a \vDash A \rightarrow B \qquad b \vDash A}{a \circ b \vDash B} \rightarrow\!\text{E}.$$

The application operator in theory terms represents application in the semantics.

The conjunction rules mirror the rules in the Anderson–Belnap system.

$$\frac{a \vDash A \qquad a \vDash B}{a \vDash A \wedge B} \wedge\text{I} \qquad \frac{a \vDash A \wedge B}{a \vDash A} \wedge\text{E} \qquad \frac{a \vDash A \wedge B}{a \vDash B} \wedge\text{E}$$

The conjunction introduction rule requires that the same theory contain both potential conjuncts before the rule can be applied. This makes sense: for a theory to contain a conjunction, it must contain both conjuncts.

There are two rules for t. The first tells us that we can assert (anywhere in a proof) that t holds in Σ:

$$\Sigma \vDash t.$$

This rule is an introduction rule for t. The elimination rule tells us that any theory that satisfies t can play the role of Σ in a theory term:

$$\frac{a(\Sigma) \vDash A \qquad b \vDash t}{a(b) \vDash A},$$

where $\Box^{-}a$ is defined, as above, as $a \circ \Sigma$, we can derive that $a \vDash t \rightarrow A$ follows from $\Box^{-}a \vDash A$:

$$\frac{\dfrac{\Box^{-}a \vDash A}{a \circ \Sigma \vDash A} \qquad [x \vDash t]^{1}}{\dfrac{a \circ x \vDash A}{a \vDash t \rightarrow A}\,1}.$$

And this is a proof of the converse:

$$\frac{a \vdash t \rightarrow A \qquad \Sigma \vDash t}{\dfrac{a \circ \Sigma \vDash A}{\Box^{-}a \vDash A}}.$$

I sometimes use the interderivability of $a \vDash t \rightarrow A$ and $\Box^{-}a \vDash A$ as a rule in proof processes rather than perform these derivations each time.

With this derivation method in hand as a way of representing proofs of theorems and proof processes, in Chapter 8, I will indicate the shortcomings of GTL and argue for the turn to a stronger logic – the logic E of relevant entailment.

Part III

The Logic E of Relevant Entailment

8   The Logic of Entailment

A Guide to Part III

In Part II, I develop a model theory for the relevant logic GTL. In Part III, this model theory is extended to treat all of the logic E, including disjunction, negation, and the quantifiers. The main task of Chapter 8 is to present and motivate the postulates for the semantic postulates that need to be added to the model theory to characterise the entailment and conjunction fragment of E. Chapter 9 adds the mechanisms for negation and disjunction, while Chapter 10 adds the semantics for quantification.

The core theory, as set out in Chapter 8, contains conjunction and entailment as its only connectives. The addition of the other usual logical connectives and quantifiers helps to characterise a class of theories that one might think of as standard, as opposed to those like constructive mathematical theories, or quantum theories, that are not closed under rules such as double negation elimination or the distribution of conjunction over disjunction. In order to treat such theories, clearly, one needs to adopt logics with a non-standard negation or disjunction. In such cases, one can retrench to the core logic and add different axioms for these connectives. In Chapters 9 and 10, I set out only the logic and semantics that deal with standard theories.

Taken together, Chapter 7 from Part II and Chapters 8 through 10 from Part III present and motivate the semantic theory. This semantic theory is a streamlined version of Kit Fine's model theory for relevant logics together with the semantics that I developed with Robert Goldblatt for quantified relevant logic [61, 134].

In Chapter 11, I reflect on the logic and semantics of entailment and its uses in a theory of inference. I examine their relationship to theories that are closed under the principles of classical logic and look at their relationship to the logic and semantic of the logic R, as I interpret and defend it in my book *Relevant Logic* [129].

8.1   Moving beyond GTL

In Chapter 7, I motivate the postulates of GTL by looking at the consequence operators determined by theories. In that chapter, I support the adoption of the semantic postulates on the basis of a particular view of what shape the models should take and the intuition that every theory should determine a closure operator. In this chapter, I motivate the extension of the definition of a model with further postulates that characterise the logic E. In part, these postulates are motivated by the fact that they deliver the logic E, which is in turn motivated by its elegant natural deduction system presented in Chapter 4. However, these postulates can also be motivated by the way in which they make theories interact.

For example, suppose a logician is looking at a theory T which contains some entailments. She wants to prove that C holds in T and knows that $A \to (B \to C)$ is a theorem of the logic that she accepts. She then proves A is in T. But she knows that B is not a theorem of her logic, but it does hold in T. It would seem that she has the following proof process in mind:

$$\frac{\dfrac{\dfrac{\Sigma \vDash A \to (B \to C) \qquad T \vDash A}{\Sigma \circ T \vDash B \to C}}{T \vDash B \to C} \qquad T \vDash B}{\dfrac{T \circ T \vDash C}{T \vDash C}\,?}$$

In GTL the last step of this proof process is invalid. It appeals to the following rule:

$$\frac{a \circ a \vDash A}{a \vDash A}\,.$$

This rule is sometimes called 'WI', but I call it 'weak contraction'. Full contraction is the postulate that we can deduce $a \circ b \vDash A$ from $(a \circ b) \circ b \vDash A$. Weak contraction, as the name suggests, can be derived from full contraction. Full contraction is valid in the semantics of E, but even weak contraction is invalid in GTL. The ability to apply entailment theories to themselves seems an important aspect of our reasoning about theories that is not captured well by GTL.

A feature of our reasoning that is difficult to represent in GTL has to do with the way in which mathematicians, logicians, and others *stage* proofs. In a proof, there are often stages. These are usually represented by lemmas. One often thinks, in setting out a proof plan, something like "If I can prove A, then if I can prove B, then I will be able to show C". It is natural to think of the agent as having in mind the proof process given above. After the agent detaches A from the original proof plan, she is left with a new proof plan, which is given by $T \vDash B \to C$. This is a new stage in the proof. In GTL, however, '$T \vDash B \to C$' does not tell her that what she has to do is prove $T \vDash B$ in order to conclude that $T \vDash C$. Rather, it gives her no direction at all. This does not seem right. The addition of weak contraction does give real sense to this stage in the proof process: it allows the agent to reach her goal by proving that $T \vDash B$. Thus, it seems that we should adopt a logic stronger than GTL.

8.2   Prefixing

Throughout this book, I use the suffixing thesis – $(A \to B) \to ((B \to C) \to (A \to C))$ – as an example of an intuitive and common proof plan. In this section, I discuss a different but very closely related thesis, prefixing:

$$(B \to C) \to ((A \to B) \to (A \to C)).$$

Prefixing is also valid in E and is often stated as an axiom. In Section 8.5, I show that the semantic postulate for suffixing can be derived from the postulate for prefixing and two other postulates. In this section, I motivate the postulate for prefixing.

The postulate corresponding to prefixing is the associativity condition:

$$(\text{Associativity}) \; a \circ (b \circ c) \leq (a \circ b) \circ c.$$

In order to understand the associativity condition, let us think of theories in terms of the consequence operators that they determine. In this idiom, associativity says:

$$C_a(C_b(c)) \leq C_{a \circ b}(c).$$

In a theory expression, '$a \circ b$', I call the theory a the "background theory" and b the "object theory". What the associativity postulate allows us to do is to take a theory expression in which there is a sequential application of theories to a given theory (c, in the statement of the associativity condition) and construct a single background theory $a \circ b$ that is applied to c as the object theory. The hereditariness lemma (Lemma 7.1) together with the associativity postulate shows that $(a \circ b) \circ c$ satisfies everything that holds in $a \circ (b \circ c)$. This construction of background theories is rather natural given the way in the framework of thinking of entailment as a theory of theory closure. When we have a theory that we want to investigate (the object theory), we construct a background theory of entailment as a tool to investigate it.

Here is a proof that the associativity postulate satisfies prefixing.

$$
\cfrac{
[x \vDash B \to C]^3 \quad
\cfrac{
\cfrac{[y \vDash A \to B]^2 \quad [z \vDash A]^1}{y \circ z \vDash B}
}{
\cfrac{
\cfrac{
\cfrac{x \circ (y \circ z) \vDash C}{(x \circ y) \circ z \vDash C} \; \text{Assoc}
}{x \circ y \vDash A \to C} \; 1
}{
\cfrac{x \vDash (A \to B) \to (A \to C)}{
\cfrac{\Sigma \circ x \vDash (A \to B) \to (A \to C)}{\Sigma \vDash (B \to C) \to ((A \to B) \to (A \to C))} \; 3
} \; 2
}
}
}{}
$$

The associativity postulate has important consequences for the theory of necessity in the model. It forces the model to satisfy a postulate that corresponds to the 4 axiom from model logic, viz.,

$$\Box A \to \Box \Box A.$$

As I say in Section 7.6, the corresponding semantic postulate is

$$\Box^- a \leq \Box^- \Box^- a.$$

Here is a little proof that associativity implies the postulate for the 4 axiom:

1.	$\Box^- a$	$= \;\; a \circ \Sigma$	Def. $\Box^-$
2.		$= \;\; a \circ (\Sigma \circ \Sigma)$	1, semantic postulate 1
3.		$\leq \;\; (a \circ \Sigma) \circ \Sigma$	2, Associativity
4.		$\leq \;\; \Box^- \Box^- a$	3, Def. $\Box^-$

We can show that associativity also implies half of the App condition (discussed in Section 7.7) that $a \circ b \leq \Box^- a \circ b$.

$$
\begin{array}{lll}
1. & a \circ b \; = \; a \circ (\Sigma \circ b) & \text{semantic postulate 1} \\
2. & \qquad\quad \leq \; (a \circ \Sigma) \circ b & \text{1, Assoc} \\
3. & \qquad\quad \leq \; \Box^- a \circ b & \text{2, Def. } \Box^-
\end{array}
$$

Thus, with associativity, the semantics makes valid,

$$(A \rightarrow B) \rightarrow \Box(A \rightarrow B).$$

8.3   Contraction

In Section 8.1, I motivate the postulate of weak contraction. Weak contraction says that every theory is closed under its own entailments. Formally, it can be written as $a \circ a \leq a$ or (Weak Contraction) $C_a(a) \leq a$. This principle is a form of self-coherence constraint on theories. It says that all theories obey their own laws of entailment. If $A \rightarrow B$ and A are both in a, then B is also in a.

The postulate of contraction generalises weak contraction. Contraction can be written as either

$$(a \circ b) \circ b \leq a \circ b$$

or

$$C_a(b) \circ b \leq C_a(b).$$

The schema that corresponds to contraction is also called 'contraction', and it is

$$(A \rightarrow (A \rightarrow B)) \rightarrow (A \rightarrow B).$$

The semantic postulate for contraction tells us that after applying a theory to another theory, applying it to the same argument again does not produce any new results. It says that we can appeal to the formulas satisfied by b over and over again as antecedents when the entailments of a are applied to them. So, for example, if $A \rightarrow (B \rightarrow C)$ is in a and A and B are both in b, we can derive that C is in $C_a(b)$. Without contraction, we can only infer that $B \rightarrow C$ is in $C_a(b)$. What contraction does is tell us that for all theories a and b, $C_a(b)$ is the *completion* of b by the entailments in b. That there are no consequents of entailments in $C_a(b)$ left to detach by formulas in b after a is applied to b.

The addition of contraction introduces a pleasing symmetry into the semantics. Recall the postulate RC,

$$(a \circ b) \circ b \leq a \circ b.$$

'RC' stands for 'reverse contraction', and now we can see why this name is used for this postulate. Instead of having the repeated theory at the end (as in contraction), RC has it at the start. The motivation for RC is that it turns all the consequence operators of a model into closure operators. Contraction does something similar.

In order to describe the similarity in more detail, I introduce the notion of a *d-theory*. It is easier to think about this syntactically. Consider sets of formulas Γ and Δ. Δ is a Γ theory if $C_\Gamma(\Delta) \subseteq \Delta$, that is, if Δ is closed under all the entailments of Γ. But we can equally talk about Γ's containing the results of the application of its own entailments to the formulas in Δ. If this is so, then

$$C_\Gamma(\Delta) \subseteq \Gamma.$$

If this holds, then I say that Γ is a "detachment theory" or "d-theory" of Δ. In the models, a is a d-theory of b if $a \circ b \leq a$. Just as reverse contraction tells us that $a \circ b$ is always a theory of a, contraction tells us that $a \circ b$ is always a d-theory of b.

Here is a proof of weak contraction ($a \circ a \leq a$) using contraction:

$$
\begin{array}{llll}
1. & a \circ a & = (\Sigma \circ a) \circ a & \text{Postulate 1} \\
2. & & \leq \ \Sigma \circ a & \text{1, Contraction} \\
3. & & \leq \ a & \text{2, Postulate 1}
\end{array}
$$

This little proof shows the sense in which contraction is a generalisation of weak contraction.

Here is a scheme that corresponds to the contraction postulate:

$$(A \to (B \to C)) \to ((A \wedge B) \to C).$$

This scheme is sometimes called *importation*, because B is imported from the right-hand side of the entailment to the left-hand side. Here is a proof of importation using the contraction postulate:

$$
\dfrac{
  \dfrac{
    [x \vDash A \to (B \to C)]^2 \qquad \dfrac{[y \vDash A \wedge B]^1}{y \vDash A}
  }{x \circ y \vDash B \to C} \qquad \dfrac{[y \vDash A \wedge B]^1}{y \vDash B}
}{}
$$

$$
\cfrac{
\cfrac{
\cfrac{
\cfrac{(x \circ y) \circ y \vDash C}{x \circ y \vDash C}\ \text{Contraction}
}{x \vDash (A \wedge B) \to C}\ 1
}{\Sigma \circ x \vDash (A \wedge B) \to C}
}{\Sigma \vDash (A \to (B \to C)) \to ((A \wedge B) \to C)}\ 2
$$

Using importation, we can prove pseudo modus ponens, $((A \to B) \wedge A) \to B$. Pseudo modus ponens directly follows from $(A \to B) \to (A \to B)$ and importation (and real modus ponens).

Weak contraction together with associativity allows us to prove reverse contraction ($a \circ (a \circ b) \leq a \circ b$), and so we do not need to include reverse contraction in the semantic postulates for E. Here is a little proof of this fact:

$$
\begin{array}{llll}
1. & a \circ (a \circ b) & \leq (a \circ a) \circ b & \text{Associativity} \\
2. & & \leq \ a \circ b & \text{1, Weak Contraction and Monotonicity}
\end{array}
$$

As I show in Chapter 7, we can use reverse contraction to prove conjunctive syllogism $(((A \rightarrow B) \wedge (B \rightarrow C)) \rightarrow (A \rightarrow C))$. We can also prove conjunctive syllogism from prefixing and importation.

8.4   Problems concerning Contraction

At the start of this chapter, I motivate the move beyond GTL by claiming that it is natural to want at least weak contraction as a rule. However, contraction (both weak and strong) has been blamed by relevant logicians for some of the so-called failings of E. In particular, contraction has been blamed both for making E undecidable and for making it susceptible to Curry-style paradoxes.

With regard to undecidability, it is unclear whether contraction alone is responsible. Whereas dropping contraction from R yields the decidable logic RW, it is not known whether EW is decidable (see [23]). What is known is that certain contraction-free subsystems of E are decidable, and no subsystem of E with contraction is decidable. Thus, it is a good bet that contraction has something to do with the undecidability of E.

It might be nice to have a logic of entailment such that if any issue arose as to whether a proof plan is valid, we could merely ask an automated theorem prover/falsifier whether it is so and be certain to get an answer. This fantasy, however, requires more than just a decidable logic. The decision procedure of the logic would have to be fairly *feasible*. We are all used to infeasible decision procedures, like truth tables for classical propositional logic. Given a formula with enough propositional variables, this procedure becomes unusable. So, decidability does not guarantee that the fantasy will come true.

On the other hand, there are for many undecidable logics, including E, proof and falsification procedures that deal well with a great many formulas of the sort that represent actual or plausible proof plans. John Slaney's program Matrix Generator for Implication Connectives (MaGIC), for example, generates logical matrices that can be used to falsify formulas in whatever logic is specified by the user. It works extremely well for formulas with five propositional variables or fewer.[1] So, when the fantasy is toned down a bit to fit what mathematicians and logicians really need a logic of entailment to do for them, the undecidability of E does not actually prove to be a real stumbling block.

The issue concerning Curry's paradox, I think, is somewhat more serious. Curry's paradox was first used by Haskell Curry to show that the original lambda calculus was trivial, in the sense that every expression in it could be proven. The paradox derives from the ability to formulate in one's logical language a sentence that in English means 'If this sentence is true, then p' for an arbitrary formula p. In order to derive the paradox, I need to add names for each formula to the language. Here, I use the standard notation and have $\ulcorner A \urcorner$ be the name of A. We also need to add to the language a truth

---

[1] At the time of writing, MaGIC could be found at https://users.cecs.anu.edu.au/~jks/magic.html.

predicate and add to the logic all the instances of the Truth-scheme formulated using entailment, that is,

$$T(\ulcorner A \urcorner) \leftrightarrow A.$$

The proof also assumes that we have some device in the language and logic that allows us to prove a fixed-point theorem, that is, for any formula of the form $A(x)$, there is a sentence C such that

$$C \leftrightarrow A(\ulcorner C \urcorner).$$

The Curry sentence is

$$T(x) \rightarrow p.$$

Given the fixed-point theorem, there is a sentence C such that

$$C \leftrightarrow (T(\ulcorner C \urcorner) \rightarrow p).$$

We can read C as saying 'If this sentence is true, then p'.

Now I can derive the paradox in the labelled deduction system. I break the proof into two subproofs. Here is the first subproof:

$$
\cfrac{
\cfrac{
\cfrac{
\cfrac{[x \vDash T(\ulcorner C \urcorner)]^1}{x \vDash C}}{T(\ulcorner C \urcorner) \rightarrow p} \quad [x \vDash T(\ulcorner C \urcorner)]^1
}{
\cfrac{\cfrac{x \circ x \vDash p}{x \vDash p}\ !}{\Sigma \circ x \vDash p}
}
}{\Sigma \vDash T(\ulcorner C \urcorner) \rightarrow p}\ 1
$$

Now that I have established that $\Sigma \vDash T(\ulcorner C \urcorner) \rightarrow p$, I can use it in the second subproof:

$$
\cfrac{
\cfrac{\cfrac{\Sigma \vDash T(\ulcorner C \urcorner) \rightarrow p}{\Sigma \vDash C}}{\Sigma \vDash T(\ulcorner C \urcorner)} \quad \Sigma \vDash T(\ulcorner C \urcorner) \rightarrow p
}{
\cfrac{\Sigma \circ \Sigma \vDash p}{\Sigma \vDash p}
}
$$

Thus, we have shown that p holds in the master theory for arbitrary p. This is an intolerable conclusion.

The exclamation mark in the first subproof indicates the step at which contraction is used. In fact, it is weak contraction that is employed here. In Chapter 10, I have an extended discussion of Curry's paradox and extended Curry's paradox. I summarise the way that I avoid the paradox there. Simply put, I do not think that a logic of entailment should contain either the Truth-scheme (in which bi-entailment is the main connective) nor do I think that it should contain the mechanisms that allow for the proof of a fixed-point theorem. Given these rejections, Curry's paradox is avoided.

8.5   Necessity in the Semantics of E

As I say in Section 8.1, it is natural that every theory is taken to be closed under its own entailments. If a theory says that B follows from A and A holds in the theory, our intuitive notion of coherence dictates that B is also true in the theory. Something similar seems to hold of necessity. If a theory says that A is necessary, then it should also say that A is true. In terms of the model theory, this can be enforced by adding the semantic postulate T:

$$\Box^- a \leq a.$$

This postulate, of course, makes valid the relevant form of the T axiom:

$$\Box A \to A.$$

Adding the T postulate also allows us to prove the other half of the App postulate, viz.,

$$\Box^- a \circ b \leq a \circ b,$$

and to turn the 4 postulate into an equality, that is,

$$\Box^- a = \Box^- \Box^- a.$$

In the resulting model theory, the following principles are sound:

$$(A \to B) \leftrightarrow \Box(A \to B)$$

$$\Box A \leftrightarrow \Box\Box A$$

Adding the T postulate also enables us to derive the following useful aggregation condition:

$$\Box^- a \circ \Box^- b \leq \Box^-(a \circ b).$$

Here is a derivation:

1.	$\Box^- a \circ \Box^- b$	$= \Box^- a \circ \Box^- \Box^- b$	postulates T and 4
2.		$= \Box^- a \circ (\Box^- b \circ \Sigma)$	1, Def. $\Box^-$
3.		$\leq (\Box^- a \circ \Box^- b) \circ \Sigma$	2, Associate
4.		$\leq \Box^-(\Box^- a \circ \Box^- b)$	3, Def. $\Box^-$
5.		$\leq \Box^-(a \circ b)$	4, T, monotonicity

The aggregation condition allows us to prove that necessity distributes over entailment:

$$(A \to B) \to (\Box A \to \Box B).$$

This scheme is the relevant version of the characteristic axiom of S3 (see Chapter 2). Here is a proof:

$$\dfrac{[x \vDash A \to B]^2 \qquad \dfrac{[y \vDash \Box A]^1}{\Box^- y \vDash A}}{\dfrac{\dfrac{\dfrac{\dfrac{\dfrac{x \circ \Box^- y \vDash B}{\Box^- x \circ \Box^- y \vDash B} \text{ App}}{\Box^- (x \circ y) \vDash B} \text{ Agg}}{x \circ y \vDash \Box B}}{x \vDash \Box A \to \Box B} {}^1}{\dfrac{\Sigma \circ x \vDash \Box A \to \Box B}{\Sigma \vDash (A \to B) \to \Box(A \to B)} {}^2}}$$

There is another semantic postulate governing necessity that is needed to characterise the logic E. To understand this postulate, let us suppose that the entailment $A \to B$ holds in a. Then, according to the logic E, a tells us that every theory is closed under $A \to B$ in two ways. First, $A \to B$ is a closure principle in the sense that I discuss in Chapters 6 and 7. This is to say that, according to a, if A holds in a theory b then so does B. Formally, this means that if A is in b, then B is in $a \circ b$. If a is a correct theory of entailment, then $a \circ b$ is just b. Otherwise, a makes some mistakes about entailment. However, E also says that, according to a, $A \to B$ can be used as the minor premise in a modus ponens with regard to the contents of any theory. This means that a says that if a formula of the form $(A \to B) \to C$ is in b, then so is C. In the semantics, this means that if $(A \to B) \to C$ is in b, then C is in $a \circ b$.

This is true of all necessitives in E. A necessitive, such as $\Box A$, is a formula that is available to use in inferences about all theories, not just the theory that contains it. Both of these schemes are theorems of E:

$$(A \to B) \to (((A \to B) \to C) \to C)$$

$$\Box A \to ((A \to B) \to B)$$

In order to make these schemes valid, I add a restricted commutativity postulate (RComm) to the semantics:

$$\Box^- a \circ \Box^- b = \Box^- b \circ \Box^- a.$$

Below is a proof of $\Box A \to ((A \to B) \to B)$ using RComm. This proof assumes that $y \vDash \Box(A \to B)$ can be proven from $y \vDash A \to B$, a proof of which is provided in Section 8.2.

$$\dfrac{\dfrac{[x \vDash \Box A]^2}{\Box^- x \vDash A} \qquad \dfrac{\dfrac{[y \vDash A \to B]^1}{y \vDash \Box(A \to B)}}{\Box^- y \vDash A \to B}}{\dfrac{\dfrac{\dfrac{\Box^- y \circ \Box^- x \vDash B}{\Box^- x \circ \Box^- y \vDash B}}{\dfrac{\dfrac{x \circ y \vDash B}{x \vDash (A \to B) \to B} {}^1}{\dfrac{\Sigma \circ x \vDash (A \to B) \to B}{\Sigma \vDash \Box A \to ((A \to B) \to B)} {}^2}}}{\qquad \Box^- x \le x \qquad \Box^- y \le y}}$$

With the addition of restricted commutativity, the suffixing thesis is valid in the semantics:

$$(A \rightarrow B) \rightarrow ((B \rightarrow C) \rightarrow (A \rightarrow C))$$

$$\dfrac{\dfrac{[x \vDash A \rightarrow B]^3 \qquad [y \vDash A]^1}{x \circ y \vDash B} \qquad [z \vDash B \rightarrow C]^2}{\dfrac{\dfrac{z \circ (x \circ y) \vDash C}{\dfrac{\Box^- z \circ (\Box^- x \circ y) \vDash C}{\dfrac{(\Box^- z \circ \Box^- x) \circ y \vDash C}{\dfrac{(\Box^- x \circ \Box^- z) \circ y) \vDash C}{\dfrac{(x \circ z) \circ y \vDash C}{\dfrac{x \circ z \vDash A \rightarrow C}{\dfrac{x \vDash (B \rightarrow C) \rightarrow (A \rightarrow C)}{\dfrac{\Sigma \circ x \vDash (B \rightarrow C) \rightarrow (A \rightarrow C)}{\Sigma \vDash (A \rightarrow B) \rightarrow ((B \rightarrow C) \rightarrow (A \rightarrow C))}}}}}}}}}}$$

(Proof annotations: App, Assoc, RComm, 1, 2, 3)

Suffixing, however, is a rather controversial logical principle. C. I. Lewis rejected it and so did Brady. In Section 8.6, I look at Wilhelm Ackermann's support for suffixing and reply to Lewis's argument against it.

8.6   Transitivity: Ackermann versus Lewis

Ackermann supported suffixing in the following passage:

[T]he formulas $A \rightarrow (B \rightarrow (A \wedge B))$, $A \rightarrow (\neg A \rightarrow B)$, and $A \rightarrow ((A \rightarrow B) \rightarrow B)$ cannot be considered universally true. The same holds for $B \rightarrow (A \rightarrow A)$, since the correctness of $A \rightarrow A$ is independent of the truth of B. In the rejection of the last formula, the implication I employ differs from strict implication, since the notion of a proposition that contains (as in $(A \wedge \neg A) \rightarrow B$) or is contained by every proposition is not the notion of a logical connection between two statements. A formula like $(A \rightarrow B) \rightarrow ((B \rightarrow C) \rightarrow (A \rightarrow C))$, on the other hand, we can assume to be universally true because under the assumption that $A \rightarrow B$, $A \rightarrow C$ follows logically from $B \rightarrow C$. [1, p 113]

The key to Ackermann's motivation for suffixing, I suggest, is his interpretation of $\rightarrow$ as "logically follows from". For both Lewis and Ackermann, if it is provable in their respective logics that $A \rightarrow B$, then, under the assumption that $B \rightarrow C$, we can derive that $A \rightarrow C$. In other words, the following rule is valid, either when $\rightarrow$ is understood as relevant entailment or as Lewis's strict implication:

$$\dfrac{\vdash A \rightarrow B}{\vdash (B \rightarrow C) \rightarrow (A \rightarrow C)}.$$

Let us read this rule as 'if it is a theorem that $A \rightarrow B$, then it is a theorem that $(B \rightarrow C) \rightarrow (A \rightarrow C)$'. This can, in turn, be read as 'if B logically follows from A, then $A \rightarrow C$ logically follows from $B \rightarrow C$'. The 'if … then' in this reading is stronger than the usual English language conditional. Ackermann, I think, would read it as entailment. Making both 'logically follows from' and 'if … then' into entailment in the sense of →, we get the following Dana Scott-style horizontalisation (see Chapter 5):

$$(A \to B) \to ((B \to C) \to (A \to C)).$$

As I interpret him, in the passage given above, Ackermann gave a brief and somewhat unsystematic Scott-style reflexivity argument for suffixing.

I think that Lewis, however, would be unmoved by Ackermann's argument. Lewis took the rule form of suffixing as what Francesco Paoli and I called an "external rule" (see Section 5.8). An external rule is a rule that governs inferences from valid formulas to valid formulas. This is different from an "internal rule", which is a rule that governs inferences from assumptions. As such, Lewis would not read the 'if ... then' in the English reading of the rule as 'logically follows' or 'entails'; rather, it is some weaker conditional and the move to horizontalisation fails.

Here is the passage in Lewis in which he explicitly rejected the thesis form of suffixing:

Both Dr. Wajsberg and Dr. Parry have proved that the principle

$$(p \dashv q) \dashv ((q \dashv r) \dashv (p \dashv r))$$

is deducible [in S3]. I doubt whether this proposition should be regarded as a valid principle of deduction: it would never lead to any inference $p \dashv r$ which would be questionable when $p \dashv q$ and $q \dashv r$ are given premises; but it gives the inference $(q \dashv r) \dashv (p \dashv r)$ whenever $p \dashv q$ is a premise. Except as an elliptical statement for "$((p \dashv q) \land (q \dashv r)) \dashv (p \dashv r)$ and $p \dashv q$ is true," this inference seems dubious. [113, p 496]

At first glance, this argument looks like a mere appeal to intuition. As such, it is not very convincing to anyone who does not share Lewis's intuitions. However, I do not think the argument is just an appeal to intuition.

We can understand Lewis's position if we recall his pragmatic interpretation of necessity and strict implication (see Chapter 2). On this theory, what is necessary is determined by the meanings of statements. The linguistic meaning of a statement is the other statements one can derive from it. The sense meaning of a statement is the collection of conditions under which the statement can be confirmed. The sense meanings that one attributes to statements come from her conceptual framework. In adopting a theory, on Lewis's conceptual pragmatism, one also adopts a conceptual framework. For example, in adopting relativity theory, one adopts sense meanings for statements about simultaneity that concern light reaching the same space-time point [107]. For Lewis, strict implication concerned the containment of sense meanings in one another. An implication, $A \dashv B$ holds in a theory if and only if the sense meaning it attributes to A contains the empirical content attributed to B, that is, the confirmation conditions for A also confirm B.

Suppose that a theory said that $A \dashv B$. If Lewis were to have accepted suffixing, he would have had to accept that the theory also said that $(B \dashv C) \dashv (A \dashv C)$. This latter formula means that the confirmation conditions of $A \dashv C$ are contained in the confirmation conditions of $B \dashv C$. I suggest that for Lewis this sort of claim was to be reserved for logical theories, but not for theories like relativity theory or evolutionary theory, which are not about meaning and derivability, but about physical or living things.

This is a point at which my view diverges from Lewis's. I think that any theory that contains an entailment of any sort is a logical theory. An entailment is a statement about what is derivable from what, and any theory that contains an entailment tells us about other theories and not just about plants, animals, or other non-logical things. Thus, I do not accept Lewis's argument against suffixing.

8.7   Recap: The Semantics of E

Before I go on to discuss other technical and philosophical properties of models of E, I review the model theory itself. A frame for the conjunction-entailment fragment of E contains a set of theories Υ, a master theory Σ, an application operator $\circ$, and a hereditariness relation $\leq$. Here are the postulates on E-frames:

	Postulate	Name
1.	$\leq$ is transitive, reflexive, and antisymmetric	Partial order
2.	$\Sigma \circ a = a$	Sigma-identity
3.	$a \leq a'$ and $b \leq b'$ implies $a \circ b \leq a' \circ b'$	Monotonicity
4.	$a \circ (b \circ c) \leq (a \circ b) \circ c$	Associativity
5.	$(a \circ b) \circ b \leq a \circ b$	Contraction
6.	$\Box^- a \leq a$	T
7.	$\Box^- a \circ \Box^- b = \Box^- b \circ \Box^- a$	Restricted commutativity

$\Box^- a$ is defined as $a \circ \Sigma$.

I sometimes also use the following derived conditions on models:

Postulate	Name
$a \circ b = \Box^- a \circ b$	Application
$a \circ (b \circ c) \leq (b \circ a) \circ b$	Twisted associativity
$\Box^- a \leq \Box^- \Box^- a$	4
$\Box^- a \circ \Box^- b \leq \Box^- (a \circ b)$	Aggregation

8.8   Canonical Judgements

Consider the entailment introduction rule of the natural deduction system for proof processes:

$$[x \vDash A]$$
$$\vdots$$
$$\frac{a \circ x \vDash B}{a \vDash A \to B}$$

There is a side condition that says that x does not occur in a. Contraction helps satisfy the side condition because it helps to eliminate all but the x at the end of $a \circ x$. Associativity helps to set up entailment introductions by aiding in isolating the lone x at the end of $a \circ x$.

A theory term that is associated to the left and does not include the repetition of any terms is ready for repeated applications of the entailment introduction rule. I call a term like this a term in *canonical form*. To make this notion precise, here is a formal definition of it. A variable x is in canonical form, as is Σ and every term name T. If $a_1, \ldots, a_n$ (for $n \geq 2$) are primitive terms (either variables, names, or Σ), and there is no repetition among them, then $((a_1 \circ a_2) \circ \ldots) \circ a_n$ is in canonical form.

In many areas of mathematics, some standard form of a term is introduced and all terms are shown to be equivalent to some term in standard form. But this is not the case here. Although we cannot show that every theory term is equivalent to one in canonical form that includes the same primitive terms, we can show that for each theory term α, there is an α' in canonical form for which $\alpha \leq \alpha'$ is provable. The proof is by induction on the complexity of α. The base case is in which α is a variable or Σ. In either case, it is in canonical form. For the inductive case, let α be $\beta \circ \delta$ and that $\beta \leq (\beta' \circ b)$ and $\delta \leq (\delta' \circ d)$, where these $\beta' \circ b$ and $\delta' \circ d$ are in canonical form. By monotonicity, $\alpha \leq ((\beta' \circ b) \circ (\delta' \circ d))$. I then reason as follows:

1. $(\beta' \circ b) \circ (\delta' \circ d)$ $\quad = \quad \Box^{-}(\beta' \circ b) \circ (\Box^{-}\delta' \circ d)$ $\quad$ App

2. $\qquad\qquad\qquad\qquad \leq \quad (\Box^{-}(\beta' \circ b) \circ \Box^{-}\delta') \circ d$ $\quad$ 1, Assoc

3. $\qquad\qquad\qquad\qquad \leq \quad (\Box^{-}\delta' \circ \Box^{-}(\beta' \circ b)) \circ d$ $\quad$ 2, RComm

4. $\qquad\qquad\qquad\qquad \leq \quad (\delta' \circ (\beta' \circ b)) \circ d$ $\quad\quad\quad$ 3, T, Mono

5. $\qquad\qquad\qquad\qquad \leq \quad ((\delta' \circ \beta') \circ b) \circ d$ $\quad\quad\quad$ 4, Assoc

If b and d are the same, then we use contraction to obtain $(\beta' \circ b) \circ (\delta' \circ d) \leq (\delta' \circ \beta') \circ b$. If δ' and β' are complex terms, we continue the same process of using associativity, restricted commutativity, and contraction until we produce a term in canonical form.

By the hereditariness lemma (Lemma 7.1), if $a \leq b$ and $a \vDash A$, then $b \vDash A$. So, the idea here is that if we have a term α and can prove $\alpha \vDash A$, then we can also show that there is some term β in canonical form such that $\alpha \leq \beta$ and $\beta \vDash A$. I use this fact in Section 8.9 to demonstrate the relationship between the semantics and the Anderson–Belnap natural deduction system for E.

8.9   Theories and Natural Deduction

I have been using a Prawitz-style labelled natural deduction system to represent reasoning about theories. In this section, I show how this system can be used to understand Anderson and Belnap's natural deduction system for the logic E, that I present in Chapter 4. I prove a fairly simple theorem to make formal and precise the relationship between the two systems. This theorem also makes clear a relationship between Anderson and Belnap's system of deduction and the theory semantics.

I use a translation between lines of an Anderson and Belnap derivation to judgements in a Prawitz-style derivation. I also, however, sometimes treat this a bit loosely and discuss translations of theory terms. A hypothesis, A_n, is translated as a judgement, $x_n \vDash A$. Where $\{i, \ldots, j\}$ is non-empty, $B_{\{i,\ldots,j\}}$ is translated as $((x_i \circ x_{i+1}) \circ \ldots) \circ a_j \vDash B$. A step $C_\emptyset$ is translated as $\Sigma \vDash C$. We then add steps in the Prawitz-style proof to produce a valid proof of the theorem.

Here is an example. Let us start with a short Anderson–Belnap proof of the theorem, $(A \rightarrow C) \rightarrow ((A \wedge B) \rightarrow C)$:

1	$A \rightarrow C_{\{1\}}$	hypothesis
2	$A \wedge B_{\{2\}}$	hypothesis
3	$A_{\{2\}}$	2, $\wedge$E
4	$C_{\{1,2\}}$	3,4, $\rightarrow$E
5	$(A \wedge B) \rightarrow C_{\{1\}}$	2–5, $\rightarrow$I
6	$(A \rightarrow C) \rightarrow ((A \wedge B) \rightarrow C)_\emptyset$	1–6, $\rightarrow$I

Here is the raw translation:

$$\dfrac{\dfrac{[x_1 \vDash A \rightarrow C]^2}{x_1 \vDash A \rightarrow C} \quad \dfrac{[x_2 \vDash A \wedge B]^1}{x_2 \vDash A}}{\dfrac{\dfrac{x_1 \circ x_2 \vDash C}{x_1 \vDash (A \wedge B) \rightarrow C} \, 1}{\Sigma \vDash (A \rightarrow C) \rightarrow ((A \wedge B) \rightarrow C)} \, 2}$$

To be in the form of a valid deduction, I remove the redundant step at the top left and add a penultimate step to introduce Σ:

$$\dfrac{\dfrac{[x_1 \vDash A \rightarrow C]^2 \quad \dfrac{[x_2 \vDash A \wedge B]^1}{x_2 \vDash A}}{\dfrac{x_1 \circ x_2 \vDash C}{x_1 \vDash (A \wedge B) \rightarrow C} \, 1}}{\dfrac{\Sigma \circ x_1 \vDash (A \wedge B) \rightarrow C}{\Sigma \vDash (A \rightarrow C) \rightarrow ((A \wedge B) \rightarrow C)} \, 2}$$

In what follows, I show that we can always manipulate a translation of an Anderson–Belnap derivation into a valid Prawitz-style proof.

I begin by proving the following lemma.

LEMMA 8.1 *If a step A_α occurs in a subproof headed by an assumption H_k and $k \notin \alpha$, then $\Box^- \tau(\alpha) \vDash A$ is a valid step in the translation of the derivation, where $\tau(\alpha)$ is the translation of α into a theory term.*

The proof is by a straightforward induction on the length of the derivation of A_α.

When α is the empty set, the translation of $A_\emptyset$ is $\Sigma \vDash A$. However, if $A_\emptyset$ is a line in a valid proof, then A is a theorem of E, and by the completeness theorem for the

Prawitz-style system, $\Sigma \vDash A$ is a valid line. Since $\Sigma = \Box^- \Sigma$, $\Box^- \Sigma \vDash A$ is also a valid step in the Prawitz-style proof.

The cases in which A_α is in that subproof because it is reiterated from an earlier subproof, or in which it is proven by either of the conjunction rules, are easy. If A_α is proven by entailment introduction, then it is an entailment, that is of the form $B \rightarrow C_\alpha$. By the inductive hypothesis $\tau(\alpha) \vDash B \rightarrow C$ is a valid step, hence so is $\Box^- \tau(\alpha) \vDash B \rightarrow C$.

The case for entailment elimination is only slightly more difficult. Suppose that A_α is derived from $B \rightarrow A_\beta$ and B_γ and α is $\beta \circ \gamma$. By the inductive hypothesis, $\Box^- \tau(\gamma) \vDash B$ is provable. Thus, we have

$$
\begin{array}{cc}
\vdots & \\
\dfrac{\tau(\beta) \vDash B \rightarrow A}{\Box^- \tau(\beta) \vDash B \rightarrow A} & \vdots \\
\multicolumn{2}{c}{\dfrac{\Box^- \tau(\beta) \vDash B \rightarrow A \qquad \Box^- \tau(\gamma) \vDash B}{\Box^- \tau(\beta) \circ \Box^- \tau(\gamma) \vDash A}}
\end{array}
$$

By the inductive hypothesis, $\tau(\beta)$ and $\tau(\gamma)$ are associated to the left, that is of the form $(\ldots (x_i \circ x_{i+1}) \circ \ldots) \circ x_j$. Now I use the same reasoning as I used in Section 8.8. I replace each of these with terms of the form $\Box^- (\ldots \Box^- (\Box^- x_i \circ x_{i+1}) \circ \ldots) \circ x_j$. This will allow us to use restricted commutativity to reorder terms as required. I call these new terms $\tau'(\beta)$ and $\tau'(\gamma)$. If the last term of $\tau'(\beta)$ has a larger subscript than the last term of $\tau'(\gamma)$, I reorder as $\Box^- \tau'(\gamma) \circ \Box^- \tau'(\beta) \vDash A$. I then derive $\Box^- \tau'(\gamma) \circ \tau'(\beta) \vDash A$ and use associativity to prove $(\Box^- \tau'(\gamma) \circ \Box^- \tau'(\beta'')) \circ x_j \vDash A$, where x_j is the last term of β and β'' is β without x_j. And then I continue the process, using restricted commutativity to reorder, associativity to reassociate, and contraction to eliminate redundant terms. After this, I remove the denecessitations and obtain $\tau(\alpha)$. If the last term of $\tau(\gamma)$ has the larger subscript (or they are the same), then I do not swap the positions of $\tau(\beta)$ and $\tau(\gamma)$. But then I proceed to use application and associativity and then carry on in the same manner.

I use Lemma 8.1 to prove the following theorem:

THEOREM 8.2 *If A_α is validly proven in an Anderson and Belnap derivation from undischarged hypotheses $H_1, \ldots, H_n$, then $\tau\alpha \vDash A$ is provable in a valid Prawitz-style derivation where $x_1 \vDash H_1, \ldots, x_n \vDash H_n$ are the only undischarged hypotheses.*

The proof uses the same technique that I use to prove the entailment elimination case of Lemma 8.1. We add denecessitation operators and then reorder as needed and then use contraction to prune redundant theory terms. This technique can be extended to treat the rules for disjunction and negation that are given in Chapter 9 and so the theorem can be proven for the full deduction system for the logic E. As I say in Section 8.8, this theorem shows that all the steps in a valid Anderson and Belnap derivation can be looked at as the result of constructing a background theory and applying it to a target theory.

This result makes precise the way in which Anderson and Belnap's natural deduction system is related to the semantics. The semantics both interprets and justifies the natural deduction system. And the fact that the semantics characterises this elegant natural deduction system also justifies the semantical theory.

8.10   Aside: The Logic of Pure Theories of Entailment

An interesting logic emerges when we look at the class of frames in which every theory is a pure theory of entailment. Suppose that Υ is a set of theories from some frame. Then let $\Box^-\Upsilon$ be the subset of Υ such that every theory in $\Box^-\Upsilon$ is identical to its own denecessitation. We can then construct a frame $< \Box^-\Upsilon, \Sigma, \circ, \leq>$. The idea here is to take the application operator and the order relation from the original frame and restrict them to $\Box^-\Upsilon$. This constructed frame is also a conjunction-entailment E frame, and its application operator is commutative. This means that for any a,b in $\Box^-\Upsilon$, $a \circ b = b \circ a$.

To prove that this structure is an E frame, it is sufficient to show that for any a,b in $\Box^-\Upsilon$, $a \circ b$ is also $\Box^-\Upsilon$. In order to prove this, it is sufficient to show that in the original frame $\Box a \circ \Box^-b = \Box^-(\Box^-a \circ \Box^-b)$. Half of this is already proven. $\Box^-(\Box^-a \circ \Box^-b) \leq \Box^-a \circ \Box^-b$ is just an instance of the T postulate. The other direction, $\Box^-a \circ \Box^-b \leq \Box^-(\Box^-a \circ \Box^-b)$, is proven as follows:

1. $\Box^-a \circ \Box^-b \quad \leq \quad \Box^-\Box^-a \circ \Box^-\Box^-b$ Postulate 4, Mono
2. $\qquad\qquad\qquad \leq \quad \Box^-(\Box^-a \circ \Box^-b)$ 1, Agg

This little proof tells us that $\Box^-\Upsilon$ is closed under application – the application of one pure theory of entailment to another pure theory of entailment always results in a pure theory of entailment. It follows from this that $< \Box^-\Upsilon, \Sigma, \circ, \leq>$ satisfies all the conditions to be a conjunction-entailment E frame. In $< \Box^-\Upsilon, \Sigma, \circ, \leq>$, the application operator ($\circ$) is commutative. If we restrict the class of frames to those in which application is commutative, the logic characterised is stronger than E. In fact it is the logic R. Now, R is usually interpreted as the logic of *contingent* relevant implication. But in this context we can interpret it as being about theories of necessarily true propositions. If T is a pure theory of entailment and A is in a, then $\Box A$ is also in a, and vice versa. On the level of frame theory, this is captured by the fact that for all theories a in a frame, a is identical to $\Box^-a$.

In terms of proof theory, we can just add to the axiomatic basis of E the axiom

$$A \to \Box A.$$

This results in the logic R (or perhaps a basis that generates all and only the axioms of R). Thinking of the axioms of R in this way recalls Hughes and Cresswell's logic VER [83].[2]

---

[2] In this logic every formula is equivalent to its own necessitation. Thus, to use Lloyd Humberstone's terminology, the necessity connective is *universally representative* – every formula is equivalent to some necessitive [86, ch 9].

It is also reminiscent of intuitionist logic (IL). On one transition of IL into the modal logic S4, we take each propositional variable to its necessitation. The understanding of $\Box$ is 'is constructible that' or 'is verifiable that'. In the same way, we can think of the conjunction-implication fragment of R as a constructive version of E. As in Kripke's semantics for IL, we can treat the points of the frame as evidential situations (instead of theories). In the Brouwer–Heyting–Kolmogorov interpretation of IL, implication and proofs are treated in terms of functions. The implication '$A \rightarrow B$' says there we have a proof that will turn a proof of A into a proof of B. The proof described by '$A \rightarrow B$' is a (partial) function from proofs to proofs. In the present semantics, in $< \Box^{-}\Upsilon, \Sigma, \circ, \leq>$ every point can be considered to be its own consequence operator, and so is a function of a sort. We can think of a point as a situation that is a conglomeration of all of the proofs that are on hand in a context. Then application can be understood in terms of the application of proofs to one another.

This intuitionist interpretation of R is interesting, but in order for it to be viable it needs to be combined with treatments of negation, disjunction, and the quantifiers. I think that the semantics I give in Chapters 9 and 10 can be altered to be fit with this intuitionist interpretation, but this would take a lot of work (especially in terms of the quantifiers) and this is a book about E, not R. So I leave this suggestion as just that, a suggestion, and return to the subject of the logic E of relevant entailment.

8.11   The Core Theory of Entailment and Its Extensions

The conjunction and entailment fragment of the logic E is the system that I call the *core logic of entailment*. Clearly, a logic of entailment, by the very nature of the beast, must contain an entailment connective. As a logic of theory closure, it also needs a way of connecting together formulas to say that they belong to the same theory, hence the inclusion of conjunction. The other standard logical connectives, that is disjunction and negation, have more peripheral roles in a logic of entailment.

In terms of theories, both negation and disjunction are problematic. There are theories with very weak conjunction and negation principles. Theories of constructive mathematics reject excluded middle and the principle of double negation elimination. Quantum logic, which some have stated should be the logic of quantum theory, rejects the distribution of conjunction over disjunction. Thus the question arises: Which disjunction and negation principles (if any) should a logic of entailment contain?

The idea that a logic of entailment should be compatible with any conceivable theory threatens to make it impossible to extend beyond the core logic. Instead, in Chapter 9, I present a view of conjunction and disjunction in the logic of theories framework that I have developed in this and the previous chapter. This view makes valid all of the propositional logic E, and so provides an interpretation of that logic, which is the main aim of this book.

In order to treat weaker theories such as quantum logical theories and intuitionist theories, I suggest a policy of *retrenchment*. We can analyse them, even if we think that they are somehow defective, using the core logic. In doing so, we create an alternative model based on a weaker logic of negation and/or disjunction. The weaker model, I postulate, would have the structure of the models that I outline in this chapter, but with extra mechanisms to treat disjunction and negation.

One might wonder why, in cases in which negation or disjunction obey different rules from those of the corresponding relevant connectives, we do not merely add both the relevant and the "deviant" connectives. We have to be sure in such cases that we will produce a logic that is a conservative extension of the manner in which the deviant theories are intended to be closed. Adding Boolean negation to E, for example, does not produce a conservative extension. It is not clear to me whether adding a disjunction over which conjunction distributes to quantum logic, say, will produce a conservative extension.

One might point out, also, that there are also many theories couched in logics that are weaker in their entailment fragment than the logic E. These theories, I contend, need to be formulated with an entailment-like connective, say, $\rightsquigarrow$, and not with $\rightarrow$. The closure of such a theory will be the same as its closure under the first-degree entailments. Since no formula of this theory contains $\rightarrow$, by the properties of E, its closure will not contain any entailments either. This makes the closure under E non-ampliative, and also not harmful.

9   Negation and Disjunction

9.1   Introduction

As I say at the end of Chapter 8, the fragment of the logic E with just entailment and conjunction is the core theory of entailment. In much of our analyses of theories, however, we employ the other standard propositional connectives, namely disjunction and especially negation. In this chapter, I extend the semantics from Chapter 8 to treat these two connectives.

There is a problem of how to add negation to a logic of entailment. To see why it is a problem, I begin by describing the technique I used to understand negation in my book, *Relevant Logic* [129]. In that book, I add negation to the proof theory for R by adding a constant f. This constant was taken there to mean 'something impossible has occurred'. Negation is defined implicationally as

$$\neg A =_{df} A \Rightarrow f.$$

Without adding any extra postulates, this definition allows the derivation of several standard theses regarding negation. Since $(A \to (A \to B)) \to (A \to B)$ is a theorem of R, then so is the thesis of reductio ad absurdum:

$$(A \Rightarrow \neg A) \Rightarrow \neg A.$$

Suffixing, $(A \Rightarrow B) \Rightarrow ((B \Rightarrow C) \Rightarrow (A \Rightarrow C))$, has as an instance the thesis of contraposition:

$$(A \Rightarrow B) \Rightarrow (\neg B \Rightarrow \neg A).$$

The permutation theorem, $(A \Rightarrow (B \Rightarrow C)) \Rightarrow (B \Rightarrow (A \Rightarrow C))$, has as a substitution instance another version of contraposition:

$$(A \Rightarrow \neg B) \Rightarrow (B \Rightarrow \neg A).$$

And the thesis of assertion, $A \Rightarrow ((A \Rightarrow B) \Rightarrow B)$, has as an instance the thesis of double negation introduction (DNI):

$$A \Rightarrow \neg\neg A.$$

Thus, in the proof theory for R, all these theses about negation come for free with the adoption of that definition. The only thesis that has to be added is double negation elimination (DNE):

$$\neg\neg A \Rightarrow A.$$

Some of these theses are valid in E, when implication is replaced with entailment, but unfortunately, it is not reasonable to interpret negation in this way in E. The implication of R is a form of contingent implication, and using it to formulate negation makes sense in the semantics of R. However, E is a rather different logic.

The conditional of E is a modal connective: it says that something necessarily follows from something. Let 'f' mean 'something impossible occurs'. Then '$A \rightarrow f$' means an impossibility necessarily follows from A. In terms of the interpretation I give to E in Chapter 8, '$A \rightarrow f$' means that every theory that contains A also contains an impossibility. But negation is not like this in theories. For example, suppose someone were to think that all monotremes were extinct. Then he might formulate a zoological theory that says that no mammals lay eggs. He would not think it impossible that mammals lay eggs, because he knows that there were mammals that lay eggs, but he thinks that they no longer exist. So his theory contains 'no mammals lay eggs' but rejects the claim about other theories that hold that some mammals lay eggs as they postulate an impossibility. Thus, the conditional treatment of negation seems a bad fit for E.

What we want from a treatment of negation (as with the other connectives) is a satisfaction condition that has a clear connection to its truth condition. With regard to entailment, this was easy. The treatment of entailment in terms of theories is its truth condition. For conjunction, there is an exact parallel between its behaviour in theories and its truth condition. However, negation is not so easy. One problem is that the truth condition for negation is currently a subject of debate. In Section 9.5, I try to show how the satisfaction condition for negation (for theories) relates to two different proposed truth conditions.

The treatment of disjunction is bound up with the treatment of negation. I accept the De Morgan definition of disjunction in terms of conjunction and negation, that is,

$$A \vee B =_{df} \neg(\neg A \wedge \neg B).$$

Like negation, the way in which disjunction acts in theories can be difficult to understand. The truth condition for disjunction is quite straightforward. A disjunction is true if and only if at least one of its disjuncts is true. However, a theory may contain a disjunction without containing either disjunct. A theory may tell us, for example, that the dinosaurs died out either because of a gigantic volcanic event or because the earth was hit by an asteroid, but not tell us which it was. Thus, the satisfaction condition for disjunction in theories must be quite different from its truth condition. In Section 9.3, I adopt a satisfaction condition from Kit Fine's "Models for Entailment" [61]. According to this satisfaction condition, a disjunction is satisfied by a theory if and only if at least one of its disjuncts is satisfied by the saturated theories that extend it. The notion of saturatedness is defined in Section 9.3, but roughly a theory is saturated if and only if disjunction and negation work in the standard ways for it. Moreover, in Section 9.5, the satisfaction conditions for disjunction and negation are shown to have a clear relationship with the truth conditions for these connectives.

This chapter presents some technical material that has not been published before. Soundness and completeness theorems are proven in the Appendix to this book.

9.2   Incompatibility

I interpret negation, both model-theoretically and proof-theoretically in terms of *incompatibility*. Consider a sequent,

$$\Gamma \Vdash \neg A.$$

This sequent says that A is incompatible with Γ. A may not be incompatible with any single formula in Γ, but perhaps with a conjunction of formulas. The semantic treatment of negation that I use builds on the idea of incompatibility.

Two theories may be compatible or they may be incompatible. Contemporary physics and our best medical theories are compatible with one another, but our physical theories are incompatible with some holistic medical theories, in which crystals are said to have magical powers. Two theories are compatible if they do not contradict one another. Note that a theory may contradict itself without contradicting every other theory. In such cases, it will be incompatible with itself but compatible with some other theories.

In order to represent the idea of incompatibility in the model theory, I borrow the orthogonality ("perp") relation $\perp$ from Robert Goldblatt's semantics for orthologic [74] and add it to my models.[1] Two theories a and b are such that $a \perp b$ if a and b are incompatible with one another. A formula $\neg A$ is satisfied by a theory a if and only if every theory that satisfies A is incompatible with a. More formally,

$$a \vDash \neg A \text{ if and only if, for all } b,\ b \vDash A \text{ implies } a \perp b.$$

Thus, negative statements in a theory, like entailment statements, make claims about the relationship of that theory with other theories in the model.

In order to extend the hereditariness lemma (Lemma 7.1) to treat negation, I add the following postulate:

$$\text{If } a \leq b, \text{ then } a^\perp \subseteq b^\perp,$$

where $a^\perp$ is the set of theories that are orthogonal to a. If we extend a theory, then the extension will be incompatible with all the theories incompatible with the original theory and perhaps with some other theories.

Given this interpretation of it, the perp relation must be symmetrical. So,

$$\text{if } a \perp b, \text{ then } b \perp a.$$

This symmetry makes valid the double negation introduction thesis (DNI):

$$A \to \neg\neg A$$

and the rule form of contraposition:

$$\frac{\vdash A \to \neg B}{\vdash B \to \neg A}.$$

---

[1] The first time that perp was used to explain how negation behaves in relevant logic is perhaps in Dunn [56]. I defined a perp relation in my article [127]. I constructed my semantics at about the same time and independently of Dunn's work.

It is particularly easy to see that contraposition holds. For suppose that $A \rightarrow \neg B$ is valid. By the semantical entailment theorem (Theorem 7.2) and the satisfaction condition for negation, for any $a \vDash A$, if b satisfies B, then a is orthogonal to b. Now, suppose that $b \vDash B$. By the symmetry of perp, if $a \vDash A$, then $b \perp A$. By the satisfaction condition for negation, $b \vDash \neg A$. Generalising, by the semantical entailment theorem, $B \rightarrow \neg A$ is valid in the model. It is quite simple to derive DNI from the contraposition rule: Just use as the premise, $\vdash \neg A \rightarrow \neg A$.

To make the contraposition thesis, $(A \rightarrow \neg B) \rightarrow (B \rightarrow \neg A)$, valid I add the following postulate:

$$(a \circ b) \perp c \text{ implies } (a \circ c) \perp b.$$

This postulate looks very much like the old algebraic versions of the rule of antilogism, which tells us that if A and B together entail the negation of C, then A and C together entail the negation of B. In the relevant context, the notion of grouping together is not captured by conjunction, but by application. Contraposition shows that entailments are closure principles in two ways. An entailment $A \rightarrow \neg B$ says that any theory that contains A also contains $\neg B$, but in a logic with the contraposition thesis, it also says that any theory that contains B also contains $\neg A$. The proof of the soundness of contraposition is in the Appendix.

Another negation thesis that is in the logic E is reductio ad absurdum:

$$(A \rightarrow \neg A) \rightarrow \neg A.$$

This scheme tells us that every formula that is self-refuting is also refuted. If assuming it allows the derivation of its negation, then its negation has to be correct. This thesis, as I show in Section 9.6, is intimately connected to the law of excluded middle. The postulate that makes reductio valid is

$$a \circ b \perp b \text{ implies } a \perp b.$$

This postulate says that if a applied to b results in a theory that is incompatible with b, then there is an incompatibility between a and b. The proof of the soundness of the reductio thesis is also in the Appendix.

The most difficult axiom scheme of E to satisfy in the model theory is double negation elimination (DNE). In order to make DNE valid, I first put in place a mechanism to treat disjunction. The intimate relationship between disjunction and negation that I mention in Section 9.1 requires that there is a semantics for disjunction in order to complete the semantics for negation.

9.3   Intersecting Theories

Suppose that a is a paleontological theory that says that the dinosaurs died out because of widespread volcanic activity and that b is a theory that says that the dinosaurs died out because an asteroid struck the earth. One's first reaction might be to say that a and b do not agree at all on the cause of the demise of the dinosaurs, but this is not completely true. Both *a* and *b* agree that the dinosaurs died out either because of widespread volcanic activity or because of an asteroid strike. It makes sense to talk about the points of agreement between two or more theories.

Hiroakira Ono [156] and Ross Brady [22] have an intersection operator in their models in order to give a semantics for disjunction. Although I had done so too in an earlier draft of this book, I do not now because I can incorporate the same idea without an additional operator. In my semantics, following Fine's [61], each theory is the intersection of its *saturated* extensions. In Fine's semantics, the notion of saturatedness is a primitive and is used to give satisfaction conditions to both disjunction and negation. Fine uses the "Routley star" operator in order to treat negation. The star operator is only defined over the class of saturated theories. I use the incompatibility relation to treat negation, and it is defined over the whole class of theories. I use incompatibility also to define saturatedness, so the latter is not a primitive notion in my theory. Here is the definition of saturatedness:

DEFINITION 9.1 (Saturated Theory)   *A theory a is saturated if and only if a^c contains a unique maximal theory under $\leq$. The maximal theory of a^c is designated by a^* ('the star of a').*

Here, a^c is the set of theories that are not orthogonal to *a*. That is, if $b \in a^c$, then $a \not\perp b$. This means that a theory *a* is saturated if and only if there is a theory a^* that is maximally compatible with *a*. I call the class of saturated theories '*S*'. For an arbitrary theory *c*, I use the notation '$S(c)$' to designate the class of saturated theories that extend *c*.

The satisfaction condition for disjunction is adopted from Fine:

$$a \vDash A \vee B \text{ if and only if, for all } b \in S(a),\ b \vDash A \text{ or } b \vDash B$$

The following postulates govern *S*:

 (i) If $a \in S$, then $a^* \in S$;
 (ii) If $a \in S$, then $a^{**} = a$;
 (iii) $\cap\{b^\perp : b \in S(a)\} \subseteq a^\perp$;
 (iv) $S(a \circ b) = \cup\{S(c \circ b) : c \in S(a)\}$.

Postulates (i) and (ii) are needed to make DNE valid. Postulates (iii) and (iv) are used to prove the following lemma:

LEMMA 9.2   *For any theory a and any formula A, $a \vDash A$ if and only if, for all $b \in S(a)$, $b \vDash A$.*

The proof is by induction on the length of *A*. The proof of the negation case appeals to (ii) and the proof of the entailment case appeals to (iv). I give the complete proof of Lemma 9.2 in the Appendix. As a direct consequence of Lemma 9.2, it can be shown that $(A \vee A) \rightarrow A$.

It is easy to prove that, for any $a \in S$, $a \vDash \neg A$ if and only if $a^* \nvDash A$. The left-to-right direction follows directly from the definition of a^c and a^*. Suppose that $a^* \nvDash A$. Since a^* is the greatest theory compatible with *a*, by the hereditariness lemma, for any theory b, if $b \vDash A$, then $a \perp b$. By the satisfaction condition for negation, $a \vDash \neg A$. This fact about saturated theories together with Lemma 9.2 implies the validity of DNE: suppose that a is a saturated theory and $a \vDash \neg\neg A$. Then $a^* \nvDash \neg A$ and so $a^{**} \vDash A$. But $a^{**} = a$. Thus, by Lemma 9.2, for any theory b, if $b \vDash \neg\neg A$, then $b \vDash A$, and so, generalising, by the semantic entailment lemma, $\neg\neg A \to A$ (DNE) is valid.

At last, the model theory for the propositional logic E of relevant entailment is fully specified!

9.4   Recap: Models for Propositional E

An E-frame is a quintuple $\langle \Upsilon, \Sigma, \circ, \perp, \leq \rangle$, where Υ is a non-empty set (of theories), Σ (the master theory) is in Υ, $\circ$ is a binary operator on theories, $\perp$ is a binary relation on theories, and $\leq$ is a partial order on theories, such that the following conditions hold.

1. $\Sigma \circ a = a$;
2. If $a \leq a'$ and $b \leq b'$, then $a \circ b \leq a' \circ b'$;
3. $a \circ (b \circ c) \leq (a \circ b) \circ c$;
4. $(a \circ b) \circ b \leq a \circ b$;
5. $\Box^- a \leq a$;
6. $\Box^- a \circ \Box^- b = \Box^- b \circ \Box^- a$;
7. If $a \leq b$, then $a^\perp \subseteq b^\perp$;
8. If $a \perp b$, then $b \perp a$;
9. If $(a \circ b) \perp c$, then $(a \circ c) \perp b$;
10. If $(a \circ b) \perp b$, then $a \perp b$;
11. If $a \in S$, then $a^* \in S$;
12. If $a \in S$, then $a^{**} = a$;
13. $\cap \{b^\perp : b \in S(a)\} \subseteq a^\perp$;
14. $S(a \circ b) = \cup \{S(c \circ b) : c \in S(a)\}$.

Where $a^\perp$ is the set of theories orthogonal to a, a^c is the (classical) set-theoretic complement of $a^\perp$, S is the set of theories a such that there is a maximal element a^* in a^c, and $S(a)$ is the set of theories in S greater than a.

An E model is a pair $\langle \mathfrak{F}, \vDash \rangle$ where $\mathfrak{F}$ is an E-frame and $\vDash$ is a relation between theories in $\mathfrak{F}$ and formulas such that the following hold:

- $a \vDash p$ if and only if, for all $b \in S(a)$, $b \vDash p$;
- $a \vDash A \wedge B$ if and only if $a \vDash A$ and $a \vDash B$;
- $a \vDash A \vee B$ if and only if, for all $b \in S(a)$, $b \vDash A$ or $b \vDash B$;
- $a \vDash \neg A$ if and only if, for all $b \in \Upsilon$, if $b \vDash A$, then $a \perp b$;
- $a \vDash A \to B$ if and only if, for all $b \in \Upsilon$, if $b \vDash A$, then $a \circ b \vDash B$.

9.5   Worlds and Truth Conditions

It is essential to have a clear relationship between the satisfaction conditions of the connectives for theories and their truth conditions in worlds. When people formulate theories, they do so to talk about this world and other nearby worlds. In formulating a theory of physics, for example, one wants to describe this world and also all the other physically (and perhaps nominally) possible worlds. The meanings of one's logical words – 'and', 'or', 'not', ... – at the very least should be used in one's theories in a manner that will fit with their truth conditions in the target worlds. Unfortunately, there is no universally accepted view about the truth condition for negation. This is, at the present time, a hotly contested issue in the philosophical literature.

I set aside, however, some important areas of the debate on negation. For example, I avoid any discussion of how to integrate a treatment of vagueness into the present view. I cannot see why a supervaluational, fuzzy, or epistemic account of vagueness could not be integrated into the present theory, but it would take the discussion of the book on a tangent away from the central issues concerning entailment.

Truth conditions for the connectives set out the behaviour of the connectives at *worlds*. These are to be distinguished from the satisfaction conditions that I give for theories, which tell us the conditions under which sentences are to be thought as *included in* theories. One might call them 'inclusion conditions', although I do not use this terminology. Also truth conditions are to be distinguished from *information conditions* that tell us when a situation contains the information that particular sentences are true. But what is a world?

This is of course another subject on which there is, to say the least, an extensive literature. And to answer the question of what is a world would take us away from semantics and drop us into hardcore metaphysics. Instead of trying to answer the question about the nature of worlds, I ask a question that can be answered in the framework of the model theory: What is a world-like theory? This question asks what properties a theory needs to have to be said to accurately and completely describe some world. Moreover, I answer this question in terms of the mechanisms of the model theory.

The worlds I am interested in are logically possible worlds. These are worlds that include all the theorems of the logic. In terms of the model theory, a theory a is world-like only if $\Sigma \leq a$. Moreover, the satisfaction conditions for disjunction in a world-like theory should match the standard truth conditions for disjunction. This assertion needs some justification. There are some relevant logicians who think that the proper treatment of disjunction is as an intensional connective [175]. I discuss this issue elsewhere [129]. In general, I think that there is room in the analysis of natural language connectives for both extensional and intensional treatments of disjunction. Some uses of disjunction are intensional and others are extensional. I do not, however, think that the intensional disjunction definable in E (as $A \oplus B \ =_{df} \ \neg A \rightarrow B$) is of use in the interpretation of ordinary disjunction. And the issue of the addition of a new primitive intensional disjunction to E is beyond the scope of this book. With regard to extensional disjunction, the standard truth condition seems right. In order to force the standard truth condition for disjunction to hold at world-like theories, I claim that every world-like theory is saturated.

This is the extent to which I wish to specify world-like theories. The set of world-like theories in a model is just $S(\Sigma)$. We can show that every theory in $S(\Sigma)$ is complete, in the sense that the law of excluded middle holds at it. Here is a little semantic argument to prove this:

1.	$a \in S(\Sigma)$	Assumption
2.	$a \not\leq a^*$	Def. of $*$
3.	$a \circ a^* \not\leq a^*$	2, reductio postulate
4.	$\Sigma \leq a$	1, Def. of $S(\Sigma)$
5.	$\Sigma \circ a^* \not\leq a^*$	3,4, mono
6.	$a^* \not\leq a^*$	5, Σ-identity

Now, suppose that A does not hold in a. Then, $a^* \vDash \neg A$ and, since a^* is compatible with itself, a^* does not satisfy A. Thus, $a \vDash \neg A$. Generalising, $a \vDash A$ or $a \vDash \neg A$ and so $a \vDash A \vee \neg A$. Thus, if a is in $S(\Sigma)$, it satisfies excluded middle and is bivalent.

In the Appendix, I show that in the canonical model there are world-like theories W such that $\Box^- W = \Sigma$. This shows that it is plausible, or at the very least possible, that the theory that is complete for the actual world has the set of E-theorems as its denecessitation. The world-like theory that I construct in the Appendix is prime and consistent.

World-like theories, however, need not be consistent. If a theory in $S(\Sigma)$ and its star are not identical to one another, then it is possible for there to be a formula and its negation that are both satisfied by that theory. We could stipulate that possible worlds are consistent and hence world-like theories are all consistent. In fact, the logic is characterised by its consistent world-like theories. We could change the definition of validity to say that a formula is valid in the class of models if and only if it is satisfied in every model by every theory in $S(\Sigma)$ that is identical to its own star. Robert Meyer and Michael Dunn proved that E remains complete over this class of models [144]. There may be good metaphysical or semantical reasons to hold that all worlds are consistent, but I do not want to get into that here.

Thinking of the behaviour of negation, disjunction, and conjunction at theories in $S(\Sigma)$, we find that they operate in accordance with the truth tables of Graham Priest's Logic of Paradox (LP) [163, 166]. The following are the truth tables for LP:

$\wedge$	T	B	F		$\vee$	T	B	F		$\neg$	
$!\,T$	T	B	F		T	T	T	T		T	F
$!\,B$	B	B	F		B	T	B	B		B	B
F	F	F	F		F	T	B	F		F	T

The exclamation marks beside 'T' and 'B' in the conjunction table indicate the designated (or true-like) values. A formula is valid over these tables if it takes the value T or the value B on all rows of these tables. 'T' stands for 'true only', 'B' for 'both true and false', and 'F' for 'false only'. If we interpret '$a \vDash A$' as saying that A is at least true according to a and '$a \models \neg A$' as saying A is at least false according to a, then over $S(\Sigma)$ we find that these three connectives behave exactly as described by the LP truth tables. Thus, semantically, we can think of E as an extension of LP. This is not to say, of course, that, like LP, E provides us with a logical basis for dealing with the logical paradoxes in the sense that it allows the paradoxes to be expressed but does not trivialise. I discuss this issue in Chapter 10.

9.6 Negation, Disjunction, and Proof Processes

In order to integrate negation into the Prawitz-style natural deduction system for proof processes, I add a second form of judgement. These new judgements are *incompatibility judgements*. An incompatibility judgement is a judgement of the form $a \perp b$, where a and b are theory terms. By having incompatibility judgements in this proof theory, I can express a lot about the semantics in that proof theory.

There are two reductio rules. One of them is a negation introduction rule and the other is a negation elimination rule:

$$
\begin{array}{cc}
[x \models A] & [x \models \neg A] \\
\vdots & \vdots \\
\dfrac{a \perp x}{a \models \neg A} & \dfrac{a \perp x}{a \models A}
\end{array}
$$

And we need a second negation elimination rule:

$$
\frac{a \models \neg A \qquad b \models A}{a \perp b}.
$$

The central motivation for introducing this Prawitz-style deduction system in addition to Anderson and Belnap's Fitch-style system is to represent reasoning about arbitrary theories. The negation rules allow one to reason clearly about the incompatibility of theories. The negation introduction rule tells us that if we derive of a theory a that any theory x that satisfies A is incompatible with a, then $a \models \neg A$. This is a fairly natural way to reason. Suppose that it is a theorem that $A \to B$ and that for a particular theory T, $T \models \neg B$. Then, the following reasoning is logically valid:

$$
\cfrac{\cfrac{\Sigma \models A \to B \qquad [x \models A]^1}{\cfrac{\Sigma \circ x \models B}{x \models B}} \qquad T \models \neg B}{\cfrac{T \perp x}{T \models \neg A} \, 1}
$$

This seems quite an intuitive way to reason. We show that a theory contains a negation by assuming that another theory contains the negand and working through what the original theory entails. Now suppose that $A \to B$ is not a theorem, but is a postulate of T itself:

$$\frac{T \vDash A \to B \qquad [x \vDash A]^1}{\dfrac{\dfrac{\dfrac{\dfrac{\dfrac{T \circ x \vDash B \qquad T \vDash \neg B}{T \perp T \circ x}}{T \circ x \perp T}}{T \circ T \perp x} \, ^!}{\dfrac{T \perp x}{T \vDash \neg A} \, ^1} \, ^{!!}}}$$

The single exclamation mark indicates the step at which the contraposition rule is used and the double exclamation marks indicate the step at which the contraction rule is applied.

The disjunction introduction rule is the standard one:

$$\frac{a \vDash A}{a \vDash A \vee B} \qquad\qquad \frac{a \vDash B}{a \vDash A \vee B}$$

I can also adopt a standard Prawitz-style disjunction elimination rule:

$$\begin{array}{ccc} & [x \vDash A] & [x \vDash B] \\ & \vdots & \vdots \\ \dfrac{a \vDash A \vee B \qquad b \circ x \vDash C \qquad b \circ x \vDash C}{b \circ a \vDash C} \end{array}$$

where x does not occur in b. The problem with this rule is that together with the conjunction rules I still cannot derive the distribution scheme: $(A \wedge (B \vee C)) \to ((A \wedge B) \vee (A \wedge C))$. The following (somewhat ugly) modification of this rule, however, suffices to produce distribution:

$$\begin{array}{ccc} & [x \vDash A \wedge B] & [x \vDash A \wedge C] \\ & \vdots & \vdots \\ \dfrac{a \vDash A \wedge (B \vee C) \qquad b \circ x \vDash D \qquad b \circ x \vDash D}{b \circ a \vDash D} \end{array}$$

where x does not occur in b. This rule is ugly because it requires the use of conjunction in the disjunction elimination rule. It ruins the lovely symmetry that one normally expects in systems of natural deduction according to which each connective has its own introduction and elimination rules. I discuss this issue again in Section 9.8.

Here is a proof using negation and disjunction rules to derive the law of excluded middle:

$$\frac{\dfrac{[x \vDash \neg(A \vee \neg A)]^2 \qquad \dfrac{[y \vDash A]^1}{y \vDash A \vee \neg A}}{\dfrac{x \perp y}{\dfrac{x \vDash \neg A}{x \vDash A \vee \neg A}} \, ^1} \qquad [x \vDash \neg(A \vee \neg A)]^2}{\dfrac{\dfrac{\dfrac{x \perp x}{\Sigma \circ x \perp x}}{\Sigma \perp x} \text{ Reductio}}{\Sigma \vDash A \vee \neg A} \, ^2}$$

And here is a proof deriving one of the De Morgan laws:

$$
\cfrac{[x \vDash A \vee B]^4 \quad \cfrac{[y \vDash A]^2 \quad \cfrac{\cfrac{[w \vDash \neg A \wedge \neg B]^1}{w \vDash \neg A}}{\cfrac{w \perp y}{\cfrac{y \perp w}{y \vDash \neg(\neg A \wedge \neg B)}}} \quad \cfrac{[z \vDash B]^3 \quad \cfrac{\cfrac{[w \vDash \neg A \wedge \neg B]^1}{w \vDash \neg B}}{\cfrac{w \perp z}{\cfrac{z \perp w}{z \vDash \neg(\neg A \wedge \neg B)}}}}{z \vDash \neg(\neg A \wedge \neg B)}{}_{2,3}}{\cfrac{\cfrac{x \vDash \neg(\neg A \wedge \neg B)}{\Sigma \circ x \vDash \neg(\neg A \wedge \neg B)}}{\Sigma \vDash (A \vee B) \to \neg(\neg A \wedge \neg B)}{}_4}
$$

9.7 Negation in the Anderson–Belnap Natural Deduction System

The negation rules in Anderson and Belnap's system are rather bitsy: they each address particular desired features of negation but do not really express a view of what negation is [3, 4]. Instead I introduce an incompatibility operator into the syntax of the logic and formulate negation rules in terms of it. In this way, I get a proof system that more closely expresses the meaning of negation as it is given in the theory semantics.

The notion of incompatibility is not a standard part of our logical vocabulary, but there are some historical cases in which it was used. I borrow much of my treatment of incompatibility from E. J. Nelson's connexive logic (see Section 2.11). Nelson uses a vertical line, |, to represent incompatibility, whereas I use a perp operator to indicate a parallel with the incompatibility relation between theories. '$A \perp B$' means 'A is incompatible with B'. As I discuss in more detail in Section 9.9, the incompatibility operator is definable using the standard connectives of E, so there is no need to give incompatibility statements special satisfaction conditions.

Incompatibility formulas (formulas of the form $A \perp B$) are implicit entailments (see Section 9.9). Thus, they can cross over subproof scope lines like entailments.

The Fitch system has two negation introduction rules ($\neg$I). These can also be thought of also as perp elimination rules.

$$
\begin{array}{cc}
A \perp B_\alpha & A \perp B_\alpha \\
A_\beta & B_\beta \\
\downarrow & \downarrow \\
\neg B_{\alpha \cup \beta} & \neg A_{\alpha \cup \beta}
\end{array}
$$

To get all of the logic E, I add another form of negation introduction, that I call *SelfIncom* for 'self-incompatibility':

$$
\begin{array}{c}
A \perp A_\alpha \\
\downarrow \\
\neg A_\alpha
\end{array}
$$

This rule says that if a proposition is incompatible with itself, it is false. This allows the derivation of the reductio principle, $(A \to \neg A) \to \neg A$.

The negation elimination rule ($\neg$E) is very similar to the negation introduction rules:

$$A \perp \neg B_\alpha$$
$$A_\beta$$
$$\downarrow$$
$$B_{\alpha \cup \beta}$$

The system also has two perp introduction rules:

$$
\begin{array}{ll}
A_k & \text{hypothesis} \\
\vdots & \\
\neg B_\alpha & \\
\hline
A \perp B_{\alpha-k} & \perp\text{I}
\end{array}
$$

$$
\begin{array}{ll}
A_k & \text{hypothesis} \\
\vdots & \\
B_\alpha & \\
\hline
A \perp \neg B_{\alpha-k} & \perp\text{I}
\end{array}
$$

where $k \in \alpha$. This second rule allows the derivation of a scheme that says that every formula is incompatible with its negation:

$$A \perp \neg A.$$

Here is a derivation of one of the contraposition principles using these negation rules:

1	$A \to \neg B_1$	hypothesis
2	B_2	hypothesis
3	A_3	hypothesis
4	$A \to \neg B_{\{1\}}$	1, reiteration
5	$\neg B_{\{1,3\}}$	3,4, $\to$E
6	$A \perp B_{\{1\}}$	3–5, $\perp$I
7	$\neg A_{\{1,2\}}$	2,6, $\neg$I
8	$B \to \neg A_{\{1\}}$	2–7, $\to$I
9	$(A \to \neg B) \to (B \to \neg A)_\emptyset$	1–8, $\to$I

And here is a little derivation showing that $\perp$ is commutative:

1	$A \perp B_{\{1\}}$	hypothesis
2	$B_{\{2\}}$	hypothesis
3	$A \perp B_{\{1\}}$	1, reiteration
4	$\neg A_{\{1,2\}}$	2,3, $\neg$I
5	$B \perp A_{\{1\}}$	2–4, $\perp$I
6	$(A \perp B) \rightarrow (B \perp A)_{\emptyset}$	1–5, $\rightarrow$I

Reductio is next:

1	$A \rightarrow \neg A_{\{1\}}$	hypothesis
2	$A_{\{2\}}$	hypothesis
3	$A \rightarrow \neg A_{\{1\}}$	1, reiteration
4	$\neg A_{\{1,2\}}$	2,3, $\rightarrow$I
5	$A \perp A_{\{1\}}$	2–4, $\perp$I
6	$\neg A_{\{1\}}$	5, SelfIncom
7	$(A \rightarrow \neg A) \rightarrow \neg A_{\emptyset}$	1–6, $\rightarrow$I

Finally, here is a proof of $(A \rightarrow \neg B) \rightarrow \neg(A \wedge B)$. I use the contrapositive of this theorem to prove Lemma A.27 in the Appendix.

1	$A \rightarrow \neg B_{\{1\}}$	hypothesis
2	$A \wedge B_{\{2\}}$	hypothesis
3	$A \rightarrow \neg B_{\{1\}}$	1, reiteration
4	$A_{\{2\}}$	2, $\wedge$E
5	$\neg B_{\{1,2\}}$	3,4, $\rightarrow$E
6	$(A \wedge B) \rightarrow B_{\emptyset}$	theorem
7	$\neg B \rightarrow \neg(A \wedge B)_{\emptyset}$	6, contraposition
8	$\neg(A \wedge B)_{\{1,2\}}$	5,7, $\rightarrow$I
9	$(A \wedge B) \perp (A \wedge B)_{\{1\}}$	2–8, $\perp$I
10	$\neg(A \wedge B)_{\{1\}}$	9, SelfIncom
11	$(A \rightarrow \neg B) \rightarrow \neg(A \wedge B)_{\emptyset}$	1–10, $\rightarrow$I

9.8   Disjunction in the Anderson–Belnap System

The disjunction introduction rules are just the classical rules with the addition of subscripts:

$$A_\alpha \qquad\qquad B_\alpha$$
$$\downarrow \qquad\qquad\quad \downarrow$$
$$A \vee B_\alpha \qquad A \vee B_\alpha$$

Anderson and Belnap's elimination rule for disjunction (in effect) is the following:

$$
\begin{array}{ll}
A \vee B_\alpha & \\
\quad A_{\{k\}} & \text{hypothesis} \\
\quad \vdots & \\
\quad C_\beta & \\
\quad B_{\{k\}} & \text{hypothesis} \\
\quad \vdots & \\
\quad C_\beta & \\
C_{\alpha \cup (\beta - k)} & \vee\mathrm{E}
\end{array}
$$

Together these introduction and elimination rules are inadequate to characterise the behaviour of disjunction in its relation to the other connectives. In particular, Anderson and Belnap found it necessary to add a special rule to make the distribution scheme valid.

Since the creation of Anderson and Belnap's system, there have been various attempts to formulate rules for disjunction that allow for a less contrived proof of distribution [24, 57]. My solution to the problem is no more elegant than these, but does not require any extra mechanism nor any change to the hypothesis rule. This is my disjunction elimination rule:

$$
\begin{array}{ll}
C \wedge (A \vee B)_\alpha & \\
\quad C \wedge A_{\{k\}} & \text{hypothesis} \\
\quad \vdots & \\
\quad D_\beta & \\
\quad C \wedge B_{\{k\}} & \text{hypothesis} \\
\quad \vdots & \\
\quad D_\beta & \\
D_{\alpha \cup (\beta - k)} & \vee\mathrm{E}
\end{array}
$$

This rule is essentially the same rule I use in the Prawitz-style system (Section 9.6) and is ugly for the same reason.[2]

Here is a proof of the converse of the De Morgan law proven in Section 9.6:

1	$\neg(\neg A \wedge \neg B)_{\{1\}}$	hypothesis
2	$\neg(A \vee B)_{\{2\}}$	hypothesis
3	$A_{\{3\}}$	hypothesis
4	$A \vee B_{\{3\}}$	3, $\vee$I
5	$A \perp \neg(A \vee B)_{\emptyset}$	3–4, $\perp$I
6	$\neg A_{\{2\}}$	2,5, $\neg$I
7	$B_{\{4\}}$	hypothesis
8	$A \vee B_{\{4\}}$	$\vee$I
9	$B \perp \neg(A \vee B)_{\emptyset}$	7–8, $\perp$I
10	$\neg B_{\{2\}}$	2,9, $\neg$I
11	$\neg A \wedge \neg B_{\{2\}}$	6,10, $\wedge$I
12	$\neg(A \vee B) \perp \neg(\neg A \wedge \neg B)_{\emptyset}$	2–11, $\perp$I
13	$A \vee B_{\{1\}}$	1,12, $\neg$E
14	$\neg(\neg A \wedge \neg B) \rightarrow (A \vee B)_{\emptyset}$	1–13, $\rightarrow$I

9.9 Axiomatisation of E

$$A \leftrightarrow B =_{df} (A \rightarrow B) \wedge (B \rightarrow A)$$

Axiom Schemes

1. $A \rightarrow A$ (Identity)
2. $(A \rightarrow B) \rightarrow (((A \rightarrow B) \rightarrow C) \rightarrow C)$ (Specialised Assertion)
3. $(A \rightarrow B) \rightarrow ((B \rightarrow C) \rightarrow (A \rightarrow C))$ (Suffixing)
4. $(B \rightarrow C) \rightarrow ((A \rightarrow B) \rightarrow (A \rightarrow C))$ (Prefixing)
5. $(A \rightarrow (A \rightarrow B)) \rightarrow (A \rightarrow B)$ (Contraction)
6. $(A \wedge B) \rightarrow A; (A \wedge B) \rightarrow B$ (Simplification)
7. $((A \rightarrow B) \wedge (A \rightarrow C)) \rightarrow (A \rightarrow (B \wedge C))$ (Conjunction in the Consequent)
8. $A \rightarrow (A \vee B); B \rightarrow (A \vee B)$ (Addition)
9. $((A \rightarrow C) \wedge (B \rightarrow C)) \rightarrow ((A \vee B) \rightarrow C)$ (Constructive Dilemma)

---

[2] Perhaps a version of Urquhart's rule in which the hypotheses of the rule have the subscript α would work [207]. It is certainly a more elegant rule and seems to do exactly the same thing as my rule. I have not given this topic enough thought.

10. $(A \lor B) \leftrightarrow \neg(\neg A \land \neg B)$ (De Morgan)
11. $\neg\neg A \to A$ (DNE)
12. $(A \land (B \lor C)) \to ((A \land B) \lor (A \land C))$ (Distribution)
13. $(A \to \neg B) \to (B \to \neg A)$ (Contraposition)
14. $(A \to \neg A) \to \neg A$ (Reductio)

Rules

$$\frac{A \to B \quad A}{B} \qquad \frac{A \quad B}{A \land B}$$

I show that this logic is complete over the semantics in the Appendix. The method for showing that this axiom system is equivalent to the Fitch-style system is outlined in Chapter 4. To show that it is equivalent to the Prawitz-style system, first show that all the theorems of the axiom system are derivable in the natural deduction system (which is quite straightforward). To show that everything provable in the Prawitz-style system is provable in the axiom system is a bit more involved.

I need a way of expressing incompatibility in the language. The perp operator can be defined in the standard language:

$$A \perp B =_{df} A \to \neg B$$

I also need a way of representing the application of theories to one another. To do so I add a "fusion" connective, $\otimes$, to the language (see Chapter 4). Fusion cannot be defined in E (as it can be in R), but rather has to be added by means of extra axioms or rules. Here are two rules that are sufficient:

$$\frac{\vdash A \to (B \to C)}{\vdash (A \otimes B) \to C} \qquad \frac{\vdash (A \otimes B) \to C}{\vdash A \to (B \to C)}$$

The first of these rules is sometimes called "importation" and the second is sometimes called "exportation". To the definition of a model, I add the following satisfaction condition for fusion:

$a \vDash A \otimes B$ if and only if there are theories b and c such that $b \vDash A$, $c \vDash B$, and $b \otimes c \leq a$.

Adding fusion in this manner produces a logic that is complete over the resulting semantics and it is easy to show that the logic is conservative over E without fusion.

I do not add fusion to the language for the logic of entailment, but only for the purposes of proving that the Prawitz-style natural deduction system is correct. With fusion, we can prove $A \to (B \to (A \otimes B))$. This is a fallacy of modality, and adding it would make every non-empty theory into a theory of entailment, and this is unacceptable.

Now that the language has incompatibility and fusion, we can translate any valid derivation of a theorem in the Prawitz-style system into a proof in the axiom system. We replace any theory variable x with A, where $x \vDash A$ is a hypothesis in the derivation, we replace $a \circ b$ with $A \otimes B$ where A is the translation of a and B is the translation of B, we replace $a \vDash B$ with $A \to B$ where A is the translation of a and replace $a \perp b$

with $A \perp B$. Then we get a valid derivation (adding lines is sometimes necessary to see that this is so).

In $E^\otimes$, we can derive the following rule:

$$\frac{\vdash A \to (B \to C) \qquad \vdash A' \to B}{\vdash (A \otimes A') \to C}.$$

This allows us to represent the entailment elimination rule in the axiom system. The entailment introduction rule is quite simple. If we replace x with A when $x \vDash A$ is a hypothesis, then in the translation of a Prawitz-style derivation, an entailment introduction

$$[x \vDash A]$$

$$\vdots$$

$$\frac{c \circ x \vDash B}{c \vDash A \to B}$$

becomes

$$A \to A$$

$$\vdots$$

$$\frac{(C \otimes A) \to B}{C \to (A \to C)}$$

where C is the translation of c. The move from the second last step to the final one is clearly valid, since it is just an instance of the exportation rule. The other rules of the Prawitz-style system are similarly derivable in $E^\otimes$.

10 Quantification

10.1 Introduction

To complete my presentation of the semantics for E, I add a treatment of the quantifiers. The semantics that I employ here is based on the model theory that I developed with Robert Goldblatt [76, 130, 134]. Quantified relevant logics have received less attention than they should have. One reason for this was the perceived level of difficulty of the semantics for quantification. Routley tried to develop a standard constant-domain semantics for quantifiers,[1] but Fine proved that Anderson and Belnap's quantified relevant logics were incomplete over this semantics [3, 65]. Fine also produced an ingenious but complicated semantics over which these logics were complete.[2] Some years later, Goldblatt and I formulated an alternative semantics. I suggest that our semantics fits better with the philosophical framework of this book. In this chapter, I explain our semantics and show how it can be understood in terms of theory closure.

To facilitate the presentation of the semantics and reduce the load of technical material that I am dumping on the reader, I am sometimes a little lax in my use of notation. I do not always explicitly distinguish between cases in which one term is uniformly substituted for another and those in which some occurrences of a term are replaced. I often have an inference from $A(\tau)$ to $A(\tau')$, say, and do not state clearly that all free occurrences of τ are replaced with τ' (which are also free). I think the reader can make these distinctions for him or herself, for they are rather obvious and follow standard approaches to quantification and identity in the literature.

10.2 Theories and Quantification

The standard approach to quantification in modal logics relativises Tarski's semantics for first-order logic to worlds. In Tarski's theory, a universally quantified statement, $\forall x A(x)$, is true on an assignment f to variables if and only if for all assignments g that are just like f except perhaps in their assignment of x, $A(x)$ is true. Similarly, an existentially quantified statement, $\exists x A(x)$, is true on f if and only if $A(x)$ is true on

---

[1] For constant-domain semantics for quantified modal logic, see Chapter 3.
[2] There is a recent paper by Shay Logan explaining Fine's semantics [117].

some such g. For the sake of simplicity, let us assume that every world has the same domain of quantification. On the standard treatment, then,

$$w \vDash_f \forall x A \text{ if and only if for all } x-\text{variants of } f, \ g, \ w \vDash_g A$$

$$w \vDash_f \exists x A \text{ if and only if for some } x-\text{variants of } f, \ g, \ w \vDash_g A$$

These clauses, however, do not fit well when we move from thinking about worlds to thinking about theories.

Recall the problem with the standard treatment of disjunction. Not all disjunctions in theories are resolved in those theories. That is, we can have a disjunction $A \vee B$ in a theory without having either A or B in that theory. The same thing is true of existentially quantified statements. It is common in mathematical theories, such as set theories, that there are existential postulates or theorems that are unresolved. A theory may tell us that there is a set of a particular cardinality without completely specifying what that set is. So, the standard clause for the existential quantifier needs a good deal of modification to be included in a treatment of the logic of theories.

Similarly, the Tarskian clause for the universal quantifier is not adequate for a semantics of theory closure. Suppose that a theory says that i is φ, j is φ, and so on, for all the objects mentioned by a. Even if this list includes everything that actually exists (or even everything that could possibly exist), this does not mean that the theory states that φ holds of everything. Rather, we need more information, such as that the things mentioned in the theory are all those things that exist. Even if in every model that we attribute to that theory, everything has a particular property, we do not necessarily attribute to the theory the statement that everything has that property. Consider, for example, a biological theory. In any model of that theory, each thing in the domain is physical that is, it is made of matter. However, one would not just because of this say that the theory says that everything is made of matter.

This example illustrates Russell's famous argument concerning general propositions from his *Lectures on Logical Atomism*:

> The old plan of complete induction, which used to occur in books, which was always supposed to be quite safe and easy as opposed to ordinary induction, that plan of complete induction, unless it is accompanied by at least one general proposition, will not yield you the result that you want. Suppose, for example, that you wish to prove in that way that "All men are mortal", you are supposed to proceed by complete induction, and say "A is a man that is mortal", "B is a man that is mortal", "C is a man that is mortal", and so on until you finish. You will not be able, in that way, to arrive at the proposition "All men are mortal" unless you know when you have finished. That is to say that, in order to arrive by this road at the general proposition "All men are mortal", you must already have the general proposition "All men are among those I have enumerated". You never can arrive at a general proposition by inference from particular propositions alone. You will always have to have at least one general proposition in your premises. [187, pp 69–70]

Greg Restall and Gillian Russell call the claim that one cannot validly infer a general claim from particular sentences, "Russell's Law" [180]. What is more plausible is their weakened form of Russell's Law, according to which a general truth cannot be derived from its particular instances. Of course, there are cases in which, in classical logic, one can infer a universally quantified claim from its instances, such as when the universally quantified statement is valid itself or the particular instances are contradictions, these are marginal cases peculiar to classical logic and irrelevant to my current project.

Let Γ be a set of formulas. Put in terms of entailment, the weakened form of Russell's Law says that even if $\Gamma \Vdash A(c)$ holds for all constants c in the language $\mathcal{L}$ (where Γ is constituted of sentences of $\mathcal{L}$), we may not be able to claim that $\Gamma \Vdash \forall x A(x)$. However, I do think that there is a closely related condition that does hold for the universal quantifier. Here is the condition:

$$(\forall 1) \; \Gamma \Vdash \forall x A(x) \text{ if and only if } \Gamma \Vdash A(c)$$

for all individual constants c in all languages $\mathcal{L}'$, where $\mathcal{L}'$ is just like $\mathcal{L}$ except that $\mathcal{L}'$ perhaps includes more individual constants.

The condition $(\forall 1)$ seems intuitive, but it is difficult to see how in a single model the idea of infinitely expanding languages can be represented. But we can *simulate* $(\forall 1)$ within a single language with only a countably infinite set of constants. To see how this can be done, I will continue with my analysis of $\Gamma \Vdash \forall x A(x)$.

Let d be a constant in an $\mathcal{L}'$ that is not in $\mathcal{L}$. Since our logic is closed under uniform substitution, $(\forall 1)$ is equivalent to $(\forall 2)$:

$$(\forall 2) \; \Gamma \Vdash \forall x A(x) \text{ if and only if } \Gamma \Vdash A(d)$$

where d does not occur in any formula in Γ. Here, d is an arbitrary constant from the standpoint of Γ. Moreover, the logic of entailment, as I argue in Chapters 3 and 6, is compact. Therefore, if $\Gamma \Vdash A(d)$, then there are $G_1, \ldots, G_n$ in Γ such that $G_1, \ldots, G_n \Vdash A(d)$ for some finite and non-zero n. And, because $\Vdash$ obeys the rule of weakening, we have

$$(\forall 3) \; \Gamma \Vdash \forall x A(x) \text{ if and only if } \exists G_1 \in \Gamma \ldots \exists G_n \in \Gamma \, (G_1, \ldots, G_n \Vdash A(d))$$

where d does not occur in any of $G_1, \ldots, G_n$.

Suppose that $\mathcal{L}$ has infinitely many constants (any infinity will do). Since the formulas of $\mathcal{L}$ are themselves of finite length and include only finitely many constants and variables, $\mathcal{L}$ includes some constant d that is not in $G_1, \ldots, G_n$. Hence, by uniform substitution, we obtain

$$(\forall 4) \; \Gamma \Vdash \forall x A(x) \text{ if and only if } \exists G_1 \in \Gamma \ldots \exists G_n \in \Gamma \, (G_1, \ldots, G_n \Vdash A(e))$$

for all constants e in $\mathcal{L}$. It is $(\forall 4)$ that I implement in my model to simulate $(\forall 1)$.

A universal statement is an inference licence of a sort. The statement that to every particle there is an associated wave function, say, tells us that if we are dealing with a particle, then we can infer that there is a wave function associated with it. The theory itself does not necessarily (or usually) tell us what particles there are, but does its universal statements tell us how to deal with whatever particles there happen to be. $(\forall 4)$ captures this idea as well as the idea that even very large collections of particular statements are inadequate to licence the inference to a universal statement.

In Section 10.3, I outline a model theory that implements $(\forall 4)$.

10.3   Frames and Propositions

Goldblatt's and my model theory provides a *general semantics* for quantification (also called an "admissible set semantics"). This means that in addition to the operators and relations employed to treat propositional logic, the definition of a frame includes a set of *propositions*. A proposition is a set of theories closed upwards under the hereditariness relation, $\leq$. Not every such "upset", however, is a proposition.

When Goldblatt and I constructed this theory, I had the informational interpretation of the Routley–Meyer semantics in mind as it was presented in [129]. The idea was to characterise a piece of general information. In this theory, a piece of information consists in a set of situations (see Section 5.6) that can be understood by people as having something substantive in common. For example, the set of situations in which it is sunny in Wellington on 17 August 2020 makes up such a piece of information, but a random set of situations (even if closed upwards under $\leq$) may not be seen to represent any information at all.

A proposition in the current theory is quite similar to a piece of information in the theory of situated inference. If we collect together all the theories that say that water is in a liquid state at 20°, at sea level on Earth (with an atmosphere like that of the actual Earth), then we have a proposition. In a model, every formula expresses a proposition. However, a random set of theories (even if they are closed upwards under $\leq$) may not be seen by people as having any substantive common content.

A proposition is true according to a theory a if and only if a is in that proposition. I use the notation '$[\![A]\!]$' to denote the proposition expressed by the formula A. This proposition is simply the set of theories that satisfy A. A proposition is closed upwards under $\leq$, that is, if a is in $[\![A]\!]$ and $a \leq b$, then b is also in $[\![A]\!]$.

In our models, a theory a satisfies $\forall x A(x)$ if and only if there is some proposition that is true according to a that entails $A(x)$ for every value of x. A proposition π entails a proposition ϕ in a frame if and only if $\pi \subseteq \phi$. Let f be an assignment of variables to objects in the domain. Then π entails $A(x)$ for every value of x if $\pi \subseteq [\![A(x)]\!]_{f'}$, where f' gives the same value to every variable except perhaps x. Stated formally, the condition for the universal quantifier is:

$$a \vDash_f \forall x A(x) \text{ if and only if } \exists \pi \in \text{Prop } \forall f'(a \in \pi \wedge \pi \subseteq [\![A(x)]\!]_{f'}).$$

It is easy to see from this condition how ($\forall$1) – ($\forall$4) are simulated in models. The proposition π does the work of the finite set of formulas that entail every instance of $A(x)$.

In Section 10.4, I go into more detail about my concept of a proposition to justify its inclusion in the current theory.

10.4   Propositions: Thick and Thin

As I say in Section 10.3, propositions are elements in the semantic theory that I support. However, as I say throughout this book, I cannot include propositions in the now-standard (or "wide") sense discussed in contemporary philosophy.

On the now-standard view, a proposition includes in it whatever is analytically implied by its content. For example, older taxonomic theories used observable morphological evidence to determine which species are subspecies of other species. More modern theories are justified in accordance with the Kripke–Putnam view that natural kind terms track internal structures such as genetic structure.

According to contemporary equine zoological theory, an extinct species of zebra – *Equus quaggaquagga* – is a subspecies of the plains zebra – *Equus quagga* [160]. If the Kripke–Putnam view is correct, then '*Equus quaggaquagga* is a subspecies of *Equus quagga*' is analytic. In older theories, *Equus quaggaquagga* is not a subspecies of *Equus quagga*. On the Kripke–Putnam view, these older theories are not just false but analytically false. Let the core genetic structure (the genetic material that makes an animal a member of a particular species) of *Equus quagga* is denoted by 'X' and the core genetic structure of *Equus quaggaquagga* is denoted by 'Y'. It might be that some zebra taxonomists knew that anything of species that has Y is a subspecies of the species that has X without knowing that *Equus quaggaquagga* is a subspecies of *Equus quagga*. The fact that these species have these genetic structures, after all, is an empirical fact. Hence, it makes sense to allow theories about species without making those theories about particular genetic structures. Thus, contemporary philosophy of science needs a theory of propositions that is not as *metaphysically thick* as sets of possible worlds.

It seems that we can talk about natural kind terms having (at least) two referents at the same time. The primary referent is the actual natural kind in the actual world (in the actual location where the term is used). For example, the primary referent of 'water' is H_2O. The secondary referent of 'water' depends on the model that one is discussing. In a model of pre-seventeenth century chemistry, 'water' refers to some primitive element. In a model of modern chemistry, it refers to H_2O.

On the debate about changes in meaning for scientific terms when theories change in important ways, I take a middle road. I do not think that the primary referents of words like 'water' or 'plains zebra' change just because we adopt new theories about them. Do their meanings change? In a way they do, because their secondary referents change, but there is still some commonality. This position avoids the worst problems about the incommensurability of meaning. We can say that one theory is right and another wrong and that they are about the same things, in the sense of primary denotation.

I do not want to get involved in a serious way in the debate about natural kind terms. It would take me too far from the main aim of the current project. So these very few, rather dogmatic, remarks will have to do. In Section 10.5, I discuss how this dual theory of reference can be integrated into the present semantical framework.

10.5   Domains

To give a semantics for quantification, I need to set out a domain for models in general and for the intended model in particular. The naïve approach to domains is one in which the domain is just made up of actual and perhaps possible individuals, like cats and dogs, and chairs and tables. In the naïve approach, names refer to their ordinary referents. For example, 'Venus', 'Hesperus', and 'Phosphorus' all refer to Venus. In Section 10.4, I suggest that natural kind terms should not be given their ordinary referents (or at least not only their ordinary referents) when they occur in theories, and the same holds for proper names.

One problem that occurs with proper names is that in reality they may co-refer, whereas in a theory they do not. Some ancient theories of astronomy may have 'Hesperus' and 'Phosphorus' being distinct objects. In the models for those theories, there are two entities that play the Hesperus and Phosphorus roles, respectively. In other theories, two names that in fact do not co-refer pick out the same individual. This problem about false distinctions and false identifications extends to natural kinds as well. Here is an example from ornithology. Until 1973, the Baltimore Oriole and Bullock's Oriole were thought to be distinct species, but observations of their interbreeding led them to be identified under the name 'Northern Oriole'. Let us say for a moment that this identification was correct. Then, 'Baltimore Oriole', 'Bullock's Oriole', and 'Northern Oriole' all refer to the same thing. But the pre-1973 theory treats 'Baltimore Oriole' and 'Bullock's Oriole' as referring to distinct species (and 'Northern Oriole' does not refer to anything). If I were to have natural kind terms merely refer to their actual referents (and the post-1973 theory were correct), then I would have to treat the pre-1973 theory as saying the same things about Baltimore Orioles, Bullock's Orioles, and Northern Orioles. But this would be to do serious violence to the pre-1973 theory.

The post-1973 theory, however, was incorrect. By 1995, genetic studies led to the claim that there was very little interbreeding between Bullock's Orioles and Baltimore Orioles, and so they were distinguished again and 'Northern Oriole' no longer referred to anything. The moral of this story is that we need a fairly flexible treatment of the reference of names and kind terms in order to do justice even to a fairly narrow range of theories.

In order to give the model theory the needed flexibility to treat individual names and natural kind terms that are identified in some theories and distinguished in others, I postulate a set of *intermediate objects* that stand semantically between the terms of the language and the things that occur in models. For example, the name 'Northern Oriole' picks out an intermediate object that determines a species in models for the 1973–1995 ornithological theory. In the model of the true theory of the actual world (i.e. the actual world itself), this intermediate object does not pick out anything. The intermediate object for 'water' picks out H_2O in the actual world, and different entities in different models. And so on.

An intermediate object is a role. It picks out, say, all the entities in models that play a certain role. In a model for an early chemical theory, there is some abstract object (such as a point) playing the role of the element phlogiston. The intermediate object *phlogiston* picks out that object from that model. Thus, at least in part, the intermediate object is a function that picks out entities from models that play particular roles. Thus, we can view the intermediate object as the role itself. At least in part, we can think of an intermediate objects as *partial* functions from models to elements in those models. I say that they are partial functions, because one would not want to say that given a particular role there is an object in every model that plays that role. Consider the intermediate object that is the role of the species *Baltimore oriole*. There is no object in any model for the standard theory of the atom that plays the role of being a Baltimore oriole. Thus, this intermediate object has no value when applied to these models. Whether we need an intermediate object to be more than just a function from models to occupants of models, I do not know, but I cannot see that for my present needs that they have to be more than mere functions.[3]

Using intermediate objects, I give a semantics for atomic formulas. Suppose that P is a unary predicate and i is an individual term. Then, $\mathfrak{M}, a \vDash_f Pi$ if and only if $f(i)$ is in $[\![P]\!]^a_{\mathfrak{M}}$, where $f(i)$ is in $\mathbb{P}$ and $[\![P]\!]^a_{\mathfrak{M}}$ is a subset of $\mathbb{P}$. The referents of all names are fixed. Thus, if n is a name, then $[\![n]\!]_{\mathfrak{M}}$ is some intermediary object in $\mathbb{P}$ and $f(n) = [\![n]\!]_{\mathfrak{M}}$ for all value assignments f.

In a model, $\mathfrak{M}$, the referents of all names are fixed. As I say above, we can fix the referent of some names to actual objects. If we add secondary domain (the actual objects), $\mathbb{A}$, to the model, and have a partial function ! from $\mathbb{P}$ into $\mathbb{A}$, then we can add an existence predicate to the language such that @ satisfies En if and only if ! ($[\![n]\!]_{\mathfrak{M}}$) is defined. I do not make any use of the existence predicate, and I treat the quantifiers as having no existential import, so I do not add these complexities to my model theory.

10.6   Identity

In order to complete the story of individual names and natural kind terms that I adumbrate in Section 10.5, I need to discuss identity. I have written about identity in relevant logic in the past [125, 129, 133], but every time I return to the subject I have looked at a different range of relevant logics. In [125], I developed a semantics for identity based on Fine's semantics for quantified relevant logic [64]. This semantics was compatible with a wide variety of logics. The theory in [129] is directed only at the logic R. And the theory in [133] is directed only at weak relevant logics such as those based on GTL (see Chapter 7). Here, I sketch a theory of identity that fits with the logic E as a logic of theory closure.

Identity, considered as a logical concept, has the following two properties:

1. Identity is an equivalence relation: it is reflexive, transitive, and symmetrical.
2. Identity warrants substitutions. A true identity $i = j$ justifies the inference from a statement A to a statement A' in which one or more instances of i are replaced with instances of j.

These two properties seem to sum up the logical behaviour of identity, but both properties have to be given further interpretation and a more precise formulation in order for them to be understood properly.

---

[3] Intermediate objects are rather like Carnap's intensional objects – which are functions from state descriptions to names [33]. But, I think, his intensional objects are total functions, whereas mine are partial.

Let us start with the properties that make a relation into an equivalence. First, let us look at reflexivity. The reflexivity postulate is, of course, $\forall x(x = x)$. Adopting reflexivity means only making it true, for every value assignment, in the master theory of every frame. I do not want reflexivity to hold in every theory. If we were to do so, given the semantics for implication, we would have $A \rightarrow \forall x(x = x)$ and $A \rightarrow i = i$ for an arbitrary formula A and name i.

With regard to transitivity, there are two alternatives: $\forall x \forall y \forall z(x = y \rightarrow (y = z \rightarrow x = z))$ and $\forall x \forall y \forall z((x = y \wedge y = z) \rightarrow x = z)$. I call the first of these 'strong transitivity' and the second 'weak transitivity', since the first entails the second in E. Suppose that we accept strong transitivity. Then, suppose that $m = e$ is true at the actual world, hence is true in the theory @. Thus, $m = e \rightarrow (e = e \rightarrow m = e)$ is true in @ and true in. But $\Box$@ is Σ, hence for any theory in which $e = e$, $m = e$. As I say in Section 10.5, this consequence is unacceptable. Weak transitivity does not have this issue, and I suggest that it be accepted as the principle of transitivity for identity.

As far as I can tell, there is nothing interesting to say about symmetry. I thus accept the standard principle of symmetry, $\forall x \forall y(x = y \rightarrow y = x)$.

The real problem of identity for relevant logics in general and for the logic of entailment in particular concerns substitution. Suppose that we were to add second-order quantifiers to our language. If we did, we could define identity in the standard manner as

$$i = j =_{df} \forall F(F(i) \leftrightarrow F(j)).$$

But this definition does not work well in the present semantics. For it entails the following schema:

$$i = j \rightarrow (A \rightarrow A).$$

This means that if $i = j$ is in a theory, then so is $A \rightarrow A$ for every formula A. And this implies that we cannot have an identity in a theory without its becoming a theory of entailment. So, I reject this definition of identity. What I want is a more subtle form of substitution.

A more relevant-looking form of substitution is given by the following full substitution axiom:

$$(i = j \wedge A(i)) \rightarrow A(j).$$

Unfortunately, this axiom is too strong. Suppose that n and m are names that actually co-refer. Then, $n = m$ is in the complete theory of the actual world, @. For any formula $A(i)$, $A(i) \rightarrow A(i)$ is also in @. Hence, by the full substitution axiom, $A(i) \rightarrow A(j)$ is in @ as well. Since the master theory, Σ, is just $\Box^-$@, $A(i) \rightarrow A(j)$ is in Σ. Therefore, if $A(i)$ is an arbitrary theory a, then so is $A(j)$. In other words, whatever a says about i it also says about j. This is exactly why I reject having the actual (or possible) objects as the domain of the model, and why I have replaced them with much metaphysically thinner objects. So, I reject full substitution.

The axiom that I think we should adopt is a weak substitution axiom:

$$(i = j \wedge A(i)) \rightarrow A(j),$$

where $A(i)$ does not contain any instances of $\rightarrow$. This axiom demands only that non-entailments are closed under substitution for identicals.[4]

Although the semantics for this logic has, in a sense, models with constant domains, there is also a sense in which domains are variable in this model. A domain for a theory can be taken to be the set of sets of intermediate objects that are said by the theory to be identical to one another. Let $[i]_a$ be the set of names j such that $i = j$ is satisfied by a. Then, the set of these (non-empty) equivalence classes could be treated as the domain of a. Or, if we wish to think of the domain as extra-linguistic, then it could be the set of intermediate objects that have the same value on a (i.e. for every model determined by a, if κ and θ have the same value, then κ is identical to θ according to a).

We can also add an existence predicate, E, to the language so that Ei satisfied by a if and only if $\exists x(i = x)$ is satisfied by T. Not all self-identities (statements of the form $i = i$) are in every theory. By defining existence this way, we can say that oxygen does not exist in the phlogiston theory and phlogiston does not exist in modern chemistry, since 'oxygen=oxygen' is not in the phlogiston theory, and 'phlogiston=phlogiston' is not in modern chemistry. Of course, we could merely take existence to be a primitive predicate, but defining it in terms of identity allows the identification of existence and the treatment of a set of intermediate objects as a single member of the domain.

10.7   Truth and Paradox

One use of this theory of domains and identity is in a treatment of the logical paradoxes. In Section 8.4, I discuss the paradoxes – in particular the Curry paradox – briefly. However, here I discuss them in more depth. The derivation of these paradoxes often depend on an inference of the following form. Where n is the name of a sentence,

$$\text{'}n\text{' refers to } A$$

to infer

$$T(n) \leftrightarrow A,$$

where $T(x)$ is a truth predicate. For example, let n be a name for the sentence $\neg T(n)$. In theories, such as those that include arithmetic and a truth predicate, we can prove that there are such theories. Thus, by the rule given above,

$$T(n) \leftrightarrow \neg T(n).$$

This rule, then, in an appropriately complicated theory, delivers a version of the liar paradox.

---

[4] In [133] I accept the principle $(i = j \wedge (A(i) \rightarrow B(i))) \rightarrow (A(j) \rightarrow A(j))$, which I reject here. The idea is that this principle helps to create a parallel between the treatment of identity and the treatment of bi-implication in relevant logics. But that is not a goal here and this principle creates problems with regard to the closure of theories.

As I have said in Sections 10.5 and 10.6, that just because a name happens actually (or even necessarily) to refer to a particular entity e, does not mean that it refers to that entity in all theories. I maintain the same line for sentences, and this means that I hold that this derivation is invalid. Just because a name actually refers to a sentence does not imply that in any theory that the truth predicate applies to the name the sentence holds.

This view, I think, is quite intuitive when it comes to theories in which names of sentences are codes of some sort, such as Gödel numbers. The weak view of identity that I adopt prevents the actual coordination of a number and a sentence from infecting other theories. Suppose that n and m are two names for a particular sentence A and that $n = m$ is true in Σ. If I were to adopt the principle FullSub or even the full substitution rule, we would be able to prove that $T(n) \leftrightarrow T(x)$. However, we cannot prove this from the weak substitution principle that I do adopt.

The issue, however, becomes more difficult if the language includes some sort of quotation device. I use corner quotes, $\ulcorner A \urcorner$, to produce a name of the sentence A. It would seem that $\ulcorner A \urcorner$ must refer to A. What else could it denote? The answer I think is that it could fail to denote. Not all theories are about language. Consider the following statement from Newton's physics.

(NL3) the mutual actions of two bodies upon one another is always equal, and direct to contrary parts.

NL3 is one of the formulations of Newton's third law of motion in his *Principia* [153, p 13]. The statement of the truth of NL3 is in some way implied by NL3 but has a different subject matter. In addition to being about the physical universe, it is also about the truth of a sentence in eighteenth century English. I claim that whereas NL3 is in the theory set out in Newton's *Principia*, the statement 'NL3 is true' is not.[5]

For this reason, I reject one half of Tarski's Truth-scheme,[6] that is, I reject:

$$A \to T(\ulcorner A \urcorner).$$

Note that what I am claiming is that the Truth-scheme is false when it is formulated using co-entailment as its main connective. I am not claiming that it is false when formulated using some weaker conditional. The total truths at the actual world, @, almost certainly includes a theory of truth. This theory might include a version of the Truth-scheme. Clearly the choice of such a theory must be compatible with the adoption of E as the theory of entailment. E has the law of excluded middle as a theorem, and so the theory of truth that is chosen must be compatible with excluded

---

[5] I think that even the instance of the law of excluded middle, $T(NL3) \vee \neg T(NL3)$, fails to hold in the *Principia*, since 'NL3' or other names for that law are not among the things that that book is about. So, that book does not really describe a theory of classical logic. It can be understood classically only if we restrict the language to contain only terms that do refer to things that book is about.

[6] This is usually called the "T-scheme" for both 'truth' and 'Tarski', but since I call the scheme $\Box A \to A$ the "T-scheme", I need another name for Tarski's famous principle.

middle. This excludes theories of truth based on three-valued logic, such as Kripke's theory [94], and theories based on four-valued logic, such as J. M. Dunn's view [51] and Albert Visser's theory [213]. There are various other theories that are compatible with excluded middle, such as the Gupta–Belnap revision theory [78]. It would be interesting to try to formulate a theory of truth based on Graham Priest's inconsistent theory of truth [163], which uses LP as its base but is extended with E (but that also rejects transparency). I am not certain whether such a theory could be made coherent, although I cannot at the moment see why not.

The issue concerning the other direction of the Truth-scheme is more complicated. Where n is the code of $A \rightarrow B$, it would seem that

$$T(n) \rightarrow (A \rightarrow B)$$

is a violation of both the variable sharing criterion and a fallacy of modality. Thus, it would seem best rejected in a theory of entailment. But, when we consider a quoted name such as $\ulcorner A \rightarrow B \urcorner$, it is less obvious what to do. Although, strictly speaking, $T(\ulcorner A \rightarrow B \urcorner)$ is not an entailment, it does seem to be about entailment, and so

$$T(\ulcorner A \rightarrow B \urcorner) \rightarrow (A \rightarrow B)$$

does not seem to be a fallacy of entailment in the same sense. Similarly, the propositional variables in the antecedent of the entailment may be within quotation marks, but they are there, and so the violation of relevance does not seem as heinous as in the previous example.

There is, however, still a problem with this direction of the Truth-scheme. If we reject both directions of the Truth-scheme for names other than quotational names and right-to-left direction for all names, then it would seem that the status of the Truth-scheme (formulated with entailment) is seriously in doubt. The support even for the one direction of the scheme with one sort of name seems quite weak, and I think we should reject it. This, again, is not to say that some version of the Truth-scheme might be a logical truth, but not formulated as a co-entailment or even as an entailment.

The same sort of reasoning that allows my view to avoid the liar paradox also enables the view to avoid the Curry and v-Curry paradoxes. In order to explain why this is so, I give a brief explanation of the Curry paradox.

Consider the following sentence:

If this sentence is true, then the moon is made of green cheese.

Here, 'this sentence' refers to the entire conditional (not just its antecedent). We can formalise it as

$$(C)\ T(\ulcorner C \urcorner) \rightarrow m.$$

If we accept all instances of the Truth-scheme, then we can derive that the moon is made of green cheese:

1	$T(\ulcorner C\urcorner)_{\{1\}}$	hypothesis
2	$T(\ulcorner C\urcorner) \leftrightarrow C_\emptyset$	Truth-scheme
3	$C_{\{1\}}$	1,2, $\leftrightarrow$E
4	$C \leftrightarrow (T(\ulcorner C\urcorner) \to m)_\emptyset$	definition of C
5	$T(\ulcorner C\urcorner) \to m$	3,4, $\leftrightarrow$E
6	$m_{\{1\}}$	1,5, $\to$E
7	$T(\ulcorner C\urcorner) \to m_\emptyset$	1–6, $\to$I
8	$C \leftrightarrow (T(\ulcorner C\urcorner) \to m)_\emptyset$	definition of C
9	$C_\emptyset$	7,8, $\leftrightarrow$E
10	$T(\ulcorner C\urcorner) \leftrightarrow C_\emptyset$	Truth-scheme
11	$T(\ulcorner C\urcorner)_\emptyset$	9,10, $\leftrightarrow$E
12	$m_\emptyset$	7,11, $\to$E

My rejection of the Truth-scheme blocks its use in lines 2 and 10 of this proof.

Jc Beall and Julien Murzi, however, have produced a version of the Curry para-dox that does not appeal to the Truth-scheme for its derivation [12].[7] They added to the language a binary predicate, Val. '$Val(\ulcorner A\urcorner, \ulcorner B\urcorner)$' means that B can be validly proven from the single premise A. They also introduced the sentence π such that

$$(\pi)\ Val(\ulcorner \pi\urcorner, \ulcorner m\urcorner).$$

Again, 'm' is an abbreviation for 'the moon is made of green cheese'. Here is the proof of the validity-Curry (or 'v-Curry') paradox:

1	$\pi_{\{1\}}$	hypothesis
2	$\pi \leftrightarrow Val(\ulcorner \pi\urcorner, \ulcorner m\urcorner)_\emptyset$	definition of π
3	$Val(\ulcorner \pi\urcorner, \ulcorner m\urcorner)_{\{1\}}$	1,2, $\leftrightarrow$E
4	$m_{\{1\}}$	1,3, ValE
5	$Val(\ulcorner \pi\urcorner, \ulcorner m\urcorner)_\emptyset$	1–4, ValI
6	$\pi \leftrightarrow Val(\ulcorner \pi\urcorner, \ulcorner m\urcorner)_\emptyset$	definition of π
7	$\pi_\emptyset$	5,6, $\leftrightarrow$E
8	$m_\emptyset$	5,7, ValE

This proof forces me to deny that there is a sentence π that holds in exactly those theories that contain $Val(\ulcorner \pi\urcorner, \ulcorner m\urcorner)$. But this is not difficult for me to deny. I hold

---

[7] Another version of their paradox is discussed in Section 5.2.

that the mechanisms that one would use to prove that such a π exists, even if there is a validity predicate of this sort are not available in the logic of entailment (or even in the true theory of the actual world). The weaker thesis,

$$\pi \equiv Val(\ulcorner\pi\urcorner, \ulcorner m\urcorner),$$

might be provable (in the complete theory of the actual world), where $\equiv$ is material equivalence, but not the stronger thesis in which the material biconditional is replaced with a co-entailment. In the present semantics, this proof is blocked by the rejection of lines 2 and 6.[8]

10.8   The Barcan Formula

The admissible set semantics for quantification, with no added conditions, characterises a system that is rather weak. In particular, it does not contain the intensional confinement scheme:

$$\forall x(A \rightarrow B(x)) \rightarrow (A \rightarrow \forall x B(x)),$$

where x is not free in A. In the semantics for the logic R, we can derive intensional confinement. In R, the fusion connective ($\otimes$) is definable, that is, $A \otimes B =_{df} \neg(A \Rightarrow \neg B)$. Given this definition, the following is provable in R:

$$((A \otimes B) \Rightarrow C) \Leftrightarrow (A \Rightarrow (B \Rightarrow C))$$

in both E and R, it is a rule that the universal quantifier distributes over the conditional of the logic. Thus, we have this derivation in R:

$$\frac{\vdash \forall x(A \Rightarrow B(x)) \Rightarrow (A \Rightarrow B(x))}{\dfrac{\vdash (\forall x(A \Rightarrow B(x)) \otimes A) \Rightarrow B(x)}{\dfrac{\vdash (\forall x(A \Rightarrow B(x)) \otimes A) \Rightarrow \forall x B(x)}{\vdash \forall x(A \Rightarrow B(x)) \Rightarrow (A \Rightarrow \forall x B(x))}}}$$

However, fusion is not definable in E, and I give reasons in Chapter 9 not to add it. In particular, if it is added to E, it allows us to derive

$$A \rightarrow (B \rightarrow (A \otimes B)).$$

This is a fallacy of modality and, more importantly, its inclusion in the system would force every non-empty theory to include at least one entailment. Thus, every theory would be a theory of entailment, and this is very counterintuitive.

There are, however, reasons to accept intensional confinement. Intensional confinement has an instance in the following scheme:

$$\forall x(t \rightarrow B(x)) \rightarrow (t \rightarrow \forall x B(x)).$$

---

[8] A referee commented that it would be better if the exact same solution was used for the v-Curry paradox as for the Curry paradox. I agree that a uniform solution is almost always better. But I think that in a sense the same solution is being used here. Banning the mechanisms that give rise to the paradoxes from the logic of entailment is a good motivation for rejecting the Truth-scheme.

Appealing to the definition of logical necessity, this can be rewritten as

$$\forall x \,\Box\, B(x) \rightarrow \Box \forall x \, B(x).$$

This latter formula is immediately recognisable to most philosophers of logic as the Barcan formula. In Chapter 5, I briefly discuss the Barcan formula. It is sometimes suggested as a horizontalisation of the rule of universal generalisation, viz.,

$$\frac{\vdash B(x)}{\vdash \forall x\, B(x)}.$$

The premise of the rule says that an arbitrary individual x is proven to be B. This is a proof technique that mathematicians and logicians often use. In modal logic, one proves something holds in an arbitrary world, say, then concludes that it holds in all worlds. In suggesting that the Barcan formula be taken to be the horizontalisation of universal generalisation, one is suggesting that 'an arbitrary thing is proven to be B' is adequately represented by 'everything is logically necessary B'. I do not think that the universal quantifier (whatever its scope) really captures the reasoning here. It is the arbitrariness of the thing that is important. What we need to capture this reasoning is a more nuanced theory of arbitrary object (such as Fine's [62]). I do not produce such a theory here, but I do reject the idea that the Barcan formula adequately captures universal generalisation. Moreover, I cannot see how intensional confinement (or the Barcan formula) can be motivated in terms of the intuitive properties of theories. Thus, I reject intensional confinement as a principle of entailment.

I do, however, accept a *rule* of intensional confinement, that is,

$$\frac{\vdash A \rightarrow B(x)}{\vdash A \rightarrow \forall x\, B(x)}$$

where x is not free in A. This rule is motivated by the satisfaction condition for the universal quantifier. It can be proven valid in the class of models without any added conditions.

10.9   The Universal Quantifier and Natural Deduction

Kit Fine's book, *Reasoning with Arbitrary Objects*, contains a deep analysis of the treatment of the universal quantifier in natural deduction systems. In the introduction, he gives a rather concise statement of the reasoning that allows us to draw universal conclusions:

As is well known, there exist certain informal procedures for arguing to a universal conclusion and from an existential premiss. We may establish that all objects of a certain kind have a given property by showing that an arbitrary object of that kind has that property .... [62, p 1]

Fine claims that the standard natural deduction universal quantifier introduction rule ($\forall$I) implements this form of reasoning. The universal quantifier introduction rule is the following:

$$A_\alpha$$

$$x \mid A_{\{k\}} \qquad \text{hypothesis}$$

$$\vdots$$

$$B(x)_{\{k\}}$$

$$\forall x\, B(x)_\alpha \qquad \forall\text{I}$$

where x does not occur free in A. The subproof is strict in two senses. First, no formula with a free x can be reiterated from a superior proof into that subproof, and second, only entailments can be reiterated from superior proofs into the subproof. What the rule says is that if a formula A is in a theory, and A entails $B(x)$ for an arbitrary variable x, then $\forall x\, B(x)$ is also in that theory. Thus, this rule is an implementation in the natural deduction system of the inclusion condition for the universal quantifier.

The Anderson and Belnap system has a different, and stronger, introduction rule for the universal quantifier. This rule is the following:

$$x \mid \vdots$$

$$B(x)_\alpha$$

$$\forall x\, B(x)_\alpha \qquad \forall\text{I}$$

The subproof labelled with 'x' is only strict in my second sense. That is, no formula with a free x can be reiterated into it. But any other formula (even non-entailments) can be reiterated into such subproofs. The Anderson and Belnap rule allows the proof of the thesis of intensional confinement:

1	$\forall x(A \rightarrow B(x))_{\{1\}}$	hypothesis
2	$A_{\{2\}}$	hypothesis
3	$x \mid \forall x(A \rightarrow B(x))_{\{1\}}$	1, reiteration
4	$A \rightarrow B(x)_{\{1\}}$	3, $\forall$E
5	$A_{\{2\}}$	2, reiteration
6	$B(x)_{\{1,2\}}$	4,5, $\rightarrow$E
7	$\forall x\, B(x)_{\{1,2\}}$	3–6, $\forall$I
8	$A \rightarrow \forall x\, B(x)_{\{1\}}$	2–7, $\rightarrow$I
9	$\forall x(A \rightarrow B(x)) \rightarrow (A \rightarrow \forall x\, B(x))_\emptyset$	1–8, $\rightarrow$I

If one wants to include intensional confinement in the logic, perhaps the most straightforward way to do so is by adding the condition to the model theory that the set of propositions in a frame is closed under fusion. This characterises exactly the logic with intensional confinement taken as the only additional axiom. I have, however, argued in Chapter 7 that we should not include fusion in the language, since its inclusion results in the provability of some fallacies of modality. Thus, we have another reason to reject intensional confinement (and the Barcan formula, and Anderson and Belnap's ∀I rule).[9]

Despite the rejection of the thesis of intensional confinement, a rule of intensional confinement is derivable in the natural deduction system (and in the semantics):

$$\frac{A \to B(x)}{A \to \forall x B(x)}$$

where x is not free in A. Suppose that $A \to B(x)$ is a theorem. Here is a proof of $A \to \forall x B(x)$:

1	$A_{\{1\}}$	hypothesis
2	x $A_{\{2\}}$	hypothesis
3	$A \to B(x)_\emptyset$	theorem
4	$B(x)_{\{2\}}$	2,3, →E
5	$\forall x B(x)_{\{1\}}$	1,2–4, ∀I
6	$A \to \forall x B(x)_\emptyset$	1–5, →I

Note that the theorem $A \to B(x)$ can be brought into a subproof that is labelled with x despite the fact that x is free in that formula. It is a theorem of the logic that $(\forall x(A \to B(x)) \wedge A) \to B(x)$, and hence, by the rule of intensional confinement,

$$(\forall x(A \to B(x)) \wedge A) \to \forall x B(x)$$

is also a theorem. This theorem is closely related to, but slightly weaker than, the intensional confinement scheme.

In the Prawitz-style system, the rules for the universal quantifier are quite straightforward. The introduction rule is:

$$\frac{\Sigma \vDash A \to B(x) \qquad a \vDash A}{a \vDash \forall x B(x)}$$

where x does not occur in A. The elimination rule, of course, is

$$\frac{a \vDash \forall x A(x)}{a \vDash A(\tau)}$$

where x is free for τ in $A(x)$. To illustrate the way in which these rules work, here is a proof of the aggregation thesis, which is an axiom of quantified E (see Section 10.11). I use 'x_0' as the individual variable to avoid confusion between theory variables and individual variables.

---

[9] We could add fusion on the semantic level without adding it in the language. Then, I think, we would have to understand propositions not as the content of theories (because some propositions would not be expressed by statements in theories) but rather as functional roles that theories play in a frame. This is an interesting reading of propositions, but does not go along with the other claims I have made for propositions in my interpretation of the semantics.

$$\dfrac{\dfrac{[x \vDash \forall x_0 A(x_0) \wedge \forall x_0 B(x_0)]^1}{x \vDash \forall x_0 A(x_0)}}{x \vDash A(x_0)} \qquad \dfrac{\dfrac{[x \vDash \forall x_0 A(x_0) \wedge \forall x_0 B(x_0)]^1}{x \vDash \forall x_0 B(x_0)}}{x \vDash B(x_0)}$$

$$\dfrac{\dfrac{x \vDash A(x_0) \wedge B(x_0)}{\Sigma \circ x \vDash A(x_0) \wedge B(x_0)}}{\dfrac{\Sigma \vDash (\forall x_0 A(x_0) \wedge \forall x_0 B(x_0)) \to (A(x_0) \wedge B(x_0))}{}}{}^1 \qquad [y \vDash \forall x_0 A(x_0) \wedge \forall x_0 B(x_0)]^2$$

$$\dfrac{\dfrac{y \vDash \forall x_0 (A(x_0) \wedge B(x_0))}{\Sigma \circ y \vDash \forall x_0 (A(x_0) \wedge B(x_0))}}{\Sigma \vDash (\forall x_0 A(x_0) \wedge \forall x_0 B(x_0)) \to \forall x_0 (A(x_0) \wedge B(x_0))} {}^2$$

10.10   The Existential Quantifier

The standard truth condition for the existential quantifier makes $\exists x A(x)$ true if and only if $A(x)$ is true for some valuation of x. However, the corresponding satisfaction condition is not reasonable as a closure condition for theories. A theory of the ascendency of mammals on Earth from little shrew-like animals might include the claim that some catastrophe destroyed the larger dinosaurs, but it might not specify an exact event. Cases of indeterminacy of this kind are common in theories.

In order to motivate a satisfaction condition, consider what it means for an existentially quantified formula to entail a formula:

$$(\exists 1) \quad \exists x A(x) \Vdash B \text{ if and only if } A(c) \Vdash B$$

where c does not occur in $A(x)$ or in B. From ($\exists 1$) and the transitivity of $\Vdash$, we can derive the following conditional:

$$(\exists 2) \quad \Gamma \Vdash \exists x A(x) \text{ and } A(c) \Vdash B \text{ only if } \Gamma \Vdash B,$$

where c does not occur in $A(x)$ or in B. Also,

$$(\exists 3) \quad A(c) \Vdash \exists x A(x).$$

Putting together ($\exists 2$) and ($\exists 3$), and generalising, we get

$$(\exists 4) \quad \Gamma \Vdash \exists x A(x) \text{ if and only if } \Gamma \Vdash B$$

for all B such that there is some variable c and $A(c) \Vdash B$, where c is not in $A(x)$ or in B.

In order to implement ($\exists 4$) in the model theory, I say that a theory a in a model satisfies $\exists x A(x)$ if and only if a is in every proposition entailed by every instance of $A(x)$. More formally, $a \vDash_f \exists x A(x)$ if and only if, for every g (where g differs only from f perhaps in its assignment of x), if $[\![A(x)]\!]_g \subseteq \pi$, then a is in π.

In the proof theory, the existential quantifier is given the usual definition in terms of the universal quantifier and negation:

$$\exists x A(x) =_{df} \neg \forall x \neg A(x).$$

This definition, together with the principles of double negation, makes the existential quantifier and the universal quantifier as *negation duals* of one another. This means that

$$Q_1 \neg A \leftrightarrow \neg Q_2 A,$$

where Q_1 can be either $\forall x$ or $\exists x$ as long as Q_2 is the other quantifier. Conjunction and disjunction are negation duals as well.

The duality of the quantifiers and conjunction and disjunction goes well beyond their relations in terms of De Morgan's laws. Their duality can be seen in comparing their proof rules in the natural deduction system.

$$
\begin{array}{|l}
\exists x \, A(x)_\alpha \\[4pt]
\quad \begin{array}{|l}
A(c)_{\{k\}} \qquad \text{hypothesis} \\[4pt]
\vdots \\[4pt]
B_{\{k\}}
\end{array} \\[4pt]
B_\alpha
\end{array}
$$

where c does not occur before the subproof and c does not occur in B. The existential quantifier introduction rule is, of course,

$$A(\tau)_\alpha$$
$$\downarrow$$
$$\exists x \, A(x)_\alpha$$

where τ is free for x in $A(\tau)$.

The existential quantifier rules in the Prawitz-style system are, like the universal quantifier rules, quite straightforward. The existential quantifier elimination rule is

$$\frac{\Sigma \vDash A(x) \rightarrow B \qquad a \vDash \exists x \, A(x)}{a \vDash B}$$

where x does not occur free in B. The introduction rule is

$$\frac{a \vDash B(\tau)}{a \vDash \exists x \, B(x)}$$

where τ is free for x in $B(\tau)$.

10.11   The Logic QE

The language extends the language for propositional logic. It includes predicate constants for relations of every finite arity written as 'P_m^n' to indicate the mth predicate of the nth arity, individual constants, $c_0, c_1, \ldots$, individual variables, $x_0, x_1, \ldots$, the quantifier $\forall$, as well as parentheses and the propositional connectives. The existential quantifier is defined in the usual manner:

$$\exists x \, A \;=_{df}\; \neg \forall x \neg A.$$

To obtain the logic QE, I add the following schemes and rule to the axiomatic basis for E (see Section 9.9):

UI $\forall x \, A(x) \to A(\tau)$, where x is free for τ in $A(x)$ (Universal Instantiation)
Agg $(\forall x \, A \wedge \forall x \, B) \to \forall x (A \wedge B)$ (Aggregation)

The rule is the intensional confinement rule:

$$\frac{\vdash A \to B}{\vdash A \to \forall x \, B}$$

where x is not free in A. The simple rule of universal generalisation is derivable from universal instantiation and the intensional confinement rule:

$$\frac{\dfrac{\dfrac{\dfrac{\vdash A}{\vdash t \to A}}{\vdash t \to \forall x \, A} \qquad \vdash t}{\vdash \forall x \, A}}{}$$

The principle of aggregation $((\forall x \, A(x) \wedge \forall x \, B(x)) \to \forall x (A(x) \wedge B(x)))$ is also derivable:

$$\frac{\dfrac{\dfrac{\vdash \forall x \, A(x) \to A(x)}{\vdash \forall x (A(x) \wedge \forall x \, B(x)) \to A(x)} \qquad \dfrac{\vdash \forall x \, B(x) \to B(x)}{\vdash \forall x (A(x) \wedge \forall x \, B(x)) \to B(x)}}{\dfrac{\vdash \forall x (A(x) \wedge \forall x \, B(x)) \to (A(x) \wedge B(x))}{\vdash \forall x (A(x) \wedge \forall x \, B(x)) \to \forall x (A(x) \wedge B(x))}}}{}$$

10.12   Model Theory for QE

A QE frame is a septuple $\langle \Upsilon, \Sigma, \circ, \perp, \leq, D, \mathsf{Prop} \rangle$ such that $\langle \Upsilon, \Sigma, \circ, \perp, \leq \rangle$ is an E frame, D is a non-empty set (the domain of quantification), and Prop is a set of sets of theories that are closed upwards under $\leq$ and if, for all $a \in \Upsilon$ and $\Phi \in \mathsf{Prop}$, if $S(a) \subseteq \Phi$, then $a \in \Phi$.

Before I define the concept of a QE-model, I need a few other definitions. A *value assignment* of variables is a function from the natural numbers into the domain. Thus, for example, if $f(0) = i$, then the value of x_0 according to f is i. An *interpretation V* of the language on a frame attributes to each n-place predicate letter and theory a set of n-tuples from D and to each individual name V assigns a member of D. For any term τ, $V_f(\tau)$ is $V(\tau)$ if τ is a name and $f(x_m)$ if τ is x_m.

A QE model is a pair $\langle \mathfrak{F}, V \rangle$, where $\mathfrak{F}$ is a QE frame and V is an interpretation. V determines a satisfaction relation $\vDash$ such that the following inductive clauses hold:

- $a \vDash_f F \tau_1 \ldots \tau_n$ if and only if $\langle V_f(\tau_1), \ldots, V_f(\tau_n) \rangle \in V(a, F)$
- $a \vDash_f A \wedge B$ if and only if $a \vDash_f A$ and $a \vDash_f B$
- $a \vDash_f \neg A$ if and only if for all b, if $b \vDash_f A$, then $a \perp b$

- $a \vDash_f A \to B$ if and only if for all b, if $b \vDash_f A$, then $a \circ b \vDash_f B$
- $a \vDash_f \forall x A(x)$ if and only if there is a proposition π, $a \in \pi$ and for all x variants g of f, $\pi \subseteq \bigcap_g [\![A(x)]\!]_g$

QE is not sound over the class of QE models. Rather, it is sound and complete over the set of semantically adequate QE models. A QE model is said to be *semantically adequate* if and only if for every formula A and every value assignment f, $[\![A]\!]_f$ is in Prop (where $[\![A]\!]$ is the set of theories a such that $a \vDash_f A$). In the Appendix, I sketch proofs of the soundness and completeness theorems for QE.

In the Appendix, however, I define QE frames and models slightly differently from here. In the Appendix, I follow [134] in postulating a set of propositional functions and use these in the definition of a frame to determine a class of propositions such that all models based on such frames are semantically adequate. This is more complicated and, from a philosophical point of view, no more conceptually compelling than the approach I use here. From a mathematical point of view, it is preferable to have a class of frames which characterises the logic. This we do not have in the current model theory. I present the current view here because it is much simpler to state and from a philosophical point of view it incorporates the same ideas.

In the Appendix, I extend the worlds theorem for E to hold for QE. I show that in the canonical model there is a theory W such that $\Box^- W = \Sigma$, where Σ is the set of theorems of QE and W is prime and all-complete. That is, if $A(c)$ is in W for all names c, then $\forall x A(x)$ is in W (where c is free for x in $A(c)$). This shows that it is plausible that the set of true entailments is captured exactly by the logic QE.[10]

In order to add identity to the semantics, I use a three-place relation $\approx$. This relation holds (or fails to hold) between two members of the domain and a theory. '$i \approx_a j$' means that i is identical to j according to a. The binary relation $\approx_a$ is symmetrical and transitive (the relation $\approx_\Sigma$ is also reflexive). Conditions need to be added to make the substitution axiom valid. I construct such conditions in the context of Fine's semantics for quantified relevant logic [125]. Goldblatt [76], Shawn Standefer [198], and Nicholas Ferenz [60] do so in the framework of Goldblatt's and my semantics.

10.13   Propositional Quantification

One natural extension of the logic is to add *propositional quantifiers*. As the name suggests, the variables in propositional quantifiers range over propositions. What is natural, with regard to a logic of entailment, of adding propositional quantifiers is that this allows the theorems of the logic to explicitly state the general logical truths. For example, with propositional quantifiers, one can write

$$\forall p(p \to p)$$

---

[10] Note that there is a difference between what I show and the stronger claim that QE is characterised by the class of its prime and all-complete extensions. I am not sure whether this is the case. Meyer and Dunn [144] proved that the logic EQ, which is QE together with the axioms $\forall x(A \lor B) \to (A \lor \forall x B)$ and $\forall x(A \to B) \to (A \to \forall x B)$ (where x is not free in A) was characterised by the class of its prime, all-complete, and consistent extensions, but I am not sure that this is true of QE.

and

$$\forall p \forall q \forall r ((p \to q) \to ((q \to r) \to (p \to r))).$$

Without propositional quantifiers, we express generality by using schemata, such as

$$A \to A$$

and

$$(A \to B) \to ((B \to C) \to (A \to C)).$$

However, these are not statements in the object language, that is, they are not in the theory of entailment itself. Thus, the theory, without propositional quantifiers, does not state that the truths of logic are themselves general. This is certainly a problem with considering a logic to be the *theory* of entailment.[11]

Adding propositional quantifiers, however, does not enable us to eliminate schemata altogether from our formulation of the logic. All the truths of the propositional logic can be represented as generalities in the logic with propositional quantifiers. We do need, however, to add schemata as well, such as

$$B(A) \to \exists p B(p),$$

where A is free for p in $B(A)$, and

$$\forall p B(p) \to B(A),$$

where p is free for A in $B(p)$. The corresponding general truths $(\forall q (B(q) \to \exists p B(p))$ and $\forall q (\forall p B(p) \to B(q))$, with the necessary side conditions) are theorems as well. Without the schemata, or some similar schemata, we cannot derive these two schemata from the general truths in the axiom system. Fortunately, the elimination of schemata is not itself a necessary goal. What we need is to be able to state the laws of logic as general truths. The fact that there are also some schemata needed to be able to do this is not a real problem, as far as I can see.[12]

In the second edition of *Principia Mathematica*, Russell constructed a logic in which all variables are bound and there are no names or parameters. One of his motivations for doing so was that he held that names do not belong in a *pure* logic. The idea is that logic should only have a subject matter in the sense that it is about everything and that no theorem of the logic should be about specific things. It seems to me that a general logic of entailment should be a pure logical system. If we allow free variables or include parameters in our language, however, it seems that we can have a pure logic in this sense and also formulate principles such as existential generalisation and universal instantiation. If like Russell, we require that all variables be bound in well-formed formulas, then the closest we can come to a principle of universal instantiation is $\forall x (\forall y A(y) \to A(x))$, and the principle that allows us to infer from

---

[11] Anderson, Belnap, and Dunn [3], hold that the extension of E to include propositional quantifiers is even more natural than its extension with individual quantifiers for this reason.

[12] We can of course eliminate schemata by using proper axioms and a rule of uniform substitution, but the rule is itself schematic, and so this tactic does not count as a true removal of schemata.

$\forall y A(y)$ to any particular thing's being A is an extra-logical principle. If we allow parameters or free variables, the extra-logical device that we need is the assignment of parameters and free variables to things in our domain. This, in my opinion, is something any theory of language needs, and hence cannot be thought to make impure a logic of entailment.

The semantics for relevant logics with propositional quantifiers has been given by Goldblatt and our student, Michael Kane [77]. This semantics is an admissible semantics just like the semantics for individual quantifiers. A value assignment is a function f from propositional variables into the set of propositions. A formula $\forall p A(p)$ is satisfied by a theory a according to a value assignment g if and only if there is some proposition in a that entails every proposition $[\![A(p)]\!]_g$, where g is just like f except perhaps in its assignment of p.

10.14   Second-Order Quantification

For both propositional and second-order quantification, the class of full models is not characterised by a finite set of axioms. In addition, the compactness theorem fails for the logic of full models. So, it is not appropriate as the logic of entailment. Nevertheless, there are interesting questions about full second-order relevant logic that have not yet been explored.

Alonzo Church makes a point about second-order logic that is similar to Tarski's argument concerning models for first-order arithmetic (see Chapter 6):

It is true that the non-effective notion of consequence, as we have introduced it in theoretical syntax, presupposes a certain absolute notion of ALL propositional functions of individuals. But this is presupposed also in classical mathematics, especially classical analysis, and objections against it lead to such modifications of classical mathematics as mathematical intuitionism ... or the partial intuitionism of Herman Weyl's *Das Kontinuum*. [38, p 326n]

In analysis, mathematicians often talk about all sets of real numbers. For example, the Dedekind completeness of the real numbers says that every set of real numbers that has an upper bound has a least upper bound. This is standardly interpreted as saying that every set of real numbers is such that if it has an upper bound, it has a least upper bound, and so this important property of the reals is stated by quantifying over every set of real numbers. If classical analysis is to be reconstructed in a logical language, and (as in *Principia Mathematica*) sets of reals are constructed from propositional functions, then it is essential to allow variables that range over all propositional functions of reals. This, however, creates a theory that is not axiomatisable. As Church says, this logic has a "non-effective notion of consequence". In this book, I am not advancing a revision of classical mathematics, but rather I think of the logic of entailment as a tool to investigate any theory, including classical and non-classical mathematical theories alike (see Chapter 11). The question, then, is how I am to advance an axiomatic theory that can do this.

The answer has to do with the separation of the theory of entailment from the theories that it investigates. Full second-order logic is a theory. It is unaxiomatisable.

However, it may be just one of the theories in the model of entailment. Its unaxiomatisability does not imply that the theory of entailment is unaxiomatisable. Recall that the objects in the domain of the model of entailment are intermediary objects. All the intermediary objects that are in the domain of the theory of full second-order logic are in the domain of the model of entailment. But, these objects do not have the properties in the model as a whole that they are given in the theory of second-order logic and so they do not force the unaxiomatisability of the theory of entailment.

As Philip Kremer has shown, the class of full models of relevant logics with propositional quantifiers characterises a logic that cannot be axiomatised [91]. It seems that the same proof goes through for second-order relevant logics. Thus, the model as a whole cannot quantify over all subsets of its domain. The model theory for second-order entailment that meets the needs of my position must be a general model in Henkin's sense.

11 Entailment and Reasoning

11.1 How to Study Theories Using Entailment Logic

There is now an extensive literature on the use of relevant logics to formulate mathematical theories, and a much smaller literature on the use of relevant logic to study scientific theories. However, this literature largely looks at what happens when we formulate a theory, such as arithmetic or set theory, using a relevant logic as a base [25, 52, 69]. To use a logic as a base, one adds to the axiomatic basis of a relevant logic the axioms for the mathematical or scientific theory in question. Although the creation of relevant mathematical theories is a very important topic for the study of relevant logic, I am not concerned with this sort of project in the present book. What I am interested in is using a relevant entailment logic to study theories based on a variety of logics. In the models that I discuss in Chapters 7–10, the logical theory (Σ) is separated from other theories that it can be used to study. In this chapter, I look more at the use of a relevant logic to study theories that are closed under other logics.

In Chapter 8, I briefly discuss theories based on weaker systems than E and theories that do not include all the principles of E, such as theories based on intuitionist or quantum logic. In this chapter, I am mostly interested in the use of E to examine logics that contain principles that are not in E. In particular, I am interested in studying theories based on classical logic and theories based on logics that include infinitary rules.

11.2 Internal Conditionals

Many, if not most, mathematical theorems and lemmas are or can be stated as conditionals. Here are a few examples of well-known mathematical theorems stated in conditional form.

- (Pythagoras) If a triangle with sides h, a and b is a right-angle triangle with hypotenuse h, then $h^2 = a^2 + b^2$.
- (Fundamental Theorem of Algebra) If a non-constant polynomial has complex coefficients, then it has at least one complex root.
- (Intermediate Value Theorem) If a function f is continuous on the interval $[a,b]$, then for every c between $f(a)$ and $f(b)$, there is some $x \in [a,b]$ such that $f(x) = c$.

Some of these theorems can be understood as entailments, but with additional information in the antecedents. But some mathematical conditionals express conditionals that are weaker than entailment. These are what I call "internal conditionals" because they express a connection between propositions that is internal to a theory. Entailments, as I have suggested, are about the closure conditions of theories in general. Internal conditionals are primarily about the theory in which they occur.

Internal conditionals have (at least) two important properties. They are transitive and ponable. I use the "maps to" arrow, $\mapsto$, to represent the internal conditional. The form of transitivity that we need this arrow to have is one that makes valid the following proof process:

$$\frac{a \models A \mapsto B \qquad a \models B \mapsto C}{a \models A \mapsto C}.$$

Mathematicians and logicians often prove that a series of conditions are equivalent to one another. They do so by proving the first implies the second, the second implies the third, and so on, and end by showing that the last implies the first. This is done so often that in seminars in mathematics departments, one often sees the speaker use the abbreviation "TFAE" for "the following are equivalent".

That a conditional is ponable means that it is governed by an internal rule of modus ponens (which is sometimes called "pseudo modus ponens" – see Section 8.3):

$$\frac{a \models A \mapsto B \qquad a \models A}{a \models B}.$$

Without this internal rule, the internal conditional is inferentially inert. The internal conditional is there to licence inferences internal to a theory, and it needs to be ponable to do this.

One way to represent the internal conditional in the semantics is due to Jc Beall et al. [11]. They added this connective to the semantics of relevant logic to characterise restricted quantification. The semantical framework they used is the Routley–Meyer semantics, which I discuss briefly in Section 5.6. They define a second ternary relation in order to give a satisfaction condition for $\mapsto$:

$$R_{\mapsto}abc \text{ if and only if } Rabc \text{ and } a \leq c.$$

Their satisfaction condition is

$$a \models A \mapsto B \text{ if and only if for all } b, c, (R_{\mapsto}abc \text{ and } b \models A) \text{ implies } c \models B.$$

This definition easily translates into the idiom of the present semantics:

$$R_{\mapsto}abc \text{ if and only if } a \circ b \leq c \text{ and } a \leq c.$$

Using this defined relation, I adopt the satisfaction condition from [11]. The constraint that $R_{\mapsto}abc$ only if $a \leq c$ makes a conditional about the theory in which it occurs (and its extensions). The involvement of the application operator in the definition gives the conditional a little bit of relevance (although not much).

The full integration of the internal conditional into the logic of proof processes would require an introduction rule. What this rule would look like, I am still uncertain. I just leave aside the issue of giving any complete proof theory for the internal conditional and confine myself to some remarks about why it or something like it would be a useful addition to the logic of entailment.

It is very easy to show that on this semantics, the internal conditional is ponable. By weak contraction, $a \circ a \leq a$ and because $\leq$ is reflexive, $a \leq a$. Hence, $R_{\mapsto}aaa$. This allows us to prove that it is valid that

$$((A \mapsto B) \wedge A) \to B,$$

and so every theory is closed under modus ponens for $\mapsto$.

To show that the internal conditional is transitive is slightly more involved. Suppose that $a \vDash A \mapsto B$ and $a \vDash B \mapsto C$. And assume that $R_{\mapsto}abc$ and $b \vDash A$. I show that $c \vDash C$. By the satisfaction condition for the internal conditional, $c \vDash B$. So, for any theory d, $R_{\mapsto}acd$, $d \vDash C$. By the definition of $R_{\mapsto}$, $a \leq c$, and by weak contraction, $c \circ c \leq c$. Thus, by the monotonicity postulate, $a \circ c \leq c$. Thus, $R_{\mapsto}acc$, and so $c \vDash C$. Generalising, $a \vDash A \mapsto C$. This proves that the internal conditional is transitive in the sense required.

11.3   Classical Theories and Proof Processes

I put this internal conditional to work in the analysis of theories that are based on logics stranger than E. Theories based on classical logic are closed under modus ponens for the material conditional. Taking $A \supset B$ to mean $\neg A \vee B$, modus ponens for the material conditional is a version of disjunctive syllogism:

$$\frac{\neg A \vee B \qquad A}{B}.$$

I discuss disjunctive syllogism and why relevant logics fail to contain it in Chapter 4. In Chapter 6, I suggest that one can appeal to the fact that a theory is closed under disjunctive syllogism without thereby using disjunctive syllogism as a principle of inference. In this section, I explain this idea more fully.

Suppose that one knows that $\neg A$ and $A \vee B$ are both theorems of Peano arithmetic. She also knows that PA is closed under disjunctive syllogism and so she concludes that B is a theorem of PA. Let us extend our Prawitz-style natural deduction system to include appeals to the closure of certain theories under rules like disjunctive syllogism and their inclusion of internal conditional expressions that express this. Using this new system we can interpret her proof process as follows. (Here, I use 'PA' not to refer to classical Peano arithmetic, but to denote the extended system that includes the internal conditional.)

$$\text{PA DS} \frac{PA \vDash \neg A \vee B \qquad PA \vDash A}{PA \vDash ((\neg A \vee B) \wedge A) \mapsto B} \qquad \frac{PA \vDash \neg A \vee B \qquad PA \vDash A}{PA \vDash (\neg A \vee B) \wedge A}.$$
$$PA \vDash B$$

The sub-process

$$\text{PA DS} \frac{PA \vDash \neg A \vee B \qquad PA \vDash A}{PA \vDash ((A \vee B) \wedge \neg A) \mapsto B}$$

needs some explanation. What it says is that because we know that the set of *theorems* of Peano arithmetic is closed under disjunctive syllogism and we are assuming that $\neg A \vee B$ and A are theorems of PA, then we can conclude that the internal conditional holds. This sub-process, clearly, not a logically valid inference in the sense that it is not of a valid form. Let us 'PA' with a generic 'T'. There are some theories that contain a disjunction and a negated disjunct without containing the other disjunct. These theories are inconsistent but they are theories nonetheless. This sub-process, then, is not a deductive inference but an *appeal* to a feature of Peano arithmetic – an appeal to the fact that the theorems of PA are, by stipulation, closed under disjunctive syllogism. This is why the inference is labelled with 'PA DS'.

The appeal to the closure of Peano arithmetic under disjunctive syllogism, blocks the use of other rules. In particular, it bars us from using it as a step in an entailment introduction. The following process is **not** valid:

$$\cfrac{\cfrac{\cfrac{[x \vDash (\neg A \vee B) \wedge A]^{1}}{x \vDash \neg A \vee B} \qquad \cfrac{[x \vDash (\neg A \vee B) \wedge A]^{1}}{x \vDash \neg A} \text{ x DS}}{x \vDash ((\neg A \vee B) \wedge A) \mapsto B}}{\cfrac{\Sigma \circ x \vDash ((\neg A \vee B) \wedge A) \mapsto B}{\Sigma \vDash ((\neg A \vee B) \wedge A) \rightarrow (((\neg A \vee B) \wedge A) \mapsto B)} \, 1}$$

The closure of x under disjunctive syllogism marks x out as a non-arbitrary theory, and so it cannot be used in a hypothesis for entailment introduction. Moreover, the appeal to the closure of x under disjunctive syllogism is not a premise in the proof. It is a side condition. This marks the difference between the inferential use of a rule and an appeal to that rule in a proof.

However, there is a weak sense in which the appeal is a logical inference. For now, let us add an operator '$\Box_{PA}$' to the language to mean 'is a theorem of Peano arithmetic'. This is a necessity operator, and corresponding to it in the semantics there is an operator on theories $\Box_{PA}^{-}$. In the intended model,

$$\Box_{PA}^{-}(\Sigma) = PA.$$

Thus, Σ says that a formula is a theorem of Peano arithmetic if and only if that formula is in fact a theorem of Peano arithmetic. Now let us consider the status of the scheme

$$(\Box_{PA}(\neg A \vee B) \wedge \Box_{PA}A) \rightarrow \Box_{PA}B.$$

This scheme is not valid. There are theories that satisfy the antecedent but do not satisfy the consequent for some instantiations of A and B. Now let us consider a slight modification of this scheme:

$$((\Box_{PA}(\neg A \vee B) \wedge \Box_{PA}A) \wedge t) \rightarrow \Box_{PA}B.$$

The addition of the necessary truth constant in the antecedent makes the scheme into an *enthymatic entailment*. It now tells us that if the antecedent is satisfied by an extension of Σ, then the consequent is also satisfied by it. This is still not true for all instances of A and B but it is true for all instances of A and $\neg A \vee B$ that are provable in Peano arithmetic. So there is a sense in which the rules given above can be represented in the object language. But it is clear that the inference is not an instance of an unconditionally valid form, because other instances of the same forms are not valid on the intended model.

11.4   Appeals and Infinitary Rules

In Chapter 6, I discuss Tarski's argument against his earlier syntactic treatment of consequence. He argued that, intuitively, the theory of the natural numbers is closed under the omega rule. This rule is infinitary, and this shows that our ordinary notion of logical consequence is not finitary or compact. I claim in that chapter that the omega rule, like disjunctive syllogism, is not a valid rule of inference, but rather is one to which we can appeal in specific circumstances. In this section, I examine proof processes that contain appeals to rules like the omega rule that cannot be represented in the syntax of the theory. There is no facility in my theory for representing conjunctions of infinitely many formulas, say, and so unlike disjunctive syllogism, there is an expressive barrier here to the inclusion of infinitely rules in the theory of entailment.

Let us consider an appeal to an infinitary rule. PA_ω is Peano arithmetic with the omega rule. Some natural numbers are Gödel encodings of formulas of the language and some are not. However, in either case, for an arbitrary number n, we can prove in Peano arithmetic that $\neg Pr(n, g)$, where Pr is provability in PA (not PA_ω) and g is the Gödel number of the Gödel sentence for PA. By the omega rule, then $\forall x \neg Pr(x, g)$ is provable in PA_ω.

This argument makes an appeal to the omega rule, but clearly does not use the omega rule in the sense that it does not list the premises of it. Probing a bit deeper into arguments like this, what typically happens is that one shows that she can derive a contradiction if $Pr(n, g)$ is provable for an arbitrary number n. An inference from a proposition holding for an arbitrary thing to its holding of everything is the sort of reasoning that one does when using the rule of universal generalisation. What guarantees that universal generalisation holds in a theory is the fact that the theory is omega complete. However, not all theories are omega complete. Thus, the inference from being able to prove some proposition A about an arbitrary object to everything's being A is not a logical entailment.

To return to a point I discuss in Section 10.8, the failure of universal generalisation to be a logical entailment is also another reason to reject the Barcan formula. Recall that the Barcan formula is

$$\forall x \Box A \rightarrow \Box \forall x A.$$

In Section 10.8, I say that I do not think that the antecedent of this entailment expresses the proof of A about an arbitrary object, but I set aside that doubt here. My point here is that the Barcan formula does not hold when we think of certain theories other than the logic itself. For example, consider

$$\forall x \Box_{PA} A \rightarrow \Box_{PA} \forall x A.$$

This clearly is false because of the omega incompleteness of Peano arithmetic. This fails again for PA_ω:

$$\forall x \Box_{PA_\omega} A \rightarrow \Box_{PA_\omega} \forall x A.$$

For some (false) theory a, $\Box^-_{PA_\omega}(a)$ is an omega incomplete theory. For a similar reason, the Barcan formula even fails in its enthymatic version,

$$\forall x (t \wedge \Box_{PA_\omega} A) \rightarrow \Box_{PA_\omega} \forall x A.$$

For some theory a that extends Σ, $\Box^-_{PA_\omega}(a)$ is an omega incomplete extension of PA_ω. The theory $\Box^-_{PA_\omega}(a)$ is inconsistent, but this does not stop its being used as a counter-example. The fact that the Barcan formula does not hold in these different cases, in these different senses, gives further evidence that it cannot be a logical entailment. That the logic of entailment is omega complete is true, but this does not mean that its logical completeness results from the closure of its axioms under a logical principle or an internal rule (see Section 5.8).

11.5   Separating Theories

The semantics of entailment that I put forward has theories being applied to one another. As I say at the beginning of this chapter, this is different from the way in which theories are usually constructed in logic. The standard method of constructing a theory is to begin with the theorems of a logic and add to it some non-logical axioms that are proper to the theory. The theory of entailment, as I think of it, is a theory for studying theories, rather than a part of every theory. It includes universal closure principles that govern theories.

The logic of entailment, however, has a dual nature. It acts as a background theory for other theories. We can appeal to the logic of entailment when trying to deduce what other theories say. The logic, considered as a set of theorems, is the collection of valid proof plans. We are entitled to use these theorems in order to deduce what a theory says. But the logic of entailment is also a theory of theories. Like a theory in the philosophy of science, which has scientific theories as its subject matter, a theory of entailment has as its subject matter theories in general.

Principles of entailment are supposed to be universal, that is, they apply to every theory. This gives the logic of entailment itself a central role in the epistemology of logic. The logic of entailment contains *all* the true entailments. Thus, where @ is the complete theory of the actual world, we have

$$\Box^-(@) = \Sigma$$

and, for any theory a, a is true only if

$$\Box^-(a) \leq \Sigma.$$

Thus, to determine whether an entailment claimed by a theory is true one need only determine whether the entailment is a theorem of the logic E.

Where a theory a does not contain any entailments, it is closed under the first-degree entailments of E. The logic of first-degree entailments is quite a weak logic. This means that the use of E as a background logical theory does not place much of a burden of justification on theories. In particular, for the assessment of most theories, for the most part it does not matter what theorems E has that do not contain any negations or disjunctions. These theorems largely do not affect the result of applying E to theories that do not contain any entailments.

As I have said, however, the true logic is part of the true theory of the actual world. This is what makes it the true logic. The logic E does contain certain theorems that could be considered controversial. For example, E contains double negation elimination (DNE, see Chapter 9) and excluded middle. If we find, for example, that our natural language claims are not closed under DNE or that our natural language negation is not exhaustive, then we should reject the offending logical principle. Our theories are formulated primarily in natural language (often with a lot of mathematical language thrown in), and what counts as a standard theory is in part determined by the logic of natural language. As I say in Chapter 8, I am willing to retrench and abandon relevant principles governing negation and disjunction, as long as they do not affect the core theory of entailment and conjunction.

The present semantics does not just allow us to look at the application of the one true logic to other theories. Rather, it allows us to look at the results of applying a wide range of theories to other theories. In so doing, we can test theories by trying them out on other theories. We can ask interesting counter-factual question of the form "What if $A \rightarrow B$ were a valid entailment?" and answer it by looking at how theories that satisfy $A \rightarrow B$ behave when applied to other theories. Such questions can be enlightening even when we are asking about formulas that we know are not theorems of the one true logic.

11.6   Theory Separation in Philosophy of Science

In the models that I have presented in this book so far, applying theories to other theories tracks the *logical entailments* of the former theories. However, I took the motivating idea of background theories from the philosophy of science, and in scientific inferences the background theories that are explicitly used are not logical of logical entailment, but rather empirical or mathematical theories of other sorts. I suggest, however, that we can use the same sort of modelling to treat the application of scientific theories or mathematical theories to one another.

Recall the theory of relevant predication that I explain briefly in Section 10.6. In this theory, there are certain predicates, called "relevant predicates". These predicates allow for a stronger axiom of substitution than do other predicates. Where F is a relevant predicate,

$$i = j \rightarrow (F(i) \rightarrow F(j)).$$

The idea here is to relativise the notion of a relevant predicate to a theory. Take, for example, a mathematical theory, a. I add to the language an entailment-like connective, $\rightsquigarrow$. I then use this connective to create principles that allow us to make inferences about other theories using the theorems of a. Then, I close the theses of a under the following rule:

$$\frac{i = j}{F(i) \rightsquigarrow F(j)}$$

where F is relevant according to a.

I also place in the semantics an operator $\odot$, that plays the role of an application operator that corresponds to $\rightsquigarrow$ in the object language. Then, the satisfaction condition for $\rightsquigarrow$ is

$$a \vDash A \rightsquigarrow B \text{ if and only if for all } b \vDash A, a \odot b \vDash B.$$

I also add to the semantics a function that attributes to each theory a set of relevant predicates. Some other additional mechanism is necessary to satisfy the principle of substitution for relevant predicates; this has largely been worked out by Philip Kremer [92] and Nicholas Ferenz [60].

Suppose that a theory a is a theory of arithmetic. One also needs to have a version of the induction principle that allows the application of a to other theories. For example,

$$(F(0) \wedge \forall x (F(x) \mapsto F(x+1)) \rightsquigarrow \forall x F(x),$$

where F is a relevant predicate according to a and $\mapsto$ is the internal conditional discussed above.This scheme allows us to use the ordering of the natural numbers to derive theorems in the theory that results from that application of a to another theory. Clearly, there are other rules and principles that may need to be added, but let this one suffice to illustrate the idea for the present.

To model theory application, the application operator, $\odot$, should obey the postulates of E-frames (see Chapter 8). The notion of a theory here may be much more restrictive than the notion of a standard theory. For example, we might be interested only in consistent theories or theories that satisfy all the actual laws of physics, or whatever. Further constraints may be added to the definition of a model in order to satisfy these further conditions. Whether the resulting frame theory should characterise a relevant logic is not clear, and frankly that is not really very important. But because the logic of theory application is an extension of E, the proof theory for E together with some extra rules could be used to represent inferences on scientific theories.

The hope with this theory of theory application is that it can resist untoward instances of *leakage* from one theory to another. For example, scientists use the calculus to determine the motion of the planets around the sun, but the fundamental theorem of calculus is not a part of the theory of planetary dynamics. I would like, but cannot prove that I can obtain, a method for altering theories like calculus to be applicable to theories like that of the motion of the planets without having very much of the former leak into the resulting theory. However, such an attempt will take me too far from the topic of logical entailment, which is what this book is about.

11.7   Confirmation and Disconfirmation

The separation of theories into background theory, target theory, and the result of applying the former to the latter complicates the issue of the confirmation or disconfirmation of theories. However, I think it does so in a way that is intuitive and somewhat pleasing. Consider the following schematic representation of the hypothetical deductive model of scientific inference:

$$\frac{\text{Theory} \qquad \text{Auxiliary Hypotheses} \qquad \text{Experimental Setup}}{\text{Prediction}}$$

In the traditional syntactic view of theories, the prediction is obtained by deduction from the sets of statements in the theory, the auxiliary hypotheses, and in the description of the experimental setup in the following manner:

$$\text{Theory} \cup \text{Auxiliary Hypotheses} \cup \text{Experimental Setup} \Vdash \text{Prediction}$$

A true prediction gives some confirmation to all the premises. A false prediction disconfirms some premise, but as Pierre Duhem and Quine point out, the hypothetical deductive scheme by itself does not tell us which premises to reject.

Let me change to the semantic mode of presentation in order to explain what the current view says about confirmation and disconfirmation. Here, we have a prediction deduced from a theory, auxiliary hypotheses, and an experimental setup represented as follows:

$$(\text{Auxiliary Hypotheses} \odot \text{Theory}) \odot \text{Experimental Setup} \vDash \text{Prediction}$$

The ordering and association of the premises may change from example to example. If the auxiliary hypotheses are largely about the equipment used in the experiment, then we might have

$$\text{Theory} \odot (\text{Auxiliary Hypotheses} \odot \text{Experimental Setup}) \vDash \text{Prediction}$$

instead.[1] We also may have more than one collection of auxiliary hypotheses and this could give rise to a more complicated representation. Setting the issue of the exact representations aside, what is confirmed or disconfirmed on this view is the result of applying the theory and auxiliary hypotheses to the experimental setup. A theory and the auxiliary hypotheses are only confirmed in a fairly indirect sense (as compared to the usual treatments of abduction).

This might seem like bad news for the separation of theories view, but I think it a good feature. In the standard view of abductive confirmation, the theory and auxiliary hypotheses are treated realistically. They are confirmed or disconfirmed in actual experiments. My view that allows for the separation and application of theories allows one to be both an antirealist about one or more background theories and remain

---

[1] Of course, if $\odot$ obeys the postulates of E frames, then, by associativity, we also have
Theory $\odot$ (Auxiliary Hypotheses $\odot$ Experimental Setup) $\leq$
(Auxiliary Hypotheses $\odot$ Theory) $\odot$ Experimental Setup and so,
(Auxiliary Hypotheses $\odot$ Theory) $\odot$ Experimental Setup $\vDash$ Prediction.

an abductive realist. Suppose, for example, that one is a fictionalist with regard to a mathematical background theory – that she believes that the theory will give useful results but that some of its ontology does not actually exist or that some of its statements are useful but false. This background theory may be used as an auxiliary in making empirical predictions but when those predictions come true the fictionalist need only be reinforced in her view that the theory is useful not that it is literally true.

Of course, we can recast the abductivist argument to say that theories a and b are given a degree of confirmation by an experiment e (where e is a theory describing the experiment) if

$$(a \odot b) \odot e \vDash P$$

and P is true. Moreover, a and b are given more confirmation by e than any other pair a', b' if a and b are, on balance, better theories than these other pairs given their theoretical virtues. My point, however, is that we can hold that, say, $a \odot b$ is true (in some cases) without holding that either a or b is true. In that way, we can accept a weak form of abductivism and still remain anti-realists about certain useful theories and auxiliary assumptions.

11.8   Further Epistemological Reflections

There are two approaches that are used in this book to support the logic E as the logic of entailment. First, there is a *bottom-up* method that is exemplified most at the end of Chapter 6. In the bottom-up method, a collection of entities are determined to be theories, or more accurately, standard theories. The logic is then made to fit this collection of theories. In Chapter 6, I appeal to the semantic view of theories and look briefly at the nature of models of scientific theories (as presented by van Fraassen and others) to support the idea that they are closed under the valid first-degree entailments. The logic of entailment that fits with this bottom-up perspective with very little further argument is Priest's logic N_4.

Then, in Chapters 7 and 8, I take a *top-down* approach, in which I consider how one constructs theories and manipulates them in proofs. In these chapters, I interpret and appeal to Anderson and Belnap's elegant natural deduction system, and construct a deduction system that represents how one applies a proof plan. This is the theory of proof processes. Using this top-down approach, I first motivate a simple logic of theory closure, GTL, and motivate the conjunction and entailment fragment of the logic E.

The problem that I deal with in Chapter 8 is how to bring the bottom-up and top-down approaches together. To a large extent, the top-down approach wins. I use the elegance and usefulness of the logic E to motivate the addition of postulates to the model theory that characterise this logic. To use the language of the philosophy of science, my argument appeals to the theoretical virtues to choose a theory of entailment. The particular virtues that are the most salient are explanatory strength and the ability to unify a wide range of examples of reasoning. This process of theory choice is described nicely by Gillian Russell [188]. On Russell's view, in order to determine what the basic laws of logic are, one reflects on his or her reasoning processes using a range of examples (reasoning about vague predicates, paradoxes, and so on) and uses theoretical virtues like explanatory unification, simplicity, and elegance to determine what is the best overall logical theory that fits with this data. In my opinion, the sort of data that Russell discusses indicates which theories (scientific and otherwise) are the ones that we consider to be standard and worthy of investigation.

In Russell's epistemology of logic the unit of selection is a logical theory as a whole. In treating the logical theory as a whole, one justifies the acceptance of all the axioms and rules of a theory. The testing of individual rules or axioms takes the form of asking what sort of theory they generate, taken together. In order to facilitate this sort of reasoning about logical theories, I have proposed a semantic framework. Within this framework (that characterises the basic logic GTL), I have asked about the addition of further postulates with the aim to producing and justifying both a reasonable semantic theory and an elegant proof theory.

I have focused on a fairly narrow class of reasoning processes and set aside issues such as vagueness in order to make the problems involved less complicated. I hope, however, that an appropriate view of vagueness (and other such issues) is out there and will one day be incorporated into my current semantics. My appeals to the elegance and simplicity of the Anderson and Belnap's proof theory, the intuitiveness of the notion of real use that it displays, and the intuitiveness of the semantic theory as interpreted in terms of theories and consequence operators justify the acceptance of E.

11.9   Entailment and Relevant Implication

In my book *Relevant Logic* [129], I construct an interpretation of the relevant logic R. I still believe almost everything I wrote in that book (with the exception of what I said about entailment). The question thus arises as to how E and R are related. Clearly, we should be able to use any reasonable theory of entailment to investigate a theory about implication. The proof theory and model theory for R are reasonable objects of investigation and can be included among the theories that make up the intended model for E.

But there is another question about the relationship between E and R: How are the theorems of E and R made simultaneously valid? The model theory for R is based on a set of situations, which I call here '*Sit*'. For each situation s in *Sit*, there is a theory a_s of s, that is, there is a theory in which all and only those propositions made true by s are included in a_s. These situational theories are also to be included among the theories of the model for entailment.

Things, however, get even more complicated. I also need a way of integrating facts about E into the semantics for R. On my interpretation of R, in a situation s, an implication $A \Rightarrow B$ means that A carries the information that B according to the information available in s. As I explain in Section 5.6, this notion of carrying information is modelled in the Routley and Meyer semantics by a ternary relation, R. $s \vDash A \Rightarrow B$ if and only if for all situations t and u, if $Rstu$ and $t \vDash A$, then $u \vDash B$.

My reading of the Routley–Meyer semantics is that in s there is information present that allows the deductive inference from there being a situation t in the same world as s such that $t \vDash A$ to there being a situation u in that world such that $u \vDash B$. In [129], I call this reading the *theory of situated inference*.

In order to represent an implication holding between entailments, I need to have a theory about what it means for a situation to satisfy (or fail to satisfy) an entailment. The meaning of $A \rightarrow B$ is that every theory that contains A also contains B, failing to satisfy entailments. In this book, I claim that the meaning of an entailment $A \rightarrow B$ is that every theory that contains A also contains B. Thus, it is required to have a theory according to which a situation that satisfies an entailment contains information about the closure conditions of theories. Information about closure conditions for theories can take various forms. There may be a set of theories that are deemed formally acceptable by scientists, there may be particular principles used by scientists and mathematicians to determine the contents of theories, there may be principles supported by the intuitions of logicians, and so on.

In order for the logics E and R to be true simultaneously, the actual world must constitute a situation in the intended model for R and Σ must accurately characterise the actual world (although not all of it) and capture all of the entailments that are true in this world. It is straightforward to show, using a metavaluations argument (see Appendix), that there are world-like theories that contain all the theorems of R and all and only those entailments that can be proven in E. We can construct a model around such a theory, but it would take some space and a lot of creative energy to argue for the acceptance of such a model. And I leave this as a project to be completed sometime in the future.

11.10   Concluding Remarks

I now come to the end of the main text of this book. It is time to give a summary of the structure and content of the book. The structure is quite simple: Chapters 2–5 set out problems for a logic of entailment, taken from the modern history of the subject, that the later chapters attempt to solve.

In Chapter 2, I suggest that Lewis's attempt at creating a logic of entailment was a failure because he failed to provide a clear and well-motivated semantical or proof-theoretical framework in which to test the ability of his logic to capture a vertebrate notion of deduction. In Chapters 4, 7, and 8, I examine Anderson and Belnap's proof theory and modify Fine's semantics to provide such frameworks. In Chapter 3, I argue that if an entailment is to be true when and only when the corresponding deduction is valid, then the semantical framework needs to suggest both an intended model and an actual world within that model at which all and only the true entailments hold. And in Chapters 3 and 6, I emphasise the need for a semantics that has a consequence relation that can be represented adequately in a logic. In particular, I argue that this consequence relation should be compact. In Chapters 7– 10, I construct such a semantics. In Chapter 4, I examine Anderson and Belnap's logic of entailment and claim that it needs a reasonable semantics and in particular a theory of meaning for the entailment connective before it can be accepted or rejected. The last half of this book attempts to produce such a theory.

Chapter 5 poses an especially important question for anyone trying to construct a logic of entailment. The problem is how to distinguish between a reasoning process that one needs to capture as an entailment and one which is something else. I claim that the use of free variables or parameters to reason about a thing in general cannot (or should not) be captured as an entailment. Similarly, I follow the mainstream in relevant logic and claim that disjunctive syllogism is not a valid rule of inference and its corresponding entailment is not a logical truth. Adding the entailments that capture these reasoning processes (at least in their most straightforward form) does damage to my overall theory and that is the main reason why I reject them. In Chapters 6 and 11, I use the distinction between the use of a claim or rule in an inference and an appeal to it to avoid the problem of capturing alleged rules of inference such as disjunctive syllogism and the omega rule. But I try, all things being equal, to represent intuitive reasoning processes as entailments within the logic itself. The problem is that in these particular cases all things are not equal.

In all, I think that I give a new interpretation of the logic E of relevant entailment, locate the logic and this interpretation in the history of entailment and give the logic a role to play going forward.

Appendix: Systems, Semantics, and Technical Results

A.1 Theorems regarding Modal Logic

THEOREM A.1 *Let W be an infinite set of worlds and $\mathbb{V}$ be the set of valuations on W. Then the formulas valid on W are just the theorems of S5.*

Proof First, note that for any $V \in \mathbb{V}$, $< W, V >$ is an S5 model, so every theorem of S5 is valid on $< W, V >$. Hence, the formulas valid on W include all the theorems of S5.

Second, let B be a non-theorem of S5. Then, there is a maximal S5-consistent set Γ such that $B \notin \Gamma$ (see [83, ch 6]). I translate Γ into a monadic first-order language, such that for each propositional variable p of the modal language, there is a predicate P in the first-order language. The translation τ is as follows:

- $\tau(p) = Px$;
- $\tau(A \wedge B) = \tau(A) \wedge \tau(B)$;
- $\tau(A \vee B) = \tau(A) \vee \tau(B)$;
- $\tau(\neg A) = \neg \tau(A)$;
- $\tau(\Box A) = \forall x(\tau(A))$;
- $\tau(\Diamond A) = \exists x(\tau(A))$.

Let Γ' be the set of translations of the members of Γ. Then, Γ' is a first-order theory. We know, by the downward Löwenheim–Skolem theorem, that there is a model for Γ' that is either finite or countably infinite. Let such a model be constructed such that the individuals of the model are worlds from W. The domain of the model, then, is a subset of W. Let us call this domain W'. We set a value assignment V on W' such that $w \in V(p)$ if and only if $P(x)$ is satisfied by w in the first-order model, for all propositional variables p and their corresponding predicates P. At least one world w in W' will be such that all and only the members of Γ are true at w according to V. With regard to all the worlds w' in W that are not in W', let V be such that $w' \in V(p)$ if and only if $w \in V(p)$, for all propositional variables p. The resulting model $< W, V >$ is an S5 model such that B is not valid on $< W, V >$.

Therefore, A is valid on the frame W if and only if A is a theorem of S5. ■

This theorem holds for propositional logic (as the proof shows) but may not hold for predicate logic with identity. Whether we can extend the theorem to this system depends on the semantics of identity. If we force $i = j$ to be true at a world w if and only if the semantic value of i is the same as that of j, then the theorem will not hold. If the domain has at least n individuals in it, we will be able to prove that the statement $\exists x_1 \ldots \exists x_n (x_1 \neq x_2 \wedge \ldots \wedge x_{n-1} \neq x_n)$ (i.e. there are at least n things). This is not a theorem of standard quantified S5 with identity.

If we treat identity as a normal binary predicate and allow its extension to vary from world to world, then we can prove that the logic of infinitely many worlds with an infinite domain is standard quantified S5 with the Barcan formula and its converse.

A.2   Tarski Logic

The logic TL has the following axiom and rules:

Axiom

$$A \to A$$

Rules

$$\frac{A \to C}{(A \wedge B) \to C} \qquad \frac{A \to B \qquad A \to C}{A \to (B \wedge C)}$$

$$\frac{A \to B \qquad B \to C}{A \to C} \qquad \frac{A \to B \qquad A}{B}$$

The TL-consequence operator, C_{TL}, is defined such that for any set of formulas Γ and any formula, $B \in C_{TL}(\Gamma)$ if and only if there are some $A_1, \ldots, A_n \in \Gamma$ such that $(A_1 \wedge \ldots \wedge A_n) \to B$ is a theorem of TL. C_{TL} satisfies Tarski's conditions on consequence operators, as shown in Chapter 6.

Here, I show that if for a logic L that contains conjunction and entailment, C_L satisfies the following conditions, then L is an extension of TL: for all sets of formulas Γ and Δ, (1) $\Gamma \subseteq C_L(\Gamma)$; (2) if $\Gamma \subseteq \Delta$, then $C_L(\Gamma) \subseteq C_L(\Delta)$; (3) $C_L(C_L(\Gamma)) \subseteq C_L(\Gamma)$; (4) $A \wedge B \in C_L(\Gamma)$ if and only if $A \in C_L(\Gamma)$ and $B \in C_L(\Gamma)$; (5) the set of theorems of L is an L-theory.

Proof Suppose that $\Gamma \subseteq C_L(\Gamma)$. Let $\Gamma = \{A\}$ for some formula A. Then, $\{A\} \subseteq C_L(\{A\})$, that is $A \in C_L(\{A\})$. Thus, there are some formulas in Γ such that their conjunction entails A according to L. Since there is only one formula in Γ, that is A. So, $A \to A$ is a theorem of L.

Suppose that $A \to B$ and $B \to C$ are theorems of L. Now consider the set $\{A\}$. $B \in C_L(\{A\})$ and $C \in C_L(C_L(\{A\}))$. Thus, by condition 3, $C \in C_L(\{A\})$. And so, by the definition of C_L, $A \to C$ is a theorem of L.

Suppose that $A \to B$ and $A \to C$ are theorems of L. Then, B and C are both in $C_L(\{A\})$. By condition 4, $B \wedge C \in C_L(\{A\})$. So, by the definition of C_L, $A \to (B \wedge C)$ is a theorem of L.

Suppose that $A \to C$ is a theorem of L. I show that $C \in C_L(\{A \wedge B\})$. $A \in C_L(\{A \wedge B\})$ by conditions 1 and 4. So, $C \in C_L(\{A \wedge B\})$. Therefore, $(A \wedge B) \to C$ is a theorem of L.

Suppose that $A \rightarrow B$ and A are theorems of L. Let Λ be the set of L theorems. Then, $A \in \Lambda$ and $B \in C_L(\Lambda)$. By condition 5, the set of L theorems is an L-theory and so $C_L(\Lambda) \subseteq \Lambda$ (see Chapter 6). Thus, $B \in \Lambda$. Hence, Λ is closed under modus ponens.   ∎

A.3   Models for TL

A TL model is a triple $< M, 0, v >$ such that M is a set (of scientific models), $0 \in M$ (0 is the "normal world" and contains the theory of the whole model), and v is a function from formulas to sets of worlds that satisfies the following conditions:

- $v(A \wedge B) = v(A) \cap v(B)$
- $0 \in v(A \rightarrow B)$ if and only if $v(A) \subseteq v(B)$

These conditions do not place any constraints on the truth of entailments at worlds other than 0. A formula A is valid in a model if and only if $0 \in v(A)$ for the model, and it is valid in the class of models if it is valid in every model in the class.

Soundness for TL over this class of models is quite straightforward. One only has to verify that the axiom is true at 0 in an arbitrary model and that the truths 0 are closed under every rule.

Completeness is also very easily proved. We start with a *hyper-regular* theory of TL. This is a theory of TL that contains all the theorems of TL and is closed under all the rules of TL. The members of M other than 0 are the theories of 0. Γ is a theory of 0 if and only if, if $A_1, \ldots, A_n$ are in Γ and $(A_1 \wedge \ldots \wedge A_n) \rightarrow B$ is in 0, then B is also in Γ. As usual, I set $a \in v(p)$ if and only if $p \in a$, for all propositional variable p and all 0 theories. For $a \neq 0$, I also set $a \in v(A \rightarrow B)$ if and only if $A \rightarrow B \in a$.

LEMMA A.2   *(i)* $A \wedge B \in w$ *if and only if* $A \in w$ *and* $B \in w$; *(ii)* $A \rightarrow B \in 0$ *if and only if for all* $w \in W$, *if* $A \in w$, *then* $B \in w$.

Proof   (i) Suppose that $A \wedge B$ is in w for an arbitrary 0 theory w. Since 0 is regular, $(A \wedge B) \rightarrow A$ and $(A \wedge B) \rightarrow B$ are both in 0. Hence, A and B are both in w.

Suppose now that $A \in w$ and $B \in w$. $(A \wedge B) \rightarrow (A \wedge B) \in 0$, hence $A \wedge B \in w$.

(ii) The left-to-right direction follows directly from the definition of a TL theory.

Suppose that for all $w \in W$, if $A \in w$, then $B \in w$. Let $[A]$ be the *principle 0 theory* of the formula A. This means that $[A]$ is just the set of formulas C such that $A \rightarrow C \in 0$. If $B \in [A]$, then $A \rightarrow B \in 0$. But $[A]$ is in W, so $A \rightarrow B \in 0$.   ∎

THEOREM A.3   *For all* $a \in M$ *and all formulas* A, $a \in v(A)$ *if and only if* $A \in a$.

Proof   By induction on the complexity of the formula A. The base case of the induction takes A to be a propositional variable. This case follows directly from the definition of the canonical model's value assignment.

Conjunction case: $A = B \wedge C$. By the inductive hypothesis, $w \in v(A)$ iff $A \in w$ and $w \in v(B)$ iff $B \in w$. By Lemma A.2, A and B are both in w if and only if $A \wedge B \in w$. So, $A \wedge B \in w$ if and only if $w \in v(A \wedge B)$.

Entailment case: $A = B \to C$. Subcase 1. $w \neq 0$. Then $w \in V(B \to C)$ if and only if $B \to C \in w$ by the construction of the canonical model.

Subcase 2. $w = 0$. By Lemma A.2, $B \to C \in 0$ if and only if, for all $w \in W$, if $B \in w$, then $C \in w$. By the inductive hypothesis, $B \to C \in w$ if and only if, for all $w \in W$, if $w \in v(B)$, then $w \in v(C)$. Thus, $B \to C \in w$ if and only if $w \in v(B \to C)$. ∎

A.4 N$_4$

The language now is the same as before, with the addition of the negation, $\neg$. The axiomatic system for N$_4$ results from adding the following definition and axiom to TL:

$$A \vee B =_{df} \neg(\neg A \wedge \neg B)$$

$$\text{(DN)} \ A \leftrightarrow \neg\neg A$$

It is clear that every theorem of TL is also a theorem of N$_4$, since N$_4$ has all the axioms and rules of TL. Thus, any derivation of a theorem in the axiom system for TL is a valid derivation of a theorem in N$_4$. To prove that N$_4$ is a conservative extension of TL, I appeal to the fact that we can extend the definition of a value assignment for an arbitrary TL model to be a value assignment for the language of N$_4$. That is, we take a value assignment v and construct a value assignment V such that v and V agree on propositional variables and entailment formulas that do not contain negation on non-normal worlds (i.e. worlds that are not 0).

I then set $w \in V(\neg A)$ if and only if w fails to be in $v(A)$ for all formulas A of the extended language. We can then prove in the standard way that $V(A \vee B) = V(A) \cup V(B)$. Then, we can define a four-valued value assignment, v, such that $T \in v_w(A)$ if and only if $w \in V(A)$ and $F \in v_w(A)$ if and only if $w \in V(\neg A)$ for all formulas A. Then it is obvious that $< W, 0, v >$ is an N$_4$ model and that, for all formulas, B, in the conjunction-entailment fragment of the language, $w \in v(B)$ if and only if $T \in v_w(B)$. In particular, $0 \in v(B)$ if and only if $T \in v_w(B)$. Thus, the valid formulas in the class of N$_4$ models are all valid formulas in the class of TL models. Thus, by soundness and completeness of these two logics, every theorem of N$_4$ in the conjunction-entailment language is a theorem of TL. In other words, N$_4$ is conservative with regard to TL.

A.5   GTL

Generalised Tarski Logic (GTL) is

$$A \to A$$

$$(A \land B) \to A; \quad (A \land B) \to B$$

$$((A \to B) \land (A \to C)) \to (A \to (A \land C))$$

$$((A \to B) \land (B \to C)) \to (A \to C)$$

$$\frac{\vdash A \to B \qquad \vdash C \to D}{\vdash (B \to C) \to (A \to D)} \qquad \frac{\vdash A \to B \qquad \vdash A}{\vdash B}$$

A.6   Models for GTL

A GTL frame is a quadruple $\langle \Upsilon, \Sigma, \circ, \leq \rangle$, such that Υ is a non-empty set, $\Sigma \in \Upsilon, \circ$ is a binary operator on Υ, and $\leq$ is a binary relation on Υ, that satisfies the following postulates. For all $a, b, c, d \in \Upsilon$,

1. $\leq$ is a partial order;
2. $\Sigma \circ a = a$ (Sigma-Identity);
3. If $a \leq c$ and $b \leq d$, $a \circ b \leq c \circ d$ (Monotonicity);
4. $a \circ (a \circ b) \leq a \circ b$ (Reverse Contraction).

A GTL model is a pair $\langle \mathfrak{F}, \vDash \rangle$, such that $\mathfrak{F}$ is a GTL frame and $\vDash$ is a relation between theories of $\mathfrak{F}$ and formulas, where the following conditions hold:

- If $a \vDash p$ and $a \leq b$, then $b \vDash p$;
- $a \vDash A \land B$ if and only if $a \vDash A$ and $a \vDash B$;
- $a \vDash A \to B$ if and only if, for all $b \vDash A$, $a \circ b \vDash B$;
- $a \vDash t$ if and only if $\Sigma \leq a$.

LEMMA 7.1 (Hereditariness) *If $a \leq b$ and $a \vDash A$, then $b \vDash A$.*

Proof By induction on the complexity of A.

Base case: A is a propositional variable. This follows directly from the conditions on the satisfaction relation.

Case 1. A is a conjunction, $B \land C$. This follows from the inductive hypothesis and the satisfaction condition for conjunction.

Case 2. A is an entailment, $B \to C$. Suppose that $a \vDash B \to C$ and $a \leq b$. Now suppose that $c \vDash B$. By the satisfaction condition for entailment, $a \circ c \vDash C$. By the monotonicity condition on frames, $a \circ c \leq b \circ c$. Thus, by the inductive hypothesis, $b \circ c \vDash C$. Generalising, $b \vDash B \to C$. ∎

THEOREM 7.2 (Semantic Entailment) $\Sigma \vDash A \to B$ *if and only if, for all $a \in \Upsilon$, $a \vDash A$ implies $a \vDash B$.*

Proof Suppose first that $\Sigma \vDash A \rightarrow B$ and that $a \vDash A$. Then $\Sigma \circ a \vDash B$. But $\Sigma \circ a = a$, and so $a \vDash B$.

Now suppose that, for all $a \in \Upsilon$, $a \vDash A$ implies $a \vDash B$. Since $\Sigma \circ a = a$, if $a \vDash A$, then $\Sigma \circ a \vDash B$. By the satisfaction condition for entailment, $\Sigma \vDash A \rightarrow B$. ∎

The soundness of the axioms is proven in Chapter 7. To show that the truths of Σ are closed under the affixing rule, suppose that $A \rightarrow B$ and $C \rightarrow D$ are both satisfied by Σ. Let $a \vDash B \rightarrow C$ and $b \vDash A$. Thus, $\Sigma \circ b \vDash B$ and so $b \vDash B$. Thus, $a \circ b \vDash C$ and so $\Sigma \circ (a \circ b) \vDash D$, that is $a \circ b \vDash D$. Thus, $a \vDash A \rightarrow D$. Generalising, by the semantic entailment theorem, $\Sigma \vDash (B \rightarrow C) \rightarrow (A \rightarrow D)$, and the truths of the model are closed under affixing. Closure under modus ponens follows from the fact that $\Sigma \circ \Sigma = \Sigma$.

To prove completeness, I construct a *canonical model*. The canonical model for each logic is closely related to the intended model for those logics. The idea is that the canonical model has properties that are desirable in the intended model and the existence of the canonical model makes it plausible that the real model has those properties. The canonical model is a structure, $\langle \Upsilon, \Sigma, \circ, \leq, \vDash \rangle$, such that Σ is the set of theorems of GTL, Υ is the set of theories (in the syntactic sense) of GTL, $\leq$ is the set-theoretic relation $\subseteq$, and $\circ$ is defined as follows. Where Γ and Δ are sets of formulas,

$$\Gamma \circ \Delta = \{B : \exists A_1 \ldots \exists A_n ((A_1 \wedge \ldots \wedge A_n) \rightarrow B \in \Gamma \wedge (\ldots (A_1 \wedge A_2) \ldots) \wedge A_n \in \Delta) \, n \geq 1\}.$$

For any set of formulas Γ, $\Sigma \circ \Gamma$ is a GTL theory. The relation $\vDash$ is defined as in the definition of a GTL model above, with the condition that for all propositional variables, p, and all theories a, $a \vDash p$ if and only if $p \in a$.

LEMMA A.4 *Let a be a GTL theory. Then $A \rightarrow B \in a$ if and only if for all GTL theories b, if $b \vDash A$, then $a \circ b \vDash B$.*

Proof The left-to-right direction follows from the definition of $\circ$. Thus, it suffices to show the right-to-left direction of the lemma.

Suppose that for all b such that $A \in b$, $B \in a \circ b$. Let $[A]$ be the set of formulas C such that $A \rightarrow C$ is a theorem of GTL. $[A]$ is a GTL theory. By assumption, $B \in a \circ [A]$. Thus, there are some $C_1, \ldots, C_n \in [A]$ such that $(C_1 \wedge \ldots \wedge C_n) \rightarrow B \in a$. By the definition of $[A]$, $A \rightarrow C_i$ is a theorem of GTL for all $1 \leq i \leq n$ and so $A \rightarrow (C_1 \wedge \ldots \wedge C_n)$ is a theorem of GTL. Thus, by affixing, $((C_1 \wedge \ldots \wedge C_n) \rightarrow B) \rightarrow (A \rightarrow B)$ is a theorem. Since a is a GTL theory, $A \rightarrow B \in a$. ∎

Since all GTL theories are closed under conjunction, I stop talking about conjunctions of formulas from theories and instead merely talk about single formulas.

LEMMA A.5 $a \circ (a \circ b) \leq a \circ b$

Proof Suppose that $C \in a \circ (a \circ b)$. Then there is some formula $B \in (a \circ b)$ such that $B \rightarrow C \in a$. Thus, there is an $A \in b$ such that $A \rightarrow B \in a$. So, both $B \rightarrow C$ and $A \rightarrow B$ are in a and hence $(A \rightarrow B) \wedge (B \rightarrow C) \in a$. By conjunctive syllogism, $A \rightarrow C \in a$. Thus, $C \in a \circ b$. Generalising, $a \circ (a \circ b) \subseteq a \circ b$, that is $a \circ (a \circ b) \leq a \circ b$. ∎

THEOREM A.6   *The frame of the canonical model is a GTL frame.*

THEOREM A.7 (Truth)   *For all $a \in \Upsilon$ and all formulas A, $a \vDash A$ if and only if $A \in a$.*

Proof   Base case: A is a propositional variable. This follows directly from the construction of the canonical model.

Case 1. A is a conjunction, $B \wedge C$. This follows from the inductive hypothesis and the satisfaction condition for conjunction.

Case 2. A is an entailment, $B \to C$. Suppose first that $a \vDash B \to C$. Follows by the inductive hypothesis and Lemma A.4. ∎

THEOREM A.8 (Weak Completeness)   *For any formula A, A is a theorem of GTL if $\Sigma \vDash A$ for all GTL models.*

Proof   Suppose that $\Sigma \vDash A$ for all GTL models. Then, $\Sigma \vDash A$ in the canonical model. However, Σ in the canonical model is just the set of GTL theorems. By the truth theorem, if $a \vDash A$ in the canonical model, then $A \in a$ for all theories A. Thus, if $\Sigma \vDash A$, then A is a theorem of GTL. ∎

THEOREM A.9 (Compactness for the Canonical Model)   *Where $\vDash$ is the consequence relation of the canonical model, for all sets of formulas Γ and all formulas A, $\Gamma \vDash A$ if and only if there is a finite subset Γ' of Γ such that $\Gamma' \vDash A$.*

Proof   Suppose that $\Gamma \vDash A$. This means that every theory in the canonical model that satisfies every member of Γ also satisfies A. Suppose that a satisfies every member of Γ. Then $\Sigma \circ a \vDash G$ for all $G \in \Gamma$ and by assumption, $\Sigma \circ a \vDash A$. By the truth theorem, $G \in \Sigma \circ a$ for all $G \in \Gamma$ and $A \in \Sigma \circ a$. Now, suppose that a is the smallest GTL theory that contains Γ, that is $C_{GTL}(\Gamma)$. Then, there are $G_1, \ldots, G_n \in \Gamma$ such that $(G_1 \wedge \ldots \wedge G_n) \to A$ is a theorem of GTL. Let Γ' be the set $\{G_1, \ldots, G_n\}$. Then, $\Gamma' \vDash A$. ∎

It is a corollary to the compactness theorem for the canonical model that the semantic consequence relation definable over the class of GTL models is itself compact. For suppose that $\Gamma \vDash_{GTL} A$ (where $\vDash_{GTL}$ is the consequence relation for the class of GTL models). Since the canonical model is a GTL model, $\Gamma \vDash A$ in the canonical model. Since the consequence relation in the canonical model is compact, there is a finite Γ' subset of Γ such that $\Gamma' \vDash A$. Since Γ' is finite, the conjunction of all the formulas in Γ', $\wedge\Gamma'$, is a formula, and so is $\wedge\Gamma' \to A$. This entailment is valid on the canonical model, and so it is a theorem of GTL. By soundness, it is valid in the class of GTL models and so $\Gamma' \vDash_{GTL} A$.

A.7   The Logics R and NR

The primitive vocabulary contains conjunction, negation, and implication. Disjunction is defined as

$$A \vee B =_{df} \neg(\neg A \wedge \neg B).$$

The logic R has the following axiomatic basis:

Axioms

1. $A \Rightarrow A$
2. $(B \Rightarrow C) \Rightarrow ((A \Rightarrow B) \Rightarrow (A \Rightarrow C))$
3. $(A \Rightarrow (A \Rightarrow B)) \Rightarrow (A \Rightarrow B)$
4. $A \Rightarrow ((A \Rightarrow B) \Rightarrow B)$
5. $(A \wedge B) \Rightarrow A; (A \wedge B) \Rightarrow B$
6. $((A \Rightarrow B) \wedge (A \Rightarrow C)) \Rightarrow (A \Rightarrow (B \wedge C))$
7. $(A \wedge (B \vee C)) \Rightarrow ((A \wedge B) \vee (A \wedge C))$
8. $\neg\neg A \Rightarrow A$
9. $(A \Rightarrow \neg B) \Rightarrow (B \Rightarrow \neg A)$
10. $(A \Rightarrow \neg A) \Rightarrow \neg A$

Rules

$$\frac{\vdash A \Rightarrow B \qquad \vdash A}{\vdash B} \qquad \frac{\vdash A \qquad \vdash B}{\vdash A \wedge B}$$

To obtain the logic NR, I add $\square$ to the language and the following axioms to the axiomatic basis for R:

1. $\square A \Rightarrow A$
2. $(\square A \wedge \square B) \Rightarrow \square(A \wedge B)$
3. $\square A \Rightarrow \square\square A$

And the following necessitation rule:

$$\frac{\vdash A}{\vdash \square A}$$

A.8 The Logic E

The language includes parentheses, propositional variables, entailment, negation, and conjunction. Disjunction and logical equivalence are defined as usual:

$$A \vee B =_{df} \neg(\neg A \wedge \neg B); \quad A \leftrightarrow B =_{df} (A \to B) \wedge (B \to A).$$

Axiom Schemes

1. $A \to A$ (Identity)
2. $(A \to B) \to (((A \to B) \to C) \to C)$ (Specialised Assertion)
3. $(A \to B) \to ((B \to C) \to (A \to C))$ (Suffixing)
4. $(B \to C) \to ((A \to B) \to (A \to C))$ (Prefixing)
5. $(A \to (A \to B)) \to (A \to B)$ (Contraction)
6. $(A \wedge B) \to A; (A \wedge B) \to B$ (Simplification)
7. $((A \to B) \wedge (A \to C)) \to (A \to (B \wedge C))$ (Conjunction in the Consequent)
8. $A \to (A \vee B); B \to (A \vee B)$ (Addition)
9. $((A \to C) \wedge (B \to C)) \to ((A \vee B) \to C)$ (Constructive Dilemma)

10. $\neg\neg A \rightarrow A$ (DNE)
11. $(A \wedge (B \vee C)) \rightarrow ((A \wedge B) \vee (A \wedge C))$ (Distribution)
12. $(A \rightarrow \neg B) \rightarrow (B \rightarrow \neg A)$ (Contraposition)
13. $(A \rightarrow \neg A) \rightarrow \neg A$ (Reductio)

Rules

$$\frac{A \rightarrow B \quad A}{B} \qquad \frac{A \quad B}{A \wedge B}$$

The following proof is used in the completeness argument for E.

$$\frac{\dfrac{(A \rightarrow B) \rightarrow ((A \wedge \neg B) \rightarrow (\neg A \vee B))}{(A \rightarrow B) \rightarrow ((A \wedge \neg B) \rightarrow \neg(A \wedge \neg B))} \qquad ((A \wedge \neg B) \rightarrow \neg(A \wedge \neg B)) \rightarrow \neg(A \wedge \neg B)}{(A \rightarrow B) \rightarrow \neg(A \wedge \neg B)}$$

By contraposition, $(A \wedge \neg B) \rightarrow \neg(A \rightarrow B)$.

A.9   Semantics for E

An E frame is a quintuple $\langle \Upsilon, \Sigma, \circ, \perp, \leq \rangle$ such that $\langle \Sigma, \circ, \leq \rangle$ is a GTL frame and $\perp$ is a binary relation on Υ such that the following definitions and postulates hold:

$$a^{\perp} =_{df} \{b : a \perp b\}$$

$$a^{c} =_{df} \Upsilon - a^{\perp}$$

$$S =_{df} \{a \in \Upsilon : a^{c} \text{ has a unique maximum element under } \leq \text{ (called } a^{*})\}$$

$$S(a) =_{df} \{b \in S : a \leq b\}$$

$$\Box^{-} a =_{df} a \circ \Sigma$$

Semantic Postulates

1. $\Sigma \circ a = a$;
2. If $a \leq a'$ and $b \leq b'$, then $a \circ b \leq a' \circ b'$;
3. $a \circ (b \circ c) \leq (a \circ b) \circ c$;
4. $(a \circ b) \circ b \leq a \circ b$;
5. $\Box^{-} a \leq a$;
6. $\Box^{-} a \circ \Box^{-} b = \Box^{-} b \circ \Box^{-} a$;
7. If $a \leq b$, then $a^{\perp} \subseteq b^{\perp}$;
8. If $a \perp b$, then $b \perp a$;
9. If $(a \circ b) \perp c$, then $(a \circ c) \perp b$;
10. If $(a \circ b) \perp b$, then $a \perp b$;
11. If $a \in S$, then $a^{*} \in S$;
12. If $a \in S$, then $a^{**} = a$;
13. $\cap\{b^{\perp} : b \in S(a)\} \subseteq a^{\perp}$;
14. $S(a \circ b) = \cup\{S(c \circ b) : c \in S(a)\}$.

The hereditariness lemma and semantic entailment theorem follow in much the same manner as for GTL.

LEMMA A.10   *For all $a \in \Upsilon$ and all formulas A, $a \vDash A$ if and only if $S(a) \subseteq [\![A]\!]$.*

Proof   By induction on the complexity of A.

Base case: A is a propositional variable. Follows directly from the constraints on $\vDash$.

Case 1. A is a conjunction. Follows from the inductive hypothesis and the satisfaction condition for conjunction.

Case 2. A is a negation, $\neg B$. Suppose that $a \vDash A$ and $b \in S(a)$. Then $a \le b$ and so $a^{\perp} \subseteq b^{\perp}$. Since $a \vDash \neg B$, $[\![B]\!] \subseteq a^{\perp}$, and so $[\![B]\!] \subseteq b^{\perp}$.

Suppose that for all $b \in S(a)$, $b \vDash \neg B$. Then $[\![B]\!] \subseteq \bigcap_{b \in S(a)} b^{\perp}$. But $\bigcap_{b \in S(a)} b^{\perp} \subseteq a^{\perp}$, so $[\![B]\!] \subseteq a^{\perp}$ that is $a \vDash \neg B$.

Case 3. A is an entailment, $B \to C$. Suppose that $a \vDash B \to C$ and $b \in S(a)$. Then, by hereditariness, $b \vDash B \to C$.

Suppose now that $b \vDash B \to C$ for all $b \in S(a)$. Assume that $c \vDash B$. I show that $a \circ c \vDash C$. By condition 9, $S(a \circ c)$ is the union of $S(b \circ c)$ for all $b \in S(a)$. By hereditariness, $S(b \circ c) \subseteq [\![C]\!]$. By the inductive hypothesis, $a \circ c \vDash C$.   ∎

LEMMA A.11   *For all $a \in S$, $a \vDash \neg A$ if and only if $a^* \nvDash A$.*

Proof   Suppose that $a \vDash \neg A$. Then if $b \vDash A$, $a \perp b$. But a^* is compatible with a, so $a^* \nvDash A$.

Suppose that $a^* \nvDash A$. By hereditariness, if $b \vDash A$, $b \nleq a^*$. Since a^* is the unique maximal theory compatible with a, if $b \vDash A$, then $a \perp b$. So, $a \vDash \neg A$.   ∎

LEMMA A.12   *$a \vDash A \vee B$ if and only if for all $b \in S(a)$ either $b \vDash A$ or $b \vDash B$.*

Proof   Suppose that $a \vDash A \vee B$. By Lemma A.10, for all $b \in S(a)$, $b \vDash A \vee B$. By the definition of disjunction, for all such b, $b \vDash \neg(\neg A \wedge \neg B)$. By Lemma A.11, $b^* \nvDash \neg A \wedge \neg B$ and so either $b^* \nvDash \neg A$ or $b^* \nvDash \neg B$. Since $b^{**} = b$, either $b \vDash A$ or $b \vDash B$.

Now suppose that for all $b \in S(a)$, $b \vDash A$ or $b \vDash B$. Suppose that $b \vDash A$. By Lemma A.11, $b^* \nvDash \neg A$ and so $b^* \nvDash \neg A \wedge \neg B$. Hence, by Lemma A.11, $b \vDash \neg(\neg A \wedge \neg B)$, that is $b \vDash A \vee B$. The same argument works if $b \vDash B$. Thus, for all $b \in S(a)$, $b \vDash A \vee B$. By Lemma A.10, $a \vDash A \vee B$.   ∎

LEMMA A.13   *$(A \wedge (B \vee C)) \to ((A \wedge B) \vee (A \wedge C))$ is valid in the class of E models.*

Proof   Suppose that $a \vDash A \wedge (B \vee C)$. Then, $a \vDash A$ and $a \vDash B \vee C$. By hereditariness, for every $b \in S(a)$, $b \vDash A$ and by Lemma A.12, either $b \vDash B$ or $b \vDash C$. Thus, for every $b \in S(a)$, $b \vDash A \wedge B$ or $b \vDash A \wedge C$ and, by Lemma A.12, $a \vDash (A \wedge B) \vee (A \wedge C)$. Therefore, by semantic entailment, $\Sigma \vDash (A \wedge (B \vee C)) \to ((A \wedge B) \vee (A \wedge C))$, that is the distribution axiom is valid in the class of E models.   ∎

LEMMA A.14   *$\neg\neg A \to A$ is valid in the class of E models.*

Proof   Suppose that $a \vDash \neg\neg A$. By Lemma A.10, for all $b \in S(a)$, $b \vDash \neg\neg A$. By Lemma A.11, $b^* \nvDash \neg A$ and so by $b^{**} = b$ and Lemma A.11, $b \vDash A$. Therefore, by

Lemma A.10, $a \vDash A$. Generalising, by the semantic entailment theorem, $\Sigma \vDash \neg\neg A \rightarrow A$, that is DNE is valid in the class of E models. ∎

LEMMA A.15 $(A \rightarrow \neg B) \rightarrow (B \rightarrow \neg A)$ *is valid in the class of E models.*

Proof Suppose that $a \vDash A \rightarrow \neg B$ and suppose that $b \vDash B$. I show that $a \circ b \vDash \neg A$. I take a random c such that $c \vDash A$. Then $a \circ c \vDash \neg B$. So $a \circ c \perp b$. By the semantic postulate given above, $a \circ b \perp c$. Thus, $a \circ b \perp c$. Generalising, $a \circ b \vDash \neg A$ as required. ∎

LEMMA A.16 $(A \rightarrow \neg A) \rightarrow A$ *is valid in the class of E models.*

Proof Suppose that $a \vDash A \rightarrow \neg A$. Also suppose that $b \vDash A$. I show that $a \perp b$. By the satisfaction clause for entailment, $a \circ b \vDash \neg A$. Thus, $a \circ b \perp b$. So, $a \perp b$ as required. Thus, $a \vDash \neg A$. ∎

The arguments in Chapter 8 together with the foregoing lemmas show that E is sound on the class of E models. To prove completeness, I construct a canonical model. To do so, I first require a version of the Lindenbaum lemma due to Dov Gabbay [71] and Nuel Belnap (see [53]).

A.10   The Lindenbaum Lemma for Propositional Relevant Logics

Before I state and prove the Lindenbaum lemma, I define a notion that is used often in the semantics of relevant logics and of which I make very heavy use.

DEFINITION A.17 (Coherent Pair) *Two sets of formulas* Γ, Δ *are said to be coherent if and only if* $\Gamma \nvdash A_n \vee \ldots \vee A_n$ *for any* $A_1, \ldots, A_n$ *in* Δ. *If* Γ *and* Δ *are coherent, then* (Γ, Δ) *are said to be a coherent pair.*

LEMMA A.18 (Lindenbaum) *If* (Γ, Δ) *is a coherent pair, then there is a prime theory* Γ' *extending* Γ *for which* (Γ', Δ) *is a coherent pair.*

Proof The proof begins with an enumeration of the formulas, $A_1, \ldots, A_n, \ldots$. Then, Γ' is constructed as follows:

$$\Gamma_0 = \Gamma; \; \Delta_0 = \Delta.$$

If $\Gamma_n \cup \{A_n\}$ and Δ_n are coherent, then $\Gamma_{n+1} = \Gamma_n \cup \{A_n\}$ and $\Delta_{n+1} = \Delta_n$. Otherwise, $\Gamma_{n+1} = \Gamma_n$ and $\Delta_{n+1} = \Delta_n \cup \{A_n\}$.

$$\Gamma' = \bigcup_{n \in \omega} \Gamma_n; \; \Delta' = \bigcup_{n \in \omega} \Delta_n.$$

It can be shown that (Γ', Δ') is a coherent pair and Γ' is a prime theory.

I first prove that if (Γ_n, Δ_n) is coherent, then so is $(\Gamma_{n+1}, \Delta_{n+1})$. Suppose that $(\Gamma_{n+1}, \Delta_{n+1})$ is not coherent. Then, there are some $G_1, \ldots, G_m \in \Gamma_n$ and $D_1, \ldots, D_p \in \Delta_n$ such that $\vdash (G_1 \wedge \ldots \wedge G_m \wedge A_n) \rightarrow (D_1 \vee \ldots \vee D_p)$

and $H_1, \ldots, H_q \in \Gamma_n$ and $E_1, \ldots, E_r \in \Delta_n$ such that $\vdash (H_1 \wedge \ldots \wedge H_q) \rightarrow (E_1 \vee \ldots \vee E_r \vee A_n)$. Combing these, we obtain

$$(a) \vdash (G \wedge H \wedge A_n) \rightarrow (D \vee E)$$

and

$$(b) \vdash (G \vee H) \rightarrow (D \vee E \vee A_n),$$

where G and H are the conjunctions of the G_is and H_is, respectively, and D and E are the disjunctions of the D_is and E_is, respectively. But,

$$\vdash (G \wedge H) \rightarrow (G \wedge H)$$

and so, from (b), we obtain

$$(c) \vdash (G \wedge H) \rightarrow ((D \vee E \vee A_n) \wedge (G \wedge H)).$$

From (c) and distribution, we get

$$(d) \vdash (G \wedge H) \rightarrow ((D \vee E) \vee (G \wedge H \wedge A_n)).$$

From (a) and (d), we obtain

$$(e) \vdash (G \wedge H) \rightarrow ((D \vee E) \vee (D \vee E))$$

that is

$$(f) \vdash (G \wedge H) \rightarrow (D \vee E).$$

But (f) says that (Γ_n, Δ_n) is not coherent, contrary to the assumption. Thus, if (Γ_n, Δ_n) is coherent, then so is $(\Gamma_{n+1}, \Delta_{n+1})$.

Suppose that (Γ', Δ') is not a coherent pair. Then, there are some $G_1, \ldots, G_n \in \Gamma'$ and $D_1, \ldots, D_m \in \Delta'$ such that $(G_1 \wedge \ldots \wedge G_n) \rightarrow (D_1 \vee \ldots \vee D_m)$. This means that at some stage of the enumeration (after all of $G_1, \ldots, G_n, D_1, \ldots, D_m$ are included), an incoherent pair is constructed. However, if (Γ_0, Δ_0) is coherent, then every constructed pair is coherent. So, (Γ', Δ') is a coherent pair.

I now show that Γ' is a prime theory. Suppose that $G_1, \ldots, G_n \in \Gamma'$ and $\vdash (G_1 \wedge \ldots \wedge G_n) \rightarrow A$. If A is not in Γ', then it is in Δ'. If it is in Δ', then (Γ', Δ') is incoherent. Since this pair is coherent, $A \in \Gamma'$. Thus, Γ' is an E theory.

Suppose that $A \vee B \in \Gamma'$. If neither A nor B is in Γ', they are both in Δ'. However, $\vdash (A \vee B) \rightarrow (A \vee B)$, so if neither A nor B is in Γ', then (Γ', Δ') is incoherent. Thus, at least one of A or B is in Γ'. Generalising, Γ' is prime. ∎

A.11   Canonical Model

Before I define the canonical model for propositional E, I state a definition that is used often in the rest of this appendix:

DEFINITION A.19 (Prime Theory)   *A theory a is prime if and only if, for all disjunctions in a, at least one disjunct is also in a.*

The canonical model for E is sextuple $\langle \Upsilon, \Sigma, \circ, \perp, \leq, \vDash \rangle$, where Υ is the set of E theories, Σ is the set of E theorems, $\circ$ and $\leq$ are defined as for the canonical model for GTL, and $\perp$ is defined as

$a \perp b$ if and only if there is a formula A such that $A \in b$ and $\neg A \in a$.

I also define S as the set of prime E theories, and for all $a \in S$, a^* is the set of formulas A such that $\neg A \notin a$.

I now prove that S and $*$ behave in accordance with the definitions and postulates of E frames.

LEMMA A.20   *For all $a \in S$, a^* is the unique maximal theory in a^c.*

Proof I first show that a^* is an E theory. Suppose that $A_1, \ldots, A_n \in a^*$ and $(A_1 \wedge \ldots \wedge A_n) \to B$ is a theorem of E. By definition of a^*, none of $\neg A_1, \ldots, \neg A_n$ is in a. By contraposition, $\neg B \to \neg(A_1 \wedge \ldots \wedge A_n)$ is a theorem of E and so is $\neg B \to (\neg A_1 \vee \ldots \vee \neg A_n)$. Since a is prime, and none of $\neg A_1, \ldots, \neg A_n$ is in a, $\neg A_1 \vee \ldots \vee \neg A_n$ is not in a and since a is an E theory, $\neg B$ is not in a. Thus, $B \in a^*$. Generalising, a^* is a theory.

Now suppose that b is a theory such that $a \not\perp b$. I show that $b \subseteq a^*$. Assume that $A \in b$. By the definition of $\perp$ for the canonical model, $\neg A \notin a$. So, by definition of a^*, $A \in a^*$. Thus, a^* is the unique maximal element of a^c.   ∎

LEMMA A.21   *For any E theory a, if a^c has a unique maximal element, then $a \in S$.*

Proof Suppose that a^c has a unique maximal element, a' and $A \vee B \in a$. Assume also that $A \notin a$ and $B \notin a$. Then, $[\neg A]$ and $[\neg B]$ are both compatible with a. For, if $[\neg A]$ were incompatible with a, then there would be some formula $\neg C \in a$ such that $\neg A \to C$ is a theorem of E. Then, $\neg C \to A$ would be a theorem of E and A would be in a. The same holds for $[\neg B]$. Since a' is the *unique* maximal element of a^c, then both $\neg A \in a'$ and $\neg B \in a'$ and $\neg A \wedge \neg B \in a'$. Since a' is compatible with a, $\neg A \wedge \neg B \notin a$, but then by the definition of $\vee$, $A \vee B \notin a$, contradicting the assumption of the proof. Generalising, a is prime, hence $a \in S$.   ∎

LEMMA A.22   *If $a \in S$, then $a^* \in S$.*

Proof Suppose that $a \in S$. I have already shown that a^* is a theory. I now show that a^* is prime. Assume that $A \vee B \in a^*$. Then $\neg(A \vee B) \notin a$ and so $\neg A \wedge \neg B \notin A$. Thus, either $\neg A$ or $\neg B$ is not in a and so either A or B is in a^*. Generalising, a^* is prime, that is $a^* \in S$.   ∎

LEMMA A.23   *If $a \in S$, then $a^{**} = a$.*

Proof Suppose that $a \in S$ and that $A \in a^{**}$. By Lemma A.22, a^{**} is a prime theory. By the definition of $*$, $\neg A \notin a^*$ and so $\neg\neg A \in a$. a is an E theory, so $A \in a$. Generalising, $a^{**} \subseteq a$.

Now suppose that $A \in a$. Then $\neg\neg A \in a$, so $\neg A \notin a^*$, and so $A \in a^{**}$. Generalising, $a \subseteq a^{**}$. Therefore, $a^{**} = a$.   ∎

LEMMA A.24 $\bigcap_{b \in S(a)} b^\perp \subseteq a^\perp$

Proof   Suppose that for all $b \in S(a)$, $b \perp c$. Then, for each $b \in S(a)$, there is an $A \in c$ such that $\neg A \in b$. Suppose, for the sake of a reductio, that there is no $A \in c$ such that $\neg A \notin a$. Let a^* be the set of A such that $\neg A \notin a$. a^* may not be a theory, but it is closed under disjunction. Suppose that A and B are both in a^*. Then, $\neg A$ and $\neg B$ are not in a. But $\neg(A \vee B) \to (\neg A \wedge \neg B)$ is a theorem of E, and so $\neg(A \vee B) \notin a$. Hence, $A \vee B \in a^*$. Thus, (a, a^*) is a coherent pair and there is a prime extension a' of a such that $a' \not\perp c$. Therefore, $\bigcap_{b \in S(a)} b^\perp \subseteq a^\perp$.   ∎

LEMMA A.25   *For any theory c, $S(a \circ c)$ is the union of the set of $S(b \circ c)$ for $b \in S(a)$.*

Proof   Assume $b \in S(a)$. It is obvious that if $d \in S(b \circ c)$, then $d \in S(a \circ c)$. So, it suffices to show that if $d \in S(a \circ c)$ there is some $b \in S(a)$ such that $d \in S(b \circ c)$.

Suppose that $d \in S(a \circ c)$. Then d is prime. Let α be the disjunctive closure of the set of formulas $A \to B$ such that $A \in c$ but $B \notin d$. Then (a, α) is a coherent pair. For suppose that $A \to B$ and $A' \to B'$ are both in α and $(A \to B) \vee (A' \to B') \in a$. Then $(A \wedge A') \to (B \vee B') \in a$. But A and A' are both in c, so $B \vee B' \in a \circ c$ and hence $B \vee B' \in d$. But d is prime, so either B or B' is in d contrary to the assumption. Thus, (a, α) is coherent and, by the Lindenbaum lemma, there is a prime extension b of a such that (b, α) is coherent. But then $b \circ c \leq d$, hence $d \in S(b \circ c)$ as required.   ∎

LEMMA A.26   *If $a \circ b \perp c$, then $a \circ c \perp b$.*

Proof   Suppose that $a \circ b \perp c$. Then there is a formula $B \in c$ such that $\neg B \in a \circ b$. Thus, there is an $A \in b$ such that $A \to \neg B \in a$. By contraposition, $B \to \neg A \in a$. Thus, $\neg A \in a \circ c$ and so $a \circ c \perp b$.   ∎

LEMMA A.27   *If $a \circ b \perp b$, then $a \perp b$.*

Proof   Suppose that $a \circ b \perp b$. Then there is a $B \in b$ such that $\neg B \in a \circ b$. Thus, there is an $A \in b$ and $A \to \neg B \in a$. $A \wedge B \in b$ and so $\neg(A \to \neg B) \in b$. Thus, $a \perp b$.   ∎

LEMMA A.28 $a \circ (b \circ c) \leq (a \circ b) \circ c$

Proof   Suppose that $C \in a \circ (b \circ c)$. Then there is some $B \in b \circ c$ such that $B \to C \in a$ and some $A \in c$ such that $A \to B \in b$. By prefixing, $(A \to B) \to (A \to C) \in a$ and $A \to C \in a \circ b$. Thus, $C \in (a \circ b) \circ c$.   ∎

LEMMA A.29 $(a \circ b) \circ b \leq a \circ b$

Proof   Suppose that $C \in (a \circ b) \circ b$. Then there is a $B \in b$ such that $B \to C \in a \circ b$. Thus, there is an $A \in b$ such that $A \to (B \to C) \in a$. So, $(A \wedge B) \to C \in a$ and $A \wedge B \in b$. Therefore, $C \in a \circ b$.   ∎

LEMMA A.30 $A \in \Box^- a$ *if and only if $t \to A \in a$.*

Proof   Suppose that $A \in \Box^- a$. Then $A \in a \circ \Sigma$. $\Sigma = [t]$. Hence, there is some formula B such that $B \to A \in a$ and $t \to B$ is a theorem of E. Thus, $t \to A \in a$.   ∎

LEMMA A.31   *(i) $\Box^- a \leq a$ and (ii) $\Box^- a \circ \Box^- b = \Box^- b \circ \Box^- a$.*

Proof   (i) Follows directly from $(t \rightarrow A) \rightarrow A$.

(ii) Suppose that $B \in \Box^-(a) \circ \Box^- b$. Then there is an $A \in \Box^- b$ such that $A \rightarrow B \in \Box^- a$. Thus, by Lemma A.30, $t \rightarrow A \in b$, that is $\Box A \in b$ and so $\Box\Box A \in b$. Hence, $\Box A \in \Box^- b$ and so $(\Box A \rightarrow B) \rightarrow B \in \Box^- b$. So, $B \in \Box^- b \circ \Box^- a$.   ∎

The foregoing lemmas show that the frame of the canonical model is an E frame. Thus, the canonical model is an E model. Extending the truth theorem to cover all of E is quite easy. The only additional case that we need to prove is for negation. To prove this, we assume first that $a \vDash \neg A$. Then, by the satisfaction condition for negation, if $b \vDash A$, then $a \perp b$. By the inductive hypothesis, $b \vDash A$ if and only if $A \in b$. Thus, $a \vDash \neg A$ if and only if for all b such that $A \in b$, $a \perp b$. Let us consider $[A]$. By assumption, $a \perp [A]$. There is some formula B such that $\neg B \in a$ and $A \rightarrow B$ is a theorem of E. Thus, $\neg B \rightarrow \neg A$ is a theorem and so $\neg A \in a$.

Now suppose that $\neg A \in a$. Thus, if $A \in b$, then $a \perp b$. By the inductive hypothesis, $A \in b$ if and only if $b \vDash A$. So, if $b \vDash A$, then $a \perp b$, and so $a \vDash \neg A$.

Since Σ is the set of theorems of E, by the truth theorem, all and only theorems of E are valid on the canonical model. The canonical model is an E model, and so only theorems of E are valid on the class of E models. The compactness of the consequence relation for the canonical model (and for the class of E models) follows as it does for GTL.

A.12   World-Like Theories of E

A regular theory is a theory that contains all the theorems of the logic. In this section, I prove that there is a prime, regular theory W (for "world") such that $\Box^- W$ is exactly the set of theorems of E. My proof is a simple variation on Meyer's method of metavaluations [143]. I begin with an arbitrary set of propositional variables, PV and construct W around that set.

W is defined as follows:

- For any propositional variable p, $p \in W$ if and only if $p \in PV$.
- For any formula of the form $\neg A$, it is in W if and only if $A \notin W$.
- For any formula of the form $A \wedge B$, it is in W if and only if $A \in W$ and $B \in W$.
- For any formula of the form $A \rightarrow B$, it is in W if and only if it is a theorem of E and either $A \notin W$ or $B \in W$.
- t is in W.

It follows directly from the clause for entailment that if $A \rightarrow B \in W$, then $A \rightarrow B$ is a theorem of E. I now show that every theorem of E is in W.

LEMMA A.32   *All instances of (a) $A \rightarrow A$, (b) $(B \rightarrow C) \rightarrow ((A \rightarrow B) \rightarrow (A \rightarrow C))$, and (c) $(A \rightarrow (A \rightarrow B)) \rightarrow (A \rightarrow B)$ are in W.*

Proof (a) Follows directly from the clause for entailment and the fact that $A \rightarrow A$ is a theorem.

(b) $(B \rightarrow C) \rightarrow ((A \rightarrow B) \rightarrow (A \rightarrow C))$ is a theorem. Suppose that $B \rightarrow C \in W$. By the clause for entailment, $B \rightarrow C$ is a theorem and so $(A \rightarrow B) \rightarrow (A \rightarrow C)$ is a theorem. Similarly, suppose that $A \rightarrow B \in W$. Then $A \rightarrow B$ is a theorem and so is $A \rightarrow C$. Now suppose that A is in W. $A \rightarrow B \in W$, so $B \in W$. $B \rightarrow C \in W$, so $C \in W$. Thus, $(B \rightarrow C) \rightarrow ((A \rightarrow B) \rightarrow (A \rightarrow C)) \in W$.

(c) $(A \rightarrow (A \rightarrow B)) \rightarrow (A \rightarrow B)$ is a theorem. Suppose that $A \rightarrow (A \rightarrow B) \in W$ and that $A \in W$. Then $A \rightarrow (A \rightarrow B)$ is a theorem and so is $A \rightarrow B$. Since $A \rightarrow (A \rightarrow B)$ and that A are both in W, so is $A \rightarrow B$ and so $B \in W$. Thus, $(A \rightarrow (A \rightarrow B)) \rightarrow (A \rightarrow B) \in W$. ∎

LEMMA A.33 *All instances of $(A \rightarrow B) \rightarrow (((A \rightarrow B) \rightarrow C) \rightarrow C)$ are in W.*

Proof $(A \rightarrow B) \rightarrow (((A \rightarrow B) \rightarrow C) \rightarrow C)$ is a theorem of E. Suppose that $A \rightarrow B \in W$. Then $A \rightarrow B$ is a theorem and so $((A \rightarrow B) \rightarrow C) \rightarrow C$ is a theorem. Now suppose that $(A \rightarrow B) \rightarrow C \in W$. By the clause for entailment and the assumption that $A \rightarrow B \in W$, $C \in W$. Hence, $((A \rightarrow B) \rightarrow C) \rightarrow C \in W$ and so $(A \rightarrow B) \rightarrow (((A \rightarrow B) \rightarrow C) \rightarrow C)$ is in W. ∎

LEMMA A.34 *(a) $(A \wedge B) \rightarrow A$ and $(A \wedge B) \rightarrow B$, (b) $(A \rightarrow C) \rightarrow ((A \wedge B) \rightarrow C)$, and (c) $((A \rightarrow B) \wedge (A \rightarrow C)) \rightarrow (A \rightarrow (B \wedge C))$ are all in W.*

Proof (a) Follows directly from the clauses for entailment and conjunction.

(b) $(A \rightarrow C) \rightarrow ((A \wedge B) \rightarrow C)$ is a theorem. Suppose that $A \rightarrow C$ is in W. Then it is a theorem and so is $(A \wedge B) \rightarrow C$. Now suppose that $A \wedge B \in W$. By the clause for conjunction, $A \in W$ and by the clause for entailment, $C \in W$. Thus, by the clause for entailment, $(A \wedge B) \rightarrow C \in W$ and so is $(A \rightarrow C) \rightarrow ((A \wedge B) \rightarrow C)$.

(c) $((A \rightarrow B) \wedge (A \rightarrow C)) \rightarrow (A \rightarrow (B \wedge C))$ is a theorem. Suppose that $(A \rightarrow B) \wedge (A \rightarrow C) \in W$, then by the clauses for conjunction and entailment, $A \rightarrow B$ and $A \rightarrow C$ are both in W and are theorems. Now suppose that A is in W. By the clause for entailment, B and C are in W, and so $B \wedge C \in W$. Thus, by the clause for entailment, $((A \rightarrow B) \wedge (A \rightarrow C)) \rightarrow (A \rightarrow (B \wedge C))$ is in W. ∎

LEMMA A.35 *For all A, B, $A \vee B$ is in W if and only if at least one of A or B is in W.*

Proof $A \vee B =_{df} \neg(\neg A \wedge \neg B)$. Suppose that $A \vee B \in W$, that is $\neg(\neg A \wedge \neg B) \in W$. By the clause for negation, $\neg A \wedge \neg B$ is not in W, and by the clause for conjunction at least one of $\neg A$ or $\neg B$ is not in W. By the clause for negation, this means that at least one of A or B is in W.

Now suppose that A is in W. Then $\neg A$ is not in W and neither is $\neg A \wedge \neg B$. Hence, $\neg(\neg A \wedge \neg B)$ is in W and so $A \vee B$ is in W. The case when B is in W is proved in the same way. ∎

LEMMA A.36 *All instances of (a) $((A \rightarrow C) \wedge (B \rightarrow C)) \rightarrow ((A \vee B) \rightarrow C)$ and (b) $(A \wedge (B \vee C)) \rightarrow ((A \wedge B) \vee (A \wedge C))$ are in W.*

Proof (a) $((A \rightarrow C) \wedge (B \rightarrow C)) \rightarrow ((A \vee B) \rightarrow C)$ is a theorem. Suppose that $(A \rightarrow C) \wedge (B \rightarrow C)$ and $A \vee B$ are in W. Then $A \rightarrow C$ and $B \rightarrow C$ are theorems. By Lemma A.35, at least one of A or B is in W. Suppose $A \in W$. By the clause for conjunction, $A \rightarrow C \in W$ and so $C \in W$ and similarly if $B \in W$. Thus, $(A \vee B) \rightarrow C \in W$ and $((A \rightarrow C) \wedge (B \rightarrow C)) \rightarrow ((A \vee B) \rightarrow C) \in W$.

(b) $(A \wedge (B \vee C)) \rightarrow ((A \wedge B) \vee (A \wedge C))$ is a theorem. Suppose that $A \wedge (B \vee C)$ is in W. By the clause for conjunction, $A \in W$ and $B \vee C \in W$. By Lemma A.35, $B \in W$ or $C \in W$ and so either $A \wedge B \in W$ or $A \wedge C \in W$. By Lemma A.35 again, $(A \wedge B) \vee (A \wedge C) \in W$. Therefore, by the clause for entailment, $(A \wedge (B \vee C)) \rightarrow ((A \wedge B) \vee (A \wedge C))$ is in W. ∎

LEMMA A.37   *Every instance of the following theses is in W: (a) $\neg\neg A \rightarrow A$; (b) $(A \rightarrow \neg B) \rightarrow (B \rightarrow \neg A)$; (c) $(A \rightarrow \neg A) \rightarrow \neg A$.*

Proof (a) $\neg\neg A \rightarrow A$ is a theorem. Suppose that $\neg\neg A \in W$. Then $\neg A \notin W$ and $A \in W$.

(b) $(A \rightarrow \neg B) \rightarrow (B \rightarrow \neg A)$ is a theorem. Suppose that $A \rightarrow \neg B \in W$ and suppose that $B \in W$. Then $A \rightarrow \neg B$ is a theorem and so is $B \rightarrow \neg A$. $\neg B \notin W$ and so $A \notin W$. Thus, $\neg A \in W$ and $B \rightarrow \neg A \in W$.

(c) $(A \rightarrow \neg A) \rightarrow \neg A$ is a theorem. Suppose that $A \rightarrow \neg A \in W$. Thus, $A \notin W$, for otherwise $\neg A \in W$ and according to the clause for negation this is impossible (I am using a classical inference here). So, $\neg A \in W$. ∎

LEMMA A.38   *All instances of the following theorem are in W: $(A \rightarrow B) \rightarrow (t \rightarrow (A \rightarrow B))$.*

Proof $(A \rightarrow B) \rightarrow (t \rightarrow (A \rightarrow B))$ is a theorem. Suppose that $A \rightarrow B$ is in W, then it is a theorem and so is $t \rightarrow (A \rightarrow B)$. By the clause for entailment, $t \rightarrow (A \rightarrow B) \in W$. ∎

That W is closed under modus ponens follows directly from the clause for entailment and that it is closed under adjunction follows directly from the clause for conjunction. To show that W is closed under necessitation, suppose that A is a theorem of E and that $A \in W$. Then $t \rightarrow A$ is a theorem and by the clause for entailment, $t \rightarrow A$ is in W. Thus, I can state the following theorem:

THEOREM A.39   *Every theorem of E is in W.*

Now I can prove the theorem that is the reason for all this work.

THEOREM A.40 $\square^- W = \Sigma$, *where Σ is the set of theorems of E.*

Proof Suppose that A is a theorem of E. Then $t \rightarrow A$ is in W and $A \in W \circ \Sigma$. Thus, $\Sigma \subseteq \square^- W$.

Now suppose that $A \in W \circ \Sigma$. Then there is some formula $B \in \Sigma$ such that $B \rightarrow A \in W$. But if $B \rightarrow A \in W$, $B \rightarrow A$ is a theorem that is $B \rightarrow A$ is in Σ. Since Σ is closed under modus ponens, $A \in \Sigma$. Thus, $\square^- W \subseteq \Sigma$. ∎

Thus, W is a regular, prime theory such that $\Box^{-}W$ is just the set of theorems of E. This makes it plausible to say that in fact the entailments that are true at the actual world are just the theorems of E. W is also consistent and I use that fact in the proof, but the consistency of the actual world is not a claim that I want to defend in this book. I am saying only that this proof makes it plausible that E is the complete true theory of entailment and I think this proof accomplishes that.

A.13   Semantics for QE

I assume a standard quantificational language with individual constants $(c_1, \ldots, c_n, \ldots)$ and variables $(x_0, \ldots, x_n, \ldots)$, the connectives of E together with $\forall$. The axiomatic basis for QE adds to the axioms and rules for E all instances of the following schemes,

$$(\forall x A \wedge \forall x B) \rightarrow \forall x (A \wedge B) \text{ (Aggregation)}$$

$$\forall x A \rightarrow A[\tau/x] \text{ where } x \text{ is free for } \tau \text{ in } A \text{ (Universal Instantiation)}$$

and the following rule:

$$\frac{\vdash A \rightarrow B}{\vdash A \rightarrow \forall x B}$$

where x is not free in A.

A QE frame is an octuple $\langle \Upsilon, \Sigma, \circ, \bot, \leq, D, \mathsf{Prop}, \mathsf{PropFun} \rangle$ where $\langle \Upsilon, \Sigma, \circ, \bot, \leq \rangle$ is an E frame, D is a non-empty set (the domain), Prop is a set of subsets of Υ, and $\mathsf{PropFun}$ is a set of functions from D^{ω} into Prop, that satisfies the following definitions and conditions. Where $X, Y \subseteq \Upsilon$,

$$X \Rightarrow Y = \{a \in \Upsilon : \forall x \in X \, (a \circ x \in Y)\}$$

$$-X = \{a \in \Upsilon : \forall x \in X \, (a \bot x)\}$$

Where φ and ψ are in $\mathsf{PropFun}$ and $f \in D^{\omega}$,

$$(\varphi \Rightarrow \psi)(f) = \varphi(f) \Rightarrow \psi(f)$$

$$(\varphi \cap \psi)(f) = \varphi(f) \cap \psi(f)$$

$$(-\varphi)(f) = -(\varphi(f))$$

And, where $X \subseteq \mathsf{Prop}$,

$$\bigsqcap X = \bigcup \left\{ Y \in \mathsf{Prop} : Y \subseteq \bigcap X \right\}.$$

$$(\forall_n \varphi)(f) = \bigsqcap_{i \in D} (\varphi)(f[i/n])$$

where $f[i/n]$ is the function that is just like f except that it gives i as the value of n.

Hereditariness Conditions: For all $\Phi \in$ Prop and all $a, b \in \Upsilon$, if $a \in \Phi$ and $a \leq b$, then $b \in \Phi$ and if $S(a) \subset \Phi$, then $a \in \Phi$.

Closure Conditions: Prop and PropFun are closed under $\cap$, $\Rightarrow$, and $-$. The set of all theories a such that $\Sigma \leq a$ is also in Prop. And for all $n \in \omega$ and all $\varphi \in$ PropFun, $(\forall_n)(\varphi)$ is in PropFun.

A QE model is a pair $\langle \mathfrak{F}, V \rangle$, where V is a function from names to D and from n-place predicates to functions from D^n to Prop. Where $f \in D^\omega$ and τ is either a name or a variable, $V_f(\tau)$ is $f(n)$ if τ is x_n and $V(\tau)$ if τ is a name. We then set

$$V(P\tau_1 \ldots \tau_n)(f) = V(P)(V_f(\tau_1), \ldots, V_f(\tau_n))$$

We stipulate that every atomic formula determines a propositional function. This means that where A is an atomic formula, there is some φ in PropFun such that for all $f \in D^\omega$, $V(A)(f) = \varphi(f)$. The valuation V determines a satisfaction relation $\vDash$ between theories, assignment functions, and formulas. The satisfaction clause for atomic formulas is

$$a \vDash_f P\tau_1 \ldots \tau_n \text{ if and only if } a \in V(P\tau_1 \ldots \tau_n)(f)$$

The relation $\vDash_f$ satisfies the usual clause for atomic statements and the clauses for the propositional connectives of $\vDash$ in E frames, with an additional clause for the universal quantifier:

$a \vDash_f \forall x A$ if and only if there exists $\Phi \in$ Prop, $a \in \Phi$ and for all $b \in \Phi$ and all $i \in D$ $b \vDash_{f[i\,n]} A$

The following proofs are sketches, provided to give the reader a sense of how the model theory works. The details can be found in [134].

LEMMA A.41   *Every formula A determines a propositional function.*

Proof   By induction on the complexity of A.

Base case: A is an atomic formula. Follows by stipulation.

Case 1. A is $B \wedge C$. By the inductive hypothesis, there are propositional functions φ and ψ that are determined by B and C, respectively. $\varphi \cap \psi$ is a propositional function and is determined by $B \wedge C$.

Case 2. A is $\neg B$. Then A determines $-\varphi$ where B determines φ.

Case 3. A is $B \rightarrow C$. Then A determines $\varphi \Rightarrow \psi$ where B determines φ and C determines ψ.

Case 4. A is $\forall x_n B$. Then A determines $\forall_n \varphi$ where B determines φ.   ∎

LEMMA A.42 $(\forall x_n A \wedge \forall x_n B) \rightarrow \forall x_n (A \wedge B)$ *is valid on the class of QE models.*

Proof   Suppose that $a \vDash_f \forall x_n A \wedge \forall x_n B$. Then there are propositions Φ and Ψ such that $a \in \Phi$, $a \in \Psi$, and $\Phi \subseteq \bigcap_{i \in D} |A|_{f[i/n]}$ and $\Psi \subseteq \bigcap_{i \in D} |B|_{f[i/n]}$. Prop is closed under intersection, so $\Phi \cap \Psi$ is in Prop. And, by Lemma A.41, $A \wedge B$ determines a propositional function. Thus, a is in a proposition that is a subset of $\bigcap_{i \in D} |A \wedge B|_{f[i/n]}$ and so $a \vDash_f \forall x_n (A \wedge B)$. By the semantic entailment theorem, $(\forall x_n A \wedge \forall x_n B) \rightarrow \forall x_n (A \wedge B)$ is valid on the class of QE models.   ∎

A tedious but simple induction on the complexity of formulas suffices to prove the following lemma:

LEMMA A.43   *If x_n is not free in A and $a \vDash_f A$, then $a \vDash_g A$, where g is an n-variant of f.*

LEMMA A.44   *If $A \to B$ is valid on the class of QE models and x_n is not free in A, then $A \to \forall x_n B$ is valid in the class of QE models.*

Proof   Suppose that $A \to B$ is valid on the class of QE models. And suppose that $a \vDash_f A$. Then, let g be some n-variant of f. By Lemma A.43, since x_n is not free in A, $a \vDash_g A$ and so $a \vDash_g B$. Where φ is the propositional function determined by A and ψ is the propositional function determined by B, $\varphi(f)$ is a subset of $\psi(f[i/n])$ for every $i \in D$ and $a \in \varphi(f)$. Thus, $a \vDash_f \forall x_n B$, and by the semantic entailment theorem, $A \to \forall x_n B$ is valid on the class of QE models.   ∎

It is easy to prove that the universal instantiation axiom is valid. Together with the proofs for propositional E, the soundness theorem follows.

The canonical model is very like the canonical model for propositional E. The theories of the model are just EQ theories, but which include only formulas without free variables (sentences). The domain is the set of names, which is countable. The propositions are just the set of $[\![A]\!]$s, where A is a sentence. The set of propositional functions are the $|A|$s where A is a (perhaps open) formula with at most $y_1, \ldots, y_n$ free and for every $f \in D^\omega$, $|A|(f)$ is $A[f(y_1)/y_1, \ldots, f(y_n)/y_n]$. It can then be shown that the canonical model is an EQ model. Moreover, it is straightforward to show, for every open formula A of EQ, that $\Sigma \in |A|(f)$ for all value assignments f if and only if A is a theorem of EQ. Hence, EQ is complete over the class of EQ models. Compactness follows from the completeness proof in the same way as it does for GTL.

A.14   World-Like Theories of QE

In this section, I extend the proof in Section A.12 to cover QE. That is I show that there is a regular prime theory W such that $\Box^- W = \Sigma$. I add to the definition of W a clause for the universal quantifier, viz.,

$$\forall x A \in W \text{ if and only if } A[c/x] \text{ for all names } c.$$

The proof that all the theorems of the propositional logic are in W remains the same as in Section A.12. I only need to prove the theorems for the quantificational axioms and rules here.

LEMMA A.45   *All instances of the following theses are in W: (a) $\forall x A \to A[c/x]$ and (b) $(\forall x A \land \forall x B) \to \forall x(A \land B)$.*

Proof   (a) Follows directly from the clauses for entailment and the universal quantifier.

(b) Suppose that $\forall x A \land \forall x B \in W$. By the clause for conjunction, $\forall x A \in W$ and $\forall x B \in W$. By the clause for the universal quantifier, $A[c/x]$ and $B[c/x]$ are in W for all names c. By the clause for conjunction, $(A \land B)[c/x]$ is in W for all names c. By the clause for the universal quantifier, $\forall x(A \land B) \in W$.   ∎

LEMMA A.46   *If $A \rightarrow B$ is a theorem of QE and is in W and c is free for x in B, and c does not occur in A, then $A \rightarrow \forall x B$ is in W.*

Proof   Suppose that $A \rightarrow B$ is a theorem and is in W. Then, $A \rightarrow \forall x B$ is a theorem. Suppose that $A \in W$, then $B[d/c] \in W$ for all names d, and so $\forall x B[x/c] \in W$, where c is free for x in B. Thus, $A \rightarrow \forall x B \in W$.   ∎

Therefore, W is a prime regular theory and is all-complete. Since $A \lor \neg A$ is a theorem of E, W is also world-like. The proof in Section A.12 suffices to show that $\Box^- W = \Sigma$.

A.15   The Prawitz-Style Proof Systems

Σ is the name of the master theory, $x, y, \ldots$ are variables, ranging over theories, and $a, b, \ldots$ are parameters and can stand for variables or for complex theory terms. Σ is a theory term. Every variable is a theory term. If a and b are theory terms, so is $(a \circ b)$. Where no confusion results, I drop outer parentheses, but in the real notation, these must remain.

A formula A is valid if it can be proven in a derivation that $\Sigma \vDash A$. The line $\Sigma \vDash t$ can be used at any point in a derivation. The other rules governing Σ are

$$\frac{\Sigma \circ a \vDash A}{a \vDash A} \qquad \frac{a \vDash A}{\Sigma \circ a \vDash A} \qquad \frac{a(\Sigma) \vDash A \qquad b \vDash t}{a(b) \vDash A},$$

where $a(b)$ results from replacing one or more occurrences of Σ with b in $a(\Sigma)$.

Here are the entailment and conjunction rules:

$$[x \vDash A]$$

$$\frac{a \vDash A \rightarrow B \qquad b \vDash A}{a \circ b \vDash B} \qquad \begin{array}{c} \vdots \\ a \circ x \vDash B \\ \hline a \vDash A \rightarrow B \end{array} \qquad \frac{a \vDash A \land B}{a \vDash A} \qquad \frac{a \vDash A \land B}{a \vDash B}$$

$$\frac{a \vDash A \qquad a \vDash B}{a \vDash A \land B}$$

In the $\rightarrow$ introduction rule, x cannot occur in a.

And here are the negation rules and the disjunction introduction rules:

$$[x \vDash A] \qquad [x \vDash \neg A]$$

$$\frac{a \vDash \neg A \qquad b \vDash A}{a \perp b} \qquad \begin{array}{c} \vdots \\ a \perp x \\ \hline a \vDash \neg A \\ a \vDash B \\ \hline a \vDash A \lor B \end{array} \qquad \begin{array}{c} \vdots \\ a \perp x \\ \hline a \vDash A \\ a \perp b \\ \hline b \perp a \end{array} \qquad \frac{a \vDash A}{a \vDash A \lor B}$$

And finally here is the disjunction elimination rule:

$$
\cfrac{
a \vDash A \wedge (B \vee C) \qquad
\cfrac{[x \vDash A \wedge B] \\ \vdots \\ b \circ x \vDash D}{} \qquad
\cfrac{[x \vDash A \wedge C] \\ \vdots \\ b \circ x \vDash D}{}
}{a \circ b \vDash D}
$$

In this last rule, x cannot occur in b.

To get the logics GTL, one needs to add the rule CS derived from the following chart. In the system for the logic E, all of these rules are valid. $\Box^-$ is defined as in Chapter 8, that is

$$\Box^- a =_{df} a \circ \Sigma.$$

Name	Semantic Postulate	Thesis
Prefixing	$a \circ (b \circ c) \le (a \circ b) \circ c$	$(B \to C) \to ((A \to B) \to (A \to C))$
Suffixing	$a \circ (b \circ c) \le (b \circ a) \circ c$	$(A \to B) \to ((B \to C) \to (A \to C))$
Contraction	$(a \circ b) \circ b \le a \circ b$	$(A \to (A \to B)) \to (A \to B)$
Conjunctive Syllogism (CS)	$a \circ (a \circ b) \le a \circ b$	$((A \to B) \wedge (B \to C)) \to (A \to C)$
Specialised Assertion	$\Box^- a \circ \Box^- b = \Box^- b \circ \Box^- a$	$(A \to B) \to (((A \to B) \to C) \to C)$
Entailment T	$\Box^- a \le a$	$(t \to A) \to A$

The semantic postulates are converted into rules by replacing a term a with a term b such that the postulate says that $a \le b$.

A.16   A Note on the Prawitz-Style Systems for E and QE

To prove that the Prawitz-style system for propositional E is correct, I first add the fusion connective ($\otimes$) into the language and add the following rules to the logic:

$$
\cfrac{\vdash A \to (B \to C)}{\vdash (A \otimes B) \to C} \qquad
\cfrac{\vdash (A \otimes B) \to C}{\vdash A \to (B \to C)}
$$

I also adopt a translation τ from a proof in the Prawitz-style system to a proof in E where $\tau(x)$ is just A, where $x \vDash A$ is a hypothesis of the proof, $\tau(a \circ b) = \tau(a) \otimes \tau(b)$, $\tau(\Sigma) = t$, $\tau(\Box^- a) = \tau(a) \otimes t$, and $\tau(a \vDash A) = \tau(a) \to A$. Here is an example of how the translation works. Consider the following derivation:

$$
\cfrac{
[x \vDash B \to C]^3 \qquad
\cfrac{
\cfrac{
\cfrac{
\cfrac{[y \vDash A \to B]^2 \qquad [z \vDash A]^1}{y \circ z \vDash B}
}{x \circ (y \circ z) \vDash C}
}{(x \circ y) \circ z \vDash C}
}{x \circ y \vDash A \to C} {\scriptstyle 1}
}{
\cfrac{
\cfrac{x \vDash (A \to B) \to (A \to C)}{\Sigma \circ x \vDash (A \to B) \to (A \to C)} {\scriptstyle 2}
}{\Sigma \vDash (B \to C) \to ((A \to B) \to (A \to C))} {\scriptstyle 3}
}
$$

The hypothesis lines come out as trivial: $(B \rightarrow C) \rightarrow (B \rightarrow C)$, $(A \rightarrow B) \rightarrow (A \rightarrow B)$, and $A \rightarrow A$. As such, they can be omitted. The translation then comes out as:

$$\frac{\dfrac{\dfrac{\dfrac{\dfrac{((A \rightarrow B) \otimes A) \rightarrow B}{((B \rightarrow C) \otimes ((A \rightarrow B) \otimes A)) \rightarrow C}}{(((B \rightarrow C) \otimes (A \rightarrow B)) \otimes A) \rightarrow C}}{((B \rightarrow C) \otimes (A \rightarrow B)) \rightarrow (A \rightarrow C)}}{(B \rightarrow C) \rightarrow ((A \rightarrow B) \rightarrow (A \rightarrow C))}}$$

I could continue to translate the last two lines, but this is unnecessary. The proof of prefixing is complete. To convince yourself that the proof works, just work backwards from the last line and you can see that each line from the third one down is a theorem of E with fusion. The second line can be proved in the axiom system in the following manner:

$$\frac{\dfrac{((A \rightarrow B) \otimes A) \rightarrow B}{(B \rightarrow C) \rightarrow (((A \rightarrow B) \otimes A) \rightarrow C)}}{((B \rightarrow C) \otimes ((A \rightarrow B) \otimes A)) \rightarrow C}$$

The first line follows (of both proofs) from the rules governing fusion and $(A \rightarrow B) \rightarrow (A \rightarrow B)$.

I cannot use exactly the same trick, however, for quantificational logic. It seems very unlikely that adding fusion to QE creates a conservative extension (intensional confinement would be a theorem if it were conservative and I doubt very much that it is – although I currently have no proof of this). In order to avoid this problem, I borrow a strategy from Nuel Belnap in his construction of a Display Logic for the system E [3, §62]. Belnap adds Boolean negation to E, but does not allow it to occur within the scope of any other operator. I do something similar with fusion. Like Belnap, I use the semantics as the framework in which to prove the accuracy of the proof theory. As usual, I set the truth condition for fusion as

$$a \vDash A \otimes B \text{ if and only if there are } b, c, \ b \vDash A, \ c \vDash B \text{ and } b \circ c \leq a.$$

I do *not* require Prop to be closed under fusion. The class of frames used is just the class of QE frames.

I set out three languages – language-1, language-2, and language-3. Every formula of the original language is a type-1 formula. Every type-1 formula is a type-2 formula. If A and B are type-2 formulas, then $A \otimes B$ is a type-2 formula. And if A is a type-2 formula and B is a type-1 formula, then $A \rightarrow B$ is a type-3 formula. It can then be proven that every type-3 formula that appears in the translation of a correct derivation in the Prawitz-style system is a theorem of QE. The class of valid type-1 formulas is closed under the theorems and rules of QE, but the class of valid type-3 theorems is closed under the theorems and rules of propositional E. We can, however, chain valid formulas of type-1 with type-3, as is shown in the proofs given below.

Every line of the translation of a valid deduction in the Prawitz-style system for QE is valid in the class of QE models. The proof for the propositional logic shows that the rules for the propositional connectives preserve validity. Consider then the introduction rule for the universal quantifier:

$$\frac{\Sigma \vDash A \rightarrow B(x) \qquad a \vDash A}{a \vDash \forall x\, B(x)}$$

where x does not occur in A. The translation is

$$\frac{t \rightarrow (A \rightarrow B(x)) \qquad \tau(a) \rightarrow A}{\tau(a) \rightarrow \forall x\, B(x)}$$

By the inductive hypothesis, $\tau(a) \rightarrow A$ is valid and so is $t \rightarrow (A \rightarrow B(x))$. Since t is valid, so is $A \rightarrow B(x)$. Thus, by the rule of intensional confinement, $A \rightarrow \forall x\, B(x)$ is valid and by transitivity, so is $\tau(a) \rightarrow \forall x\, B(x)$.

That the universal elimination rule and the existential introduction rule preserve validity is obvious. Now consider the existential elimination rule:

$$\frac{\Sigma \vDash A(x) \rightarrow B \qquad a \vDash \exists x\, A(x)}{a \vDash B}$$

where x does not occur free in B. The translation is

$$\frac{t \rightarrow (A(x) \rightarrow B) \qquad \tau(a) \rightarrow \exists x\, A(x)}{\tau(a) \rightarrow B}$$

Here is a proof of validity preservation:

$$\frac{\dfrac{\dfrac{\dfrac{\dfrac{\dfrac{\vDash t \rightarrow (A(x) \rightarrow B)}{\vDash A(x) \rightarrow B}}{\vDash \neg B \rightarrow \neg A(x)}}{\vDash \neg B \rightarrow \forall x \neg A(x)}}{\vDash \neg \forall x \neg x\, A(x) \rightarrow \neg\neg B}}{\vDash \exists x\, A(x) \rightarrow B} \qquad \vDash \tau(a) \rightarrow \exists x\, A(x)}{\vDash \tau(a) \rightarrow B}$$

References

[1] Wilhelm Ackermann. Begründung einer strenge Implikation. *The Journal of Symbolic Logic*, 21:113–128, 1956.

[2] Alan Anderson. Some open problems concerning the system E of entailment. *Acta Philosophica Fennica*, 16:9–18, 1963.

[3] Alan Anderson, Nuel D. Belnap, and J. M. Dunn. *Entailment: Logic of Relevance and Necessity*, volume II. Princeton University Press, Princeton, 1992.

[4] Alan Anderson and Nuel D. Belnap. *Entailment: Logic of Relevance and Necessity*, volume I. Princeton University Press, Princeton, 1975.

[5] Richard B. Angell. Deducibility, entailment and analytic containment. In J. Norman and R. Sylvan, editors, *Directions in Relevant Logic*, pages 119–143. Kluwer, Dordrecht, 1989.

[6] Aristotle. Prior analytics. In *Aristotle, Complete Works*, volume 1, pages 39–113. Princeton University Press, Princeton, 1985.

[7] Thomas Baldwin. C.I. Lewis and the analyticity debate. In Erich H. Reck, editor, *The Historical Turn in Analytic Philosophy*, pages 201–228. Palgrave Macmillan, Basingstoke, 2013.

[8] Jon Barwise and John Perry. *Situations and Attitudes*. MIT Press, Cambridge, MA, 1983.

[9] David Basin, Marcello D'Augustino, Dov Gabbay, Sean Matthews, and Luca Vigano, editors. *Labelled Deduction*. Springer, Dordrecht, 2000.

[10] Jc Beall. There is no logical negation: True, false, both, and neither. *The Australasian Journal of Logic*, 14:1–29, 2017.

[11] Jc Beall, Ross Brady, Alan Hazen, Graham Priest, and Greg Restall. Relevant restricted quantification. *Journal of Philosophical Logic*, 35:587–598, 2006.

[12] Jc Beall and Julian Murzi. Two flavors of Curry's paradox. *The Journal of Philosophy*, 110:143–165, 2013.

[13] Oskar Becker. Zur Logik der Modalitäten. *Jahrbuch für philosophie und philosophische Forschung*, 9:497–548, 1930.

[14] John L. Bell. Infinitary logic. In Edward N. Zalta, editor, *The Stanford Encyclopedia of Philosophy*. Metaphysics Research Lab, Stanford University, Stanford, CA, Winter 2016 edition, 2016.

[15] John Bigelow. *The Reality of Numbers*. Oxford University Press, Oxford, 1988.

[16] Katalin Bimbó and J. Michael Dunn. Larisa Maksimova's early contributions to relevance logic. In S. Odintsov, editor, *Larisa Maksimova on Implication, Interpolation, and Definability*. Springer Verlag, Basel, 2018.

[17] Katalin Bimbó, J. Michael Dunn, and Nicholas Ferenz. Two manuscripts: One by Routley, one by Meyer: The origins of the Routley-Meyer semantics for relevance logics. *The Australasian Journal of Logic*, 15:171–209, 2018.

[18] David Bloor. *Knowledge as Social Imagery*. University of Chicago Press, Chicago, second edition, 1991.

[19] Bernard Bolzano. *Theory of Science*. Oxford University Press, Oxford, 2014.

[20] George Boole. *An Investigation of the Laws of Thought*. Macmillan, London, 1854. I used the 1958 Dover reprint.

[21] George Boolos. *The Logic of Provability*. Cambridge University Press, Cambridge, 1993.

[22] Ross Brady. A content semantics for quantified relevant logic I. *Studia Logica*, 47: 111–127, 1988.

[23] Ross Brady. Gentzenization and decidability of some contraction-less relevant logics. *Journal of Philosophical Logic*, 20:97–117, 1991.

[24] Ross Brady. Normalized natural deduction systems for some relevant logics I: The logic DW. *The Journal of Symbolic Logic*, 71:35–66, 2006.

[25] Ross Brady. *Universal Logic*. Center for the Study of Language and Information, Stanford, CA, 2006.

[26] Bryson Brown and Graham Priest. Chunk and permeate: An inconsistent inference strategy. Part I: The infinitesimal calculus. *Journal of Philosophical Logic*, 33:379–388, 2004.

[27] John Buridan. *Treatise on Consequences*. Fordham University Press, New York, 2015. Translated with an introduction by Stephen Read and an editorial introduction by Hubert Hubien.

[28] Rudolf Carnap. Testability and meaning. *Philosophy of Science*, 3:419–471, 1936.

[29] Rudolf Carnap. *Logical Syntax of Language*. Routledge and Kegan Paul, London, 1937.

[30] Rudolf Carnap. On inductive logic. *Philosophy of Science*, 12:72–97, 1945.

[31] Rudolf Carnap. Modalities and quantification. *The Journal of Symbolic Logic*, 11:33–64, 1946.

[32] Rudolf Carnap. Meaning postulates. *Philosophical Studies*, 3:65–73, 1952.

[33] Rudolf Carnap. *Meaning and Necessity: A Study in Semantics and Modal Logic*. University of Chicago Press, Chicago, second edition, 1956.

[34] Rudolf Carnap. Carnap's intellectual autobiography. In Paul A. Schilpp, editor, *The Philosophy of Rudolf Carnap*, pages 3–86. Open Court, La Salle, IL, 1963.

[35] Rudolf Carnap. *An Introduction to the Philosophy of Science*. Dover Publications, Garden City, NY, 1995. Originally published in 1966.

[36] Brian Chellas. *Modal Logic: An Introduction*. Cambridge University Press, Cambridge and New York, 1980.

[37] Alonzo Church. The weak theory of implication. In Albert Menne and Alexander Wilhelmy, editors, *Kontrolliertes Deken, Untersuchungen zum Logikkalulkül und zur Logik der Eizelwissenschaften*, pages 22–37. Kommissions-Verlag Karl Alber, Freiburg, 1951.

[38] Alonzo Church. *Introduction to Mathematical Logic*. Princeton University Press, Princeton, 1956.

[39] Nino B. Cocchiarella. On the primary and secondary semantics of logical necessity. *Journal of Philosophical Logic*, 4:13–27, 1975.

[40] Nino B. Cocchiarella. *Logical Studies in Early Analytic Philosophy*. Ohio State University Press, Columbus, 1987.

[41] B. Jack Copeland. The genesis of possible world semantics. *Journal of Philosophical Logic*, 31:99–137, 2002.

[42] Max Cresswell. The completeness of Carnap's modal predicate logic. *The Australasian Journal of Logic*, 11:46–61, 2014.

[43] Max Cresswell. Carnap's modal predicate logic. In Max Cresswell, Edwin Mares, and Adriane Rini, editors, *Logical Modalities from Aristotle to Carnap*, pages 298–316. Cambridge University Press, Cambridge, 2016.

[44] M. J. Cresswell. The interpretation of some Lewis systems of modal logic. *Australasian Journal of Philosophy*, 45:198–206, 1967.

[45] M. J. Cresswell. S1 is not so simple. In Walter Sinnott-Armstrong, Diana Raffman, and Nicholas Asher, editors, *Modality, Morality, and Belief*, pages 29–40. Cambridge University Press, Cambridge, 1995.

[46] M. J. Cresswell. Revisiting McKinsey's 'syntactical' construction of modality. *The Australasian Journal of Logic*, 17:123–140, 2020.

[47] John N. Crossley and Lloyd Humberstone. The logic of 'actually'. *Reports on Mathematical Logic*, 3:11–29, 1977.

[48] Martin Davies and Lloyd Humberstone. Two notions of necessity. *Philosophical Studies*, 38:1–30, 1980.

[49] Kosta Došen. The first axiomatisation of relevant logic. *Journal of Philosophical Logic*, 21:339–356, 1992.

[50] Kosta Došen. Modal logic as metalogic. *Journal of Logic, Language, and Information*, 1:173–201, 1992.

[51] J. Michael Dunn. Natural versus formal languages. Given at an American Philosophical Association meeting, 1968.

[52] J. Michael Dunn. Relevant Robinson's arithmetic. *Studia Logica*, 38:407–418, 1979.

[53] J. Michael Dunn. Relevance logic and entailment. In D. M. Gabbay and F. Guenthner, editors, *Handbook of Philosophical logic*, volume III, pages 117–224. Kluwer, Dordrecht, 1984.

[54] J. Michael Dunn. Relevant predication I: The formal theory. *Journal of Philosophical Logic*, 16:347–381, 1987.

[55] J. Michael Dunn. Relevant predication II: Intrinsic properties and internal relations. *Philosophical Studies*, 60, 1990.

[56] J. Michael Dunn. Star and perp. *Philosophical Perspectives*, 7:331–357, 1993.

[57] J. Michael Dunn and Greg Restall. Relevance logic. In D. M. Gabbay and F. Guenthner, editors, *Handbook of Philosophical Logic*, volume VI, pages 1–128. Springer Verlag, Basel, second edition, 2002.

[58] Güunther Eder. Frege's 'On the foundations of geometry' and axiomatic metatheory. *Mind*, 125:5–40, 2016.

[59] John Etchemendy. *The Concept of Logical Consequence*. Harvard University Press, Cambridge, MA, 1990.

[60] Nicholas Ferenz. Identity in relevant logics: A relevant predication approach. In Martin Blicha and Igor Sedlar, editors, *The Logica Yearbook 2020*, pages 49–64. College Publications, London, 2021.

[61] Kit Fine. Models for entailment. *Journal of Philosophical Logic*, 3:347–372, 1974.

[62] Kit Fine. *Reasoning with Arbitrary Objects*. Blackwell, Oxford, 1985.

[63] Kit Fine. Analytic implication. *Notre Dame Journal of Formal Logic*, 27:169–179, 1986.

[64] Kit Fine. Semantics for quantified relevance logic. *Journal of Philosophical Logic*, 14:27–59, 1988.

[65] Kit Fine. Incompleteness for quantified relevance logics. In Richard Sylvan and Jean Norman, editors, *Directions in Relevant Logic*, pages 205–255. Springer, Dordrecht, 1989.

[66] Frederick B. Fitch. Natural deduction rules for obligation. *American Philosophical Quarterly*, 3:27–38, 1966.

[67] Gottlob Frege. *Philosophical and Mathematical Correspondence*. University of Chicago Press, Chicago, 1980.

[68] Gottlob Frege. On the foundations of geometry. In Brian McGuiness, editor, *Collected Papers on Mathematics, Logic, and Philosophy*, pages 293–340. Blackwell, 1984.

[69] Harvey Friedman and Robert K. Meyer. Whither relevant arithmetic. *The Journal of Symbolic Logic*, 57:824–831, 1992.

[70] André Fuhrmann. Models for relevant modal logics. *Studia Logica*, 49:501–514, 1990.

[71] Dov M. Gabbay. On second-order intuitionist propositional calculus with full comprehension. *Archive for Mathematical Logic*, 16:177–186, 1974.

[72] Ronald N. Giere. *Explaining Science*. University of Chicago Press, Chicago, 1988.

[73] J. D. Goheen and J. L. Mothershead, editors. *Collected Papers of Clarence Irving Lewis*. Stanford University Press, Stanford, CA, 1970.

[74] Robert Goldblatt. Semantic analysis of orthologic. *Journal of Philosophical Logic*, 3:19–35, 1974.

[75] Robert Goldblatt. Mathematical modal logic: A view of its evolution. In Dov Gabbay and John Woods, editors, *Handbook of the History of Logic*, volume 7, pages 1–98. Elsevier, Leiden, 2006.

[76] Robert Goldblatt. *Quantifiers, Propositions, and Identity*. Cambridge University Press, Cambridge, 2011.

[77] Robert Goldblatt and Michael Kane. An admissible semantics for propositionally quantified relevant logics. *Journal of Philosophical Logic*, 39:73–100, 2010.

[78] Anil Gupta and Nuel Belnap. *The Revision Theory of Truth*. MIT Press, Cambridge, MA, 1993.

[79] Jacques Herbrand. Investigations in proof theory. In Warren D. Goldfarb, editor, *Jacques Herbrand, Logical Writings*, pages 44–202. Harvard University Press, Cambridge, MA, 1971.

[80] Jaakko Hintikka. Standard vs. nonstandard logic: Higher-order, modal, and first-order logics. In E. Agazzi, editor, *Modern Logic – A Survey*, pages 283–296. D. Reidel, Dordrecht, 1980.

[81] Jaakko Hintikka and Merrill Hintikka. *The Logic of Epistemology and the Epistemology of Logic*. D. Reidel, Dordrecht, 1989.

[82] George Hughes and Max Cresswell. *An Introduction to Modal Logic*. Methuen, York, 1968.

[83] George Hughes and Max Cresswell. *A New Introduction to Modal Logic*. Routledge, London, 1996.

[84] Lloyd Humberstone. Operational semantics for positive *R*. *Notre Dame Journal of Formal Logic*, 29:61–80, 1988.

[85] Lloyd Humberstone. Smiley's distinction between rules of inference and rules of proof. In J. Lear and A. Oliver, editors, *The Force of Argument: Essays in Honour of Timothy Smiley*, pages 107–126. Routledge, London, 2010.

[86] Lloyd Humberstone. *The Connectives*. MIT Press, Cambridge, MA, 2011.

[87] Bruce Hunter. Clarence Irving Lewis. In Edward N. Zalta, editor, *The Stanford Encyclopedia of Philosophy*. Metaphysics Research Lab, Stanford University, Stanford, CA, Spring 2021 edition, 2021.

[88] Frank Jackson. *From Metaphysics to Ethics: A Defence of Conceptual Analysis*. Oxford University Press, Oxford, 1998.

[89] Bjarni Jónsson and Alfred Tarski. Boolean algebras with operators, part I. *American Journal of Mathematics*, 73:891–939, 1951.

[90] David Kaplan. Demonstratives. In Joseph Almog and John Perry, editors, *Themes from Kaplan*, pages 481–564. Oxford University Press, Oxford, 1989.

[91] Philip Kremer. Quantifying over propositions in relevance logic: The non-axiomatisability of primary interpretations of ∀p and ∃p. *The Journal of Symbolic Logic*, 58:334–349, 1993.

[92] Philip Kremer. Relevant identity. *Journal of Philosophical Logic*, 28:199–222, 1999.

[93] Saul A. Kripke. Semantical analysis of modal logic II. Non-normal modal propositional calculi. In J. W. Addison, L. Henkin, and A. Tarski, editors, *The Theory of Models*, pages 206–220. North-Holland, Amsterdam, 1965.

[94] Saul A. Kripke. An outline of a theory of truth. *Journal of Philosophy*, 72:690–716, 1975.

[95] Bruce Kuklick. *The Rise of American Philosophy*. Yale University Press, New Haven, 1977.

[96] Casimir Kuratowski. Sur l'operation $\overline{A}$ de l'Analysis Situs. *Fundamenta Mathematicae*, 3:182–199, 1922.

[97] Christine Ladd-Franklin. On the algebra of logic. In C. S. Peirce, editor, *Studies in Logic by Members of the John Hopkins University*, pages 17–71. Johns Hopkins, Baltimore, 1883.

[98] Mark Lance and Philip Kremer. The logical structure of linguistic commitment II: Systems of relevant entailment commitment. *Journal of Philosophical Logic*, 25:425–449, 1996.

[99] Saunders Mac Lane. A logical analysis of mathematical structure. *The Monist*, 45:118–130, 1935.

[100] E. J. Lemmon. New foundations for Lewis modal systems. *The Journal of Symbolic Logic*, 22:176–186, 1957.

[101] John Lemmon. *Beginning Logic*. Thomas Nelson and Sons, London, 1965.

[102] C. I. Lewis. Implication and the algebra of logic. *Mind*, 21:522–531, 1912.

[103] C. I. Lewis. The issues concerning material implication. *Journal of Philosophy, Psychology, and Scientific Methods*, 14:350–356, 1917.

[104] C. I. Lewis. *Survey of Symbolic Logic*. University of California Press, Berkeley, first edition, 1918.

[105] C. I. Lewis. Strict implication – An emendation. *The Journal of Philosophy, Psychology and Scientific Methods*, 17:300–302, 1920.

[106] C. I. Lewis. Facts, systems, and the unity of the world. *The Journal of Philosophy*, 20:141–151, 1923. Reprinted in [73, pp 383–393]. Page references are to the reprinted version.

[107] C. I. Lewis. A pragmatic conception of the *a priori*. *The Journal of Philosophy*, 20:169–177, 1923. Reprinted in [73, pp 231–239]. Page references are to the reprinted version.

[108] C. I. Lewis. *Mind and the World Order: Outline of a Theory of Knowledge*. Charles Scribner and Sons, New York, 1929.

[109] C. I. Lewis. Alternative systems of logic. *The Monist*, 17:481–507, 1932. Reprinted in [73, pp 400–419]. Page references are to the reprinted version.

[110] C. I. Lewis. *Analysis of Knowledge and Valuation*. Open Court, LaSalle, IL, 1946.

[111] C. I. Lewis. The material and conceptual in logic and philosophy. In the C. I. Lewis papers of the Stanford Archives. Call number M0174 Box 10 Folder 8, 1948.

[112] C. I. Lewis. Notes on the logic of intension. In H. H. M. Kallen and S. Langer, editors, *Structure, Method and Meaning: Essays in Honor of Henry M. Sheffer*, pages 25–34. Bobbs-Merrill, New York, 1951. Reprinted in [73, pp 420–429]. Page references are to the reprinted version.

[113] C. I. Lewis and C. H. Langford. *Symbolic Logic*. Dover, New York, first edition, 1951. Originally published in 1932.

[114] C. I. Lewis and C. H. Langford. *Symbolic Logic*. Dover, New York, second edition, 1959.

[115] David Lewis. *On the Plurality of Worlds*. Blackwell, Oxford, 1986.

[116] Casimir Lewy. *Meaning and Modality*. Cambridge University Press, Cambridge, 1976.

[117] Shay Logan. Notes on stratified semantics. *Journal of Philosophical Logic*, 48:749–786, 2019.

[118] Fraser MacBride. Truthmakers. In Edward N. Zalta, editor, *The Stanford Encyclopedia of Philosophy*. Metaphysics Research Lab, Stanford University, Stanford, CA, Fall 2021 edition, 2021.

[119] Hugh MacColl. *Symbolic Logic and Its Applications*. Longmans, Green, and Co., London, 1906.

[120] Hugh MacColl. 'If' and 'imply'. *Mind*, 17:453–455, 1908.

[121] John MacFarlane. What does it mean to say that logic is formal? PhD thesis, University of Pittsburgh, 2000.

[122] Larisa Maksimova. A semantics for the calculus E of entailment. *Bulletin of the Section of Logic*, 2:18–21, 1973.

[123] Ruth Barcan Marcus. The deduction theorem in a functional calculus of first order based on strict implication. *The Journal of Symbolic Logic*, 11:115–118, 1946.

[124] Ruth Barcan Marcus. Strict implication, deducibility, and the deduction theorem. *The Journal of Symbolic Logic*, 18:234–236, 1953.

[125] Edwin Mares. Semantics for relevance logic with identity. *Studia Logica*, 51:1–20, 1992.

[126] Edwin Mares. Classically complete modal relevant logics. *Mathematical Logic Quarterly*, 39:165–177, 1993.

[127] Edwin Mares. A star-free semantics for R. *The Journal of Symbolic Logic*, 60:579–590, 1995.

[128] Edwin Mares. CE is not a conservative extension of E. *Journal of Philosophical Logic*, 29:263–275, 2000.

[129] Edwin Mares. *Relevant Logic: A Philosophical Interpretation*. Cambridge University Press, Cambrdge, 2004.

[130] Edwin Mares. General information in relevant logic. *Synthese*, 167:343–362, 2009.

[131] Edwin Mares. Conjunction and relevance. *Journal of Logic and Computation*, 22:7–21, 2012.

[132] Edwin Mares. Relevant logic and the philosophy of mathematics. *Philosophy Compass*, 7:481–494, 2012.

[133] Edwin Mares. From iff to is: Some new thoughts on identity in relevant logic. In Can Başkent and Thomas Macalay Ferguson, editors, *Graham Priest on Dialetheism and Paraconsistency*, pages 343–363. Springer Verlag, Basel, 2019.

[134] Edwin Mares and Robert Goldblatt. An alternative semantics for quantified relevant logic. *The Journal of Symbolic Logic*, 71:163–187, 2006.

[135] Edwin Mares and Francesco Paoli. Logic consequence and the paradoxes. *Journal of Philosophical Logic*, 43:439–469, 2014.

[136] Edwin Mares and Francesco Paoli. C.I. Lewis, E.J. Nelson, and the modern origins of connexive logic. *Organon F*, 26:405–426, 2019.

[137] Edwin Mares and Shawn Standefer. The relevant logic E and some close neighbours. *The IFCoLog Journal of Logics and Their Applications*, 4:695–730, 2017.

[138] Tim Maudlin. *Philosophy of Physics: Quantum Theory*. Princeton University Press, Princeton and Oxford, 2019.

[139] Grover Maxwell. The ontological status of theoretical entitles. *Minnesota Studies in the Philosophy of Science*, 3:3–27, 1962.

[140] J. C. C. McKinsey. On the syntactic construction of systems of modal logic. *The Journal of Symbolic Logic*, 10:83–94, 1944.

[141] J. C. C. McKinsey and Alfered Tarski. The algebra of topology. *Annals of Mathematics*, 45:141–191, 1944.

[142] Robert K. Meyer. Entailment and relevant implication. *Logique et analyse*, 11:472–479, 1968.

[143] Robert K. Meyer. Metacompleteness. *Notre Dame Journal of Formal Logic*, 17:501–516, 1976.

[144] Robert K. Meyer and J. Michael Dunn. E, R, and γ. *The Journal of Symbolic Logic*, 34:460–474, 1969.

[145] Robert K. Meyer and Errol Martin. S (for syllogism) revisited. *The Australasian Journal of Logic*, 16:49, 2019.

[146] Robert K. Meyer and Chris Mortensen. Inconsistent models for relevant arithmetic. *The Journal of Symbolic Logic*, 49:917–929, 1984.

[147] G. E. Moore. External and internal relations. *Proceedings of the Aristotelian Society*, 20:40–62, 1920.

[148] Chris Mortensen. *Inconsistent Mathematics*. Springer Verlag, Dordrecht, 1995.

[149] Everett J. Nelson. Towards an intensional logic of propositions. PhD thesis, Harvard University, Cambridge, MA, 1929.

[150] Everett J. Nelson. Intensional relations. *Mind*, 39:440–453, 1930.

[151] Everett J. Nelson. On three logical principles in intension. *The Monist*, 43:268–284, 1933.

[152] Everett J. Nelson. A note on contradiction: A protest. *The Philosophical Review*, 45:505–508, 1936.

[153] Isaac Newton. *Mathematical Principles of Natural Philosophy*. University of California Press, Berkeley, 1934.

[154] Daniel Nolan. Impossible worlds: A modest approach. *Notre Dame Journal of Formal Logic*, 38:535–572, 1997.

[155] Daniel Nolan. Reflections on Routley's ultralogic program. *The Australasian Journal of Logic*, 15:407–430, 2018.

[156] Hiroakira Ono. Semantics for substructural logics. In Kosta Došen and Peter Schröder-Heister, editors, *Substructural Logics*, pages 259–291. Oxford University Press, Oxford, 1993.

[157] I. Orlov. The calculus of compatibility of propositions (in Russian). *Matematicheskii Sbornik*, 35:263–286, 1928.

[158] William T. Parry. Implication. PhD thesis, Harvard University, Cambridge, MA, 1932.

[159] William T. Parry. Analytic implication: Its history, justification and varieties. In J. Norman and R. Sylvan, editors, *Directions in Relevant Logic*, pages 101–118. Kluwer, Dordrecht, 1989.

[160] Casper-Emil T. Pederson, Anders Albrechtsen, Paul D. Etter, Eric A. Johnson, Ludovic Orlando, Louns Chikhi, Hans R. Siegismund, and Rasmus Heller. Southern African origin and cryptic structure in the highly mobile plains zebra. *Nature Ecology and Evolution*, 2:491–498, 2018.

[161] Emil Post. Introduction to a general theory of propositions. *American Journal of Mathematics*, 43:163–185, 1921.

[162] Emil L. Post. Introduction to a general theory of elementary propositions. PhD thesis, Columbia University, New York, 1921.

[163] Graham Priest. The logic of paradox. *Journal of Philosophical Logic*, 8:219–241, 1979.

[164] Graham Priest. What is a non-normal world? *Logique et analyse*, 139:291–302, 1992.

[165] Graham Priest. Inconsistent models of arithmetic, Part I. *Journal of Philosophical Logic*, 26:223–235, 1997.

[166] Graham Priest. *An Introduction to Non-classical Logic: From If to Is*. Cambridge University Press, Cambridge, second edition, 2008.

[167] Graham Priest. Mission impossible. In Yale Weiss and Romina Padro, editors, *Saul Kripke on Modal Logic*. Springer Verlag, Basel, forthcoming.

[168] W. V. O. Quine. Two dogmas of empiricism. *Philosophical Review*, 60:20–43, 1951. Reprinted in [170].

[169] W. V. O. Quine. Reply to Professor Marcus. *Synthese*, 13:323–330, 1961.

[170] W. V. O. Quine. *From a Logical Point of View*. Harvard University Press, Cambridge, MA, third edition, 1980.

[171] W. V. O. Quine. *Mathematical Logic*. Harvard University Press, Cambridge, MA, 1981.

[172] F. P. Ramsey. The foundations of mathematics. *Proceedings of the London Mathematical Society*, 25:338–384, 1925. Reprinted in [174]. Page references are to the reprinted version.

[173] F. P. Ramsey. Theories. In [174], pages 112–139. Originally written in 1929 and published posthumously in 1931.

[174] F. P. Ramsey. *Philosophical Papers*. Cambridge University Press, Cambridge, 1990. Edited by D. H. Mellor.

[175] Steven Read. *Relevant Logic: The Philosophical Interpretation of Inference*. Blackwell, Oxford, 1988.

[176] Stephen Read. Hugh MacColl and the algebra of strict implication. *Nordic Journal of Philosophical Logic*, 3:53–84, 1998.

[177] Hans Reichenbach. *Elements of Symbolic Logic*. Macmillan, London, 1947.

[178] Greg Restall. *Introduction to Substructural Logics*. Routledge, London, 2000.

[179] Greg Restall. Multiple conclusions. *Analysis and Metaphysics*, 6:14–34, 2007.

[180] Greg Restall and Gillian Russell. Barriers to implication. In Charles Pigden, editor, *Hume on Is and Ought*, pages 243–258. Palgrave Macmillan, London, 2010.

[181] David Ripley. Extending classical logic with transparent truth. *Review of Symbolic Logic*, 5:354–378, 2012.

[182] Richard Routley. *Ultralogic and Universal: The Sylvan Jungle*, volume 4. Springer Verlag, Basel, 2019. Edited by Zach Weber.

[183] Richard Routley and Robert K. Meyer. Semantics for entailment II. *Journal of Philosophical Logic*, 1:53–73, 1972.

[184] Richard Routley and Robert K. Meyer. Dialectical logic, classical logic, and the consistency of the world. *Studies in Soviet Thought*, 16:1–25, 1976.

[185] Bertrand Russell. *Introduction to Mathematical Philosophy*. George Allen and Unwin, London, 1919.

[186] Bertrand Russell. *The Analysis of Matter*. Dover, New York, 1954. Reprint of the 1927 book.

[187] Bertrand Russell. *The Philosophy of Logical Atomism*. Open Court, La Salle, IL, 1985.

[188] Gillian Russell. The justification of the basic laws of logic. *Journal of Philosophical Logic*, 44:793–803, 2015.

[189] Alexander Sandgren and Koji Tanaka. Two kinds of logical impossibility. *Noûs*, 54: 795–806, 2020.

[190] Dana Scott. Engendering an illusion of understanding. *Journal of Philosophy*, 68: 787–807, 1971.

[191] Moh Shaw-Kwei. The deduction theorem and two new logical systems. *Methodos*, 2:56–75, 1950.

[192] Gila Sher. *The Bounds of Logic: A Generalized Viewpoint*. MIT Press, Cambridge, MA, 1991.

[193] John Slaney. A general logic. *Australasian Journal of Philosophy*, 68:74–89, 1990.

[194] Timothy J. Smiley. Entailment and deducibility. *Proceedings of the Aristotelian Society*, 59:233–354, 1958–1959.

[195] Timothy J. Smiley. Relative necessity. *Journal of Symbolic Logic*, 28:113–134, 1963.

[196] Nicholas J. J. Smith. Frege's judgement stroke and the conception of logic as the study of inference not consequence. *Philosophy Compass*, 4:639–665, 2009.

[197] Robert Stalnaker. Assertion. *Syntax and Semantics*, 9:315–332, 1978.

[198] Shawn Standefer. Identity in Mares-Goldblatt models for quantified relevant logic. *Journal of Philosophical Logic*, 50:1389–1415, 2021.

[199] Frederick Suppe. The search for philosophical understanding of scientific theories. In Frederick Suppe, editor, *The Structure of Scientific Theories*, pages 3–232. University of Illinois Press, Urbana, IL, second edition, 1977.

[200] Stanisław J. Surma. The deduction theorem in certain fragments of the Lewis system S2 and the system of Feys-Von Wright. *Bulletin of the Section of Logic*, 1, 1972.

[201] Alfred Tarski. Fundamental concepts of the methodology of the deductive sciences. In [203], pages 60–109. 1930.

[202] Alfred Tarski. On the concept of logical consequence. In [203], pages 409–420. 1936.

[203] Alfred Tarski. *Logic, Semantics, Metamathematics: Papers from 1923 to 1938*. Hackett, Indianapolis, second edition, 1983.

[204] Neil Tennant. *Anti-realism and Logic: Truth as Eternal*. Oxford University Press, Oxford, 1987.

[205] Alasdair Urquhart. Semantics for relevance logics. *The Journal of Symbolic Logic*, 37:159–169, 1972.

[206] Alasdair Urquhart. Intensional languages via nominalization. *Pacific Philosophical Quarterly*, 63:186–192, 1982.

[207] Alasdair Urquhart. Relevance logic: Problems open and closed. *The Australasian Journal of Logic*, 13:11–20, 2016.

[208] Alasdair Urquhart. The story of γ. In K. Bimbó, editor, *J. Michael Dunn on Information Based Logics*, pages 93–105. Springer, Basel, 2016.

[209] Bas C. van Fraassen. *The Scientific Image*. Oxford University Press, Oxford, 1980.

[210] Bas C. van Fraassen. *Laws and Symmetry*. Oxford University Press, Oxford, 1989.

[211] Karin Verelst. Newton versus Leibniz: Intransparency versus inconsistency. *Synthese*, 191:2907–2940, 2014.

[212] Peter Vickers. *Understanding Inconsistent Science*. Oxford University Press, Oxford, 2013.

[213] Albert Visser. Four-valued semantics and the liar. *Journal of Philosophical Logic*, 13:181–212, 1984.

[214] Heinrich Wansing and Matthias Unterhuber. Connexive conditional logic, part I. *Logic and Logical Philosophy*, 28:567–610, 2019.

[215] Zach Weber. Transfinite numbers in paraconsistent set theory. *The Review of Symbolic Logic*, 3:71–92, 2010.

[216] Yale Weiss. Connexive extensions of regular conditional logic. *Logic and Logical Philosophy*, 28:611–627, 2019.

[217] Yale Weiss. New(ish) foundations for theories of entailment. In Yale Weiss and Romina Padro, editors, *Saul Kripke on Modal Logic*. Springer Verlag, Basel, forthcoming.

[218] Alfred N. Whitehead and Bertrand Russell. *Principia Mathematica*. Cambridge University Press, Cambridge, first edition, 1910–1913. (Merchant Books reprint, 2009).

[219] Ludwig Wittgenstein. *Tractatus Logico-Philosophicus*. Routledge, London, 1974.

Index

abduction/inference to the best explanation, 223, 224
Abelard, Peter, 12
Ackermann, Wilhelm, 70, 71, 89–93, 166–168
Anderson, Alan, xi, 2, 5, 9, 70, 71, 74–77, 79, 81–83, 89–91, 93, 105, 108, 111, 119, 147, 153, 169–171, 183, 185, 188, 189, 192, 206, 207, 212, 224, 226, 227
Angell, Richard, 93, 94
Aristotle, 12, 13, 15
Aristotle's Thesis, 44, 46

Bar Hillel, Yehoshua, 12
Barcan Marcus, Ruth, 43, 44, 59, 64, 103, 204, 205, 207, 219, 220, 229
Barwise, Jon, 108
Beall, Jc, 11, 203, 204
Becker, Oskar, 34
Belnap, Nuel, xi, 2, 5, 9, 70, 71, 74–77, 79, 81–83, 89–91, 93, 105, 108, 111, 119, 147, 153, 169–171, 183, 185, 188, 189, 192, 202, 206, 207, 212, 224, 226, 227, 238, 250
Bigelow, John, 8
Bimbó, Katalin, 145
Boethius's Thesis, 44–46
Brady, Ross, 91–94, 150, 166, 179
Brandom, Robert, xii
Brown, Bryson, 120
Buridan, Jean, 12

Carnap, Rudolf, 10, 12, 23, 30, 31, 53, 63, 66, 130
Chellas, Brian, 145
Church, Alonzo, 6, 70, 71, 213
Cocchiarella, Nino, 23, 62–66
Coffa, Alberto, 147
compactness, 6, 7, 14, 18, 56, 57, 60, 63, 66, 86, 120–122, 127, 129, 130, 138, 143, 149, 194, 213, 219, 227, 234, 242
consequence operator, 120–126
consequence relation, 6, 17, 18, 55–57, 60, 63, 66, 73, 85, 86, 98, 120, 123–125
Cresswell, M.J., 54, 68, 145, 172
Crossley, John, 65
Curry, Haskell, 162, 200, 202

Davies, Martin, 65
Došen, Kosta, 105
Duhem, Pierre, 223
Dunn, J.M., 7, 90, 108, 132, 145, 177, 182, 202, 211, 212

entailment
  nested, xii, 2, 7, 14, 19, 23, 32, 40, 66, 67, 70, 71, 105, 106, 112, 115, 126, 134, 136, 138, 139
Etchemendy, John, 12, 13

Ferenz, Nicholas, 145, 211
Fine, Kit, xii, 2, 47, 48, 139, 157, 176, 179, 192, 198, 205, 206, 211, 226
Fitch, Frederick, 71
Fitch-style natural deduction, 4, 71–74, 183
Frege, Gottlob, 25–28, 38

Gabbay, Dov, 151, 238
Gentzen systems, 4, 39, 103–105, 124
Giere, Ronald, 131, 132
Goldblatt, Robert, 2, 157, 177, 192, 195, 211, 213
Gupta, Anil, 202
Gödel, Kurt, 97

Herbrand, Jacques, 26, 30, 31, 42
Hintikka, Jaakko, 57–60, 62, 63, 114
horizontalization, 102, 104, 110, 111, 114, 143, 167, 205
Hughes, George, 54, 145, 172
Humberstone, Lloyd, 65, 113

identity, 198–200
implication
  analytic, 46, 48, 93–95
  relevant, xi, 70, 74–77, 80, 86, 105–109, 111, 148, 172, 225, 226
  strict, xi, 2–4, 23–26, 28, 30, 31, 34–44, 49, 51, 53, 55, 56, 66, 67, 69, 77–80, 86, 88, 91, 92, 97–99, 113, 166, 167
intuition, 3

Jónsson, Bijarni, 123

Kane, Michael, 213
Kremer, Philip, xii, 214, 222

Printed in the United States
by Baker & Taylor Publisher Services